The Behavioral Consequences of Stroke

Tom A. Schweizer • R. Loch Macdonald
Editors

The Behavioral Consequences of Stroke

Springer

Editors
Tom A. Schweizer, Ph.D.
Neuroscience Research Program
University of Toronto
St. Michael's Hospital
Toronto, ON, Canada

R. Loch Macdonald, M.D., Ph.D.
Division of Neurosurgery
University of Toronto
St. Michael's Hospital
Toronto, ON, Canada

ISBN 978-1-4614-7671-9 ISBN 978-1-4614-7672-6 (eBook)
DOI 10.1007/978-1-4614-7672-6
Springer New York Heidelberg Dordrecht London

Library of Congress Control Number: 2013943826

Printed on acid-free paper

Springer is part of Springer Science+Business Media (www.springer.com)

To my wife and children, Melanie, Evan, and Averie for their unwavering love and support and for bringing absolute joy to my life in ways that no professional pursuit can approach.

TAS

To my parents, Neil (a psychologist) and Lea who gave me the drive to understand the brain; to my wife and children, Sheilah, Iain, Robyn, and Erin, who support me unerringly; and to my patients, who I try my best to help, even if not always successfully.

RLM

Foreword

Behavioral Consequences of Stroke is the first major textbook that has appeared on this topic. This volume comes very timely at a stage when the stroke field is undergoing major changes in recognizing the true burden of cerebrovascular disease, in consolidating and putting into clinical practice what has been learned in the recent past, and in directing focus to several areas that have been neglected previously. The topic of behavioral consequences after stroke belongs to the latter category.

Until about a decade ago, the picture of the global burden of stroke was very incomplete with data mainly limited to high-income parts of the world. Since then, the true face of the scope of cerebrovascular disease has been disclosed, showing a situation that is much more grave than previously recognized. The core epidemiological data are commonly cited in almost any publication dealing with stroke and have been iterated so many times that it may have a blunting effect. Nevertheless, the data are massive, and the challenge has not declined during the last decade.

As I write this foreword, I have the *Lancet* December 15, 2012 issue in front of me, devoted to the new round of global burden of disease data, which will serve as the new landmark references for several years to come. The world demographics are rapidly changing: life expectancy has increased throughout the world from an average of 56.4 years in 1970 to 67.5 years in 2010 for males and from 61.2 to 73.3 years for females. Worldwide, stroke remains the second commonest cause of death after ischemic heart disease and ahead of chronic obstructive pulmonary disease and the third commonest cause of years of life lost, behind both of these. Since the world population has both grown and become older, the total years lived with disability has increased by one third, and stroke remains in third place in the rank order of all diseases as per Disability Adjusted Life Years (DALYs) lost.

The need for global actions to prevent stroke as well as other non-communicable diseases (NCDs) have been recognized by the WHO and the UN, and prompted the important UN high-level meeting on prevention of NCDs in September 2011, the second time that a medical topic was addressed at the UN General Assembly. The New York declaration has now been followed by the first-ever global monitoring framework that comprises nine voluntary global targets and 25 indicators to prevent and control NCDs including stroke. The combat of stroke is not only a task for medical professionals; stroke has now clearly entered the global political agenda.

Despite progress in prevention, stroke will not be eradicated but will remain with us as long as we can foresee. When I started practicing, in the early 1970s we were convinced that there would never be any effective acute therapy for stroke; analogous to cardiac arrest, the brain would suffer major irreversible damage within few minutes, and when the patients would arrive at hospital, treatment possibilities would have passed. Fortunately, this nihilistic approach of stroke (and cardiac arrest) was proven wrong. Within only a few decades, the treatment opportunities of stroke have changed dramatically, with stroke unit therapy and thrombolytic therapy as the diamonds in the crown, representing some of the most effective therapies in the acute medical field overall and some of the most shining glories of evidence-based medicine. Establishment of well-functioning acute stroke services is now a prioritized task in all regions.

Even with improvements in acute treatments, a majority of patients with stroke will not be completely cured but will have residual symptoms and disabilities. Globally more than 30 million persons live with effects after a stroke; almost half of these have moderate to severe disability, and almost one half are within the age of working life. Rehabilitation services and efficient long-term follow-up after stroke has emerged as fields in need of much strengthening, following an era when much attention has been paid to the acute phase. Just as the concept of organized stroke care has emerged for acute services, we now need to establish organized post-acute stroke services—a task of considerable challenge.

In the measurement of outcomes after stroke, clinical trials and observational studies have usually focused on death, recurrent stroke, and disability measured as dependence/need of support from others. However, the focus has gradually moved to other areas—the behavioral consequences of stroke. Had we listened more carefully to the voice of patients and caregivers in the past, these areas might have been better explored today. Patients and caregivers have repeatedly highlighted the less obvious non-motor consequences after stroke: difficulties in communication, executive functions, fatigue, depression, personality changes, and much more subtle neuropsychological changes. It has also come as some surprise that such effects may be profound even after apparently mild strokes. It took some time before we listened, and it has taken some time for this field to develop scientifically.

The present book on *Behavioral Consequences of Stroke* therefore appears most timely. Tom Schweizer and Loch Macdonald, both well-recognized leaders in this field, have gathered a strong international team of researchers to summarize the current state-of-the art and provide guidance how to incorporate current knowledge into clinical practice. The book broadly covers neurocognitive aspects of ischemic and hemorrhagic stroke, including the clinical manifestations, basic science background, diagnostic and testing issues, imaging aspects, and implications for practice.

I warmly recommend this book to scientists as well as clinicians in this important field of stroke. Let it be closely read, with pages and binding worn out by use, and let the wisdom written be transformed into reality to the benefit of patient and caregivers.

Lund University, Lund, Sweden Bo Norrving

Preface

Stroke remains the fourth most common cause of death in North America and an increasingly common cause in low- and middle-income countries. As death from some types declines due to improved care, the burden of disability increases. Much attention is paid to paralysis and the more obvious physical disabilities these patients incur, but we are becoming increasingly aware that the cognitive and neurobehavioral complications are important contributors to stroke morbidity and even to functional neurological recovery. Physicians and health care providers have been more focused on saving lives as well, but as mortality declines, the details and quality of life of the survivor are becoming more important. Furthermore, these deficits actually overshadow focal neurological impairment in some types of stroke such as subarachnoid hemorrhage.

I am a neurosurgeon and my coeditor is a cognitive neuroscientist. This book was conceived when we began working together and finding that we were teaching each other about our respective understandings of disability after stroke. We noted that there was no book that covered the broad general topic of neurobehavioral and cognitive function after stroke that would be appropriate for the wide range of health care professionals that need to know something about this field. To that end, we have assembled an up-to-date overview of the cognitive and neurobehavioral consequences of stroke.

Who needs to know about these deficits? Everyone who cares for patients with stroke should be aware of and have some understanding of these effects of stroke, and for good reason. As we learn from this book, early treatment of depression and recognition of posttraumatic stress disorder are key to understanding disability and recovery after stroke. There are many other reasons that await the reader of this book.

Given its increasing importance, and the broad audience including neurologists, neurosurgeons, physiatrists, and other medical doctors, as well as physical, speech and occupational therapists, nurses, and psychologists to name a few, a broad summary of the key issues was needed. The goal of this book is to provide a current understanding of each of the major cognitive/neurobehavioral effects of stroke. It is not meant to delve into excessive detail but to give the reader a broad and up-to-date understanding of the field. Because we hope our readers will be from varied disciplines, we include chapters on the epidemiology, general treatment, imaging, and an update on some stroke clinical trials. We solicited authors who were leaders in their respective fields and who can provide the current state of the particular topic of their chapter.

We both learned a lot by reading this book. We hope you will also, and will find it useful.

Toronto, ON, Canada

Tom A. Schweizer, Ph.D.
R. Loch Macdonald, M.D., Ph.D.

Contents

1 **Stroke Epidemiology, Etiology, and Background** 1
Deven Reddy and Robert G. Hart

2 **Treatment of Stroke** 15
Douglas J. Cook and Michael Tymianski

3 **Sensorimotor Control After Stroke** 37
W. Richard Staines, David A.E. Bolton, and William E. McIlroy

4 **Limb Apraxia: Types, Neural Correlates, and Implications for Clinical Assessment and Function in Daily Living** 51
Eric A. Roy, Sandra E. Black, Vessela Stamenova, Deborah Hebert, and David Gonzalez

5 **Spatial Neglect: Not Simply Disordered Attention** 71
James Danckert

6 **Post-stroke Aphasia** 95
Marcelo L. Berthier, Guadalupe Dávila, Natalia García-Casares, and Ignacio Moreno-Torres

7 **Disorders of Emotional Communication After Stroke** 119
Kenneth M. Heilman

8 **Dysexecutive Syndrome After Stroke** 135
Olivier Godefroy, Pierre Yves Garcia, Jean Marc Bugnicourt, Claire Leclercq, and Martine Roussel

9 **Memory Loss After Stroke** 151
Signy Sheldon and Gordon Winocur

10 **Behavior After Aneurysmal Subarachnoid Hemorrhage: Cognition and Functional Outcome** 177
Timour Al-Khindi, R. Loch Macdonald, Stephan Mayer, and Tom A. Schweizer

11 **Cognitive Dysfunction After Intracerebral Hemorrhage, Vasculitis, and Other Stroke Syndromes** 199
Eric E. Smith and José Andrés Venegas-Torres

12 **Post-stroke Depression** 227
Bradleigh D. Hayhow, Simone Brockman, and Sergio E. Starkstein

13 **Better Dead than Alive? Quality of Life After Stroke** 241
Thomas Schenk and Adam J. Noble

14 **MRI Methods Applied to Stroke** 257
Bradley J. MacIntosh and Simon J. Graham

15 **Clinical Outcomes, Stroke Trials, and Cognitive Outcome** 283
Benjamin W.Y. Lo, Filip Stojic, Julian Spears, Tom A. Schweizer, and R. Loch Macdonald

16 **Cognitive Rehabilitation and Recovery After Stroke** 315
Audrey Bowen and Emma Patchick

Index 341

Contributors

Timour Al-Khindi, B.Sc. Johns Hopkins University School of Medicine, Baltimore, MD, USA

Marcelo L. Berthier, M.D., Ph.D. Department of Medicine, Unit of Cognitive Neurology and Aphasia, Centro de Investigaciones Médico-Sanitarias (CIMES), Univesity of Malaga, Malaga, Spain

Sandra E. Black, M.D., F.R.C.P.(C) Brill Chair in Neurology, University of Toronto and Cognitive Neurology, Sunnybrook Health Sciences Centre, Toronto, ON, Canada

David A.E. Bolton, Ph.D. Department of Kinesiology, University of Waterloo, Waterloo, ON, Canada

Audrey Bowen, B.A., M.Sc., Ph.D. Vascular and Stroke Research Centre, University of Manchester (MAHSC), Salford Royal Foundation NHS Trust, Salford, UK

Simone Brockman, B.Sc.(Hons.) Department of Psychiatry, Fremantle Hospital, Fremantle, WA, Australia

Jean Marc Bugnicourt, M.D. Department of Neurology, Amiens University Hospital, Amiens, France

Douglas J. Cook, M.D. Department of Neurosurgery, Lucile Packard Children's Hospital, Stanford Hospital and Clinics, Stanford, CA, USA

James Danckert, Ph.D. Department of Psychology, University of Waterloo, Waterloo, ON, Canada

Guadalupe Dávila, Ph.D. Department of Psychobiology, University of Malaga, Malaga, Spain

Pierre Yves Garcia, Ph.D. Department of Neurology, Centre Hospitalier Compiegne, Compiegne, France

Natalia García-Casares, M.D., Ph.D. Department of Medicine, Unit of Cognitive Neurology and Aphasia, Centro de Investigaciones Médico-Sanitarias (CIMES), University of Malaga, Malaga, Spain

Olivier Godefroy, M.D., Ph.D. Department of Neurology, University Hospital, Amiens, France

David Gonzalez, Ph.D. Department of Kinesiology, University of Waterloo, Waterloo, ON, Canada

Simon J. Graham, Ph.D., P.Eng. Department of Medical Biophysics, University of Toronto, Toronto, ON, Canada

Sunnybrook Research Institute, Sunnybrook Health Sciences Centre, Toronto, ON, Canada

Heart and Stroke Foundation, Centre of Stroke Recovery, Toronto, ON, Canada

Robert G. Hart, M.D. Department of Medicine (Neurology), Hamilton General Hospital, Hamilton, ON, Canada

Bradleigh D. Hayhow, B.A.(Hons.), B.M.B.S. Department of Psychiatry, Fremantle Hospital, Fremantle, WA, Australia

Deborah Hebert, B.Sc.(OT), M.Sc.(Kin), Ph.D.(C) Toronto Rehabilitation Institute, University Health Network, Toronto, ON, Canada

Kenneth M. Heilman, M.D. Department of Neurology, University of Florida College of Medicine, Gainesville, FL, USA

Claire Leclercq, M.D. Department of Neurology, CHU Amiens—Hôpital Nord, Amiens, France

Benjamin W.Y. Lo, M.D., Ph.D.(cand.), F.R.C.S.C. Division of Neurosurgery, Department of Surgery, University of Toronto, St. Michael's Hospital, Toronto, ON, Canada

R. Loch Macdonald, M.D., Ph.D. Division of Neurosurgery, Department of Surgery, University of Toronto, St. Michael's Hospital, Toronto, ON, Canada

Bradley J. MacIntosh, Ph.D. Department of Medical Biophysics, University of Toronto, Toronto, ON, Canada

Sunnybrook Research Institute, Sunnybrook Health Sciences Centre, Toronto, ON, Canada

Heart and Stroke Foundation, Centre of Stroke Recovery, Toronto, ON, Canada

Stephan Mayer, M.D., F.C.C.M. Neurological Intensive Care Unit, Columbia University Medical Center, New York, NY, USA

William E. McIlroy, Ph.D. Department of Kinesiology, University of Waterloo, Waterloo, ON, Canada

Ignacio Moreno-Torres, Ph.D. Department of Spanish Language, Philosophy, University of Malaga, Malaga, Spain

Adam J. Noble, B.Sc., M.Sc., Ph.D. Unit of Neurology and General Practice, Department of Clinical Neuroscience, Institute of Psychiatry, King's College London, London, UK

Bo Norrving Department of Clinical Neuroscience, Lund University, Lund, Sweden

Emma Patchick, B.Sc. School of Psychological Sciences, University of Manchester, Manchester, UK

Deven Reddy, M.B.Ch.B., M.Sc., C.C.F.P. Division of Neurosurgery, Department of Surgery, McMaster University, Hamilton General Hospital, Hamilton, ON, Canada

Martine Roussel, Ph.D. Department of Neurology, University Hospital, Amiens, France

Eric A. Roy, Ph.D., C.Psych. Department of Kinesiology and Psychology, University of Waterloo, Waterloo, ON, Canada

Thomas Schenk, B.Sc., M.Sc., Ph.D. Department of Neurology, University Clinic, Friedrich-Alexander University Erlangen-Nuremberg, Erlangen, Bavaria, Germany

Tom A. Schweizer, Ph.D. Division of Neurosurgery, Department of Surgery, University of Toronto, St. Michael's Hospital, Toronto, ON, Canada

Keenan Research Centre of the Li Ka Shing Knowledge Institute, St. Michael's Hospital, Toronto, ON, Canada

Heart and Stroke, Foundation Centre for Stroke Recovery, Toronto, ON, Canada

Institute of Biomedical Engineering, University of Toronto, Toronto, ON, Canada

Signy Sheldon, Ph.D. Rotman Research Institute, Baycrest Centre for Geriatric Care, Toronto, ON, Canada

Eric E. Smith, M.D., M.P.H. Department of Clinical Neurosciences, Foothills Medical Centre, Calgary, AB, Canada

Julian Spears, M.D., S.M., F.R.C.S.C., F.A.C.S. Division of Neurosurgery, Department of Surgery, University of Toronto, St. Michael's Hospital, Toronto, ON, Canada

W. Richard Staines, Ph.D. Department of Kinesiology, University of Waterloo, Waterloo, ON, Canada

Vessela Stamenova, Ph.D. (C) The Rotman Research Institute, Baycrest Centre for Geriatric Care, Toronto, ON, Canada

Sergio E. Starkstein, M.D., Ph.D. Department of Psychiatry, Fremantle Hospital, Fremantle, WA, Australia

Filip Stojic Keenan Research Centre in the Li Ka Shing Knowledge Institute, St. Michael's Hospital, Toronto, ON, Canada

Michael Tymianski, M.D., Ph.D., F.R.C.S.C. Department of Neurosurgery, Toronto Western Hospital, Toronto, ON, Canada

José Andrés Venegas-Torres, M.D. Department of Clinical Neurosciences, Calgary Stroke Program, University of Calgary, Foothills Medical Centre, Calgary, AB, Canada

Gordon Winocur, Ph.D. Department of Psychiatry, University of Toronto, Toronto, ON, Canada

Department of Psychology, University of Toronto, Toronto, ON, Canada

The Rotman Research Institute, Baycrest, Toronto, ON, Canada

Chapter 1
Stroke Epidemiology, Etiology, and Background

Deven Reddy and Robert G. Hart

Introduction: Historical Perspective

> … it is a disease in which the functions of relation are suspended, while those of organic life continue. A fit of apoplexy, it has been observed, resembles in many respects, profound sleep. There is the same insensibility to external impressions, the same unconsciousness of everything that is passing around; the action of the heart and respiration go on in both instances, but the individual is shut out from the world, sight, hearing, smell, touch and taste being abolished for the time [1].

Throughout the ages various theories and beliefs prevailed conceptualizing physician and lay perceptions of the entity of stroke. The term "apoplexy" (from the Greek word *apoplexia*, meaning striking or hitting away) emerged in the time of Hippocrates and was used into the early part of the twentieth century [2]. Initial theories centered around the concept of blood being one of four crucial humors necessary for survival and a vital spirit. Apoplexy would result when its motion was interrupted. Galen (129–c. 200 CE) expanded on these theories, which were accepted for many centuries. In this time, apoplexy carried a grim prognosis and patients were usually advised as such. Supernatural phenomena were also a popular belief through the ages, with stroke seen as a punitive judgment at the hand of God.

In the eighteenth century, the theory of the "apoplectic habitus" was put forward:

"Those Persons, above all others, are in danger of sudden deaths, that are of an unwieldy, corpulent Body; that have short Necks, strait Chests, and are subject to hitch in their Breathing; great, large heads, with a very sanguine or pale Countenance,

D. Reddy, M.B.Ch.B., M.Sc., C.C.F.P. (✉)
Division of Neurosurgery, Department of Surgery, McMaster University, Hamilton General Hospital, Hamilton, ON, Canada
e-mail: vesh@sympatico.ca

R.G. Hart, M.D.
Department of Medicine (Neurology), Hamilton General Hospital, Hamilton, ON, Canada

T.A. Schweizer and R.L. Macdonald (eds.), *The Behavioral Consequences of Stroke*, DOI 10.1007/978-1-4614-7672-6_1, © Springer Science+Business Media New York 2014 if they indulge in a luxurious Manner of Living, seldom escape a sudden, fatal stroke" [3]. Overindulgence and a lack of moderation were regarded as key causes of stroke in the eighteenth and nineteenth centuries and people thought to be at risk were encouraged to lead a life of balance and moderation.

Bloodletting as a treatment for stroke was popularized in the seventeenth century, this as well as purges and enemas were primary treatment modalities until the early twentieth century. Bloodletting continued despite statistical evidence gathered in the nineteenth century that found it harmful, though it finally fell out of favor when the 1935 edition of Osler's *Principles and Practice of Medicine* concluded that it was not useful in the treatment of apoplexy. The importance of this historical context of stroke allows us to recognize that current practices are not necessarily the truth but simply reflect present-day understanding of this disease process and of the mechanisms at play as we transition to future preventative and therapeutic treatment interventions. It is conceivable that at least some of what we do now and how we conceptualize stroke may be regarded in an unfavorable light in the future, the knowledge of which should serve to temper our enthusiasm to cast our predecessors' efforts and beliefs in a negative light.

Epidemiology

The second half of the twentieth century saw phenomenal gains in our understanding of the pathophysiology of stroke and the care of stroke patients. The establishment of stroke care units has allowed the growth of multidisciplinary teams with a focus on stroke and has helped advance stroke research. Stroke prevention and management remains a formidable healthcare challenge. It is the second most common cause of death after ischemic heart disease and accounts for 9 % of deaths worldwide [4].

In developed countries, where there are more healthcare resources, 10–12 % of all deaths are attributed to stroke and the consequences of stroke, and of these deaths, 12 % occur in patients younger than 65 years of age [5]. It is estimated that there are 5 million stroke deaths per year, two-thirds of which occur in developing countries [6]. The incidence of stroke decreased 42 % from 163 in 1970–1979 to 94 per 100,000 person-years in 2000–2008 in high income countries [7]. In low to middle income countries the stroke incidence rate doubled from 52 to 117 per 100,000 person-years over the same time. The decline in stroke in high income countries was driven by an 11 % decrease in ischemic stroke with no change in the incidence of intracerebral or subarachnoid hemorrhage. There is limited data on stroke subtypes in low to middle income countries, but the data show a similar incidence of ischemic stroke (70 and 67 per 100,000 in high and low to middle income countries, respectively) with almost twice the incidence of intracerebral hemorrhage in these countries compared to high income countries.

Age-adjusted stroke mortality rates demonstrate significant geographic variation. Globally the stroke mortality rate averages at 100/100,000 people/year.

Table 1.1 World Health Organization 2004 estimated age standardized death rate for cerebrovascular disease [8]

Country	Age-adjusted mortality rate per 100,000 per year
Canada	25.7
USA	30.4
Australia	32.6
United Kingdom	45.6
Kyrgyzstan	249.4
India	108
China	156.5
Russian Federation	228
South Africa	141.1
Nigeria	152.8
Brazil	91.1
Pakistan	117.9

Early stroke mortality rates declined 1.1 % per year between 1970 and 2008 in high income and 0.6 % per year in low to middle income countries. There is, however, a wide variation from country to country with Canada at 25.7/100,000 people/year and Russia at 228/100,000 people/year [8]. A number of reasons have been postulated to explain the geographic variation including differences in stroke care, risk factor exposure, and various genetic factors. Case fatality rates in high and low to middle income countries were similar for ischemic stroke (13–23 %) but higher in low to middle income countries for intracerebral hemorrhage (30–48 % compared to 25–35 % in high income countries) and subarachnoid hemorrhage (40–48 % compared to 25–35 %) [7]. The reasons for these variations are likely multifactorial. What is apparent though, is that poorer, developing countries have considerably higher mortality rates than developed countries (Table 1.1; Fig. 1.1) [9].

Estimates suggest that stroke consumes 2–4 % of the total healthcare costs globally. In developed nations this proportion exceeds 4 % of healthcare costs [10]. In the United States, the estimated cost of stroke in 2008 was $34.3 billion; this includes direct medical costs of $18.8 billion [11].

United States data estimates the annual stroke incidence to be 795,000, of which approximately 610,000 are first-time strokes [12]. In the United States, stroke as a cause of death declined by 19 % from 1998 to 2008 so that it is now the fourth most common cause of death. The lifetime risk of stroke is greater for women than for men, and women are generally older at stroke onset. Racial differences have also been observed with African- and Mexican-Americans having a higher incidence of stroke than Caucasian-Americans [12].

Whilst cardio-embolic stroke and ischemic stroke incidence rates in general have been relatively stable or declining over the past few decades, the incidence of intracerebral hemorrhage has not changed and anticoagulant-associated intracerebral hemorrhage has been increasing. This trend will likely continue in developed countries as the average age of the population increases and there is increasing use of

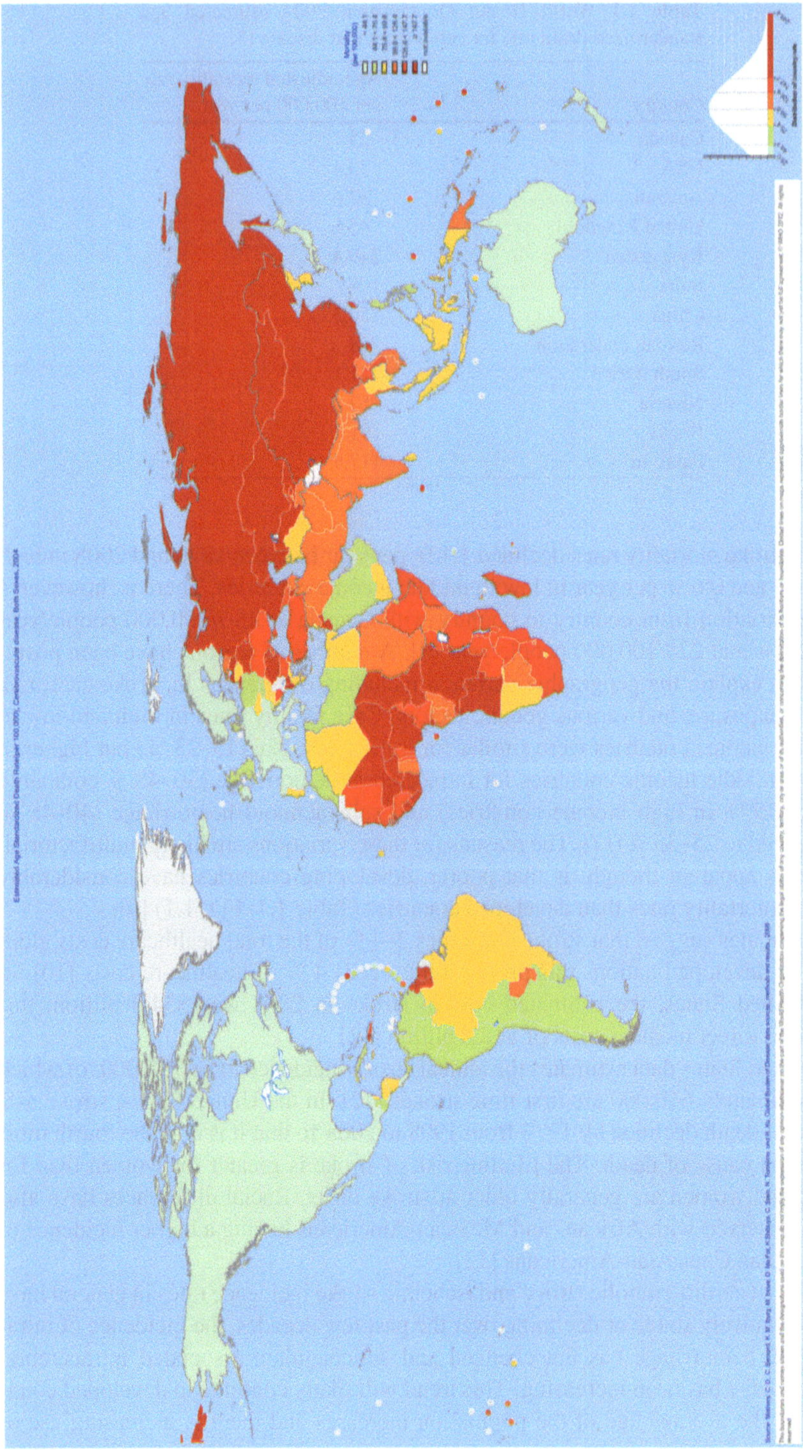

Fig. 1.1 WHO Global Infobase. Estimated age standardized death rate, cerebrovascular disease, both sexes, 2004. Reprinted with permission from the World Health Organization

anticoagulants. Whether new oral direct thrombin inhibitors such as dabigatran and oral factor Xa inhibitors such as apixaban and rivaroxaban, which may be associated with lower rates of intracranial hemorrhage, will prevent this is unclear [13].

Etiology

Stroke—by definition a loss of brain function due to disturbance in blood supply to the brain—has been classified based on the duration of symptoms, clinical presentation, etiology, vascular or anatomical area of the brain affected, and by syndrome. About 80–85 % of strokes are ischemic and 15–20 % are hemorrhagic. The causative mechanism may be related to local intracranial vascular pathologies such as atherosclerosis, a more distant source of pathology such as emboli from extracranial artery disease or the heart, or reduced cerebral perfusion in the setting of circulatory failure. These categories are further subdivided based on the causative process (Fig. 1.2).

Ischemic stroke may be caused by large artery thrombosis, cardio-embolism, small vessel occlusion, venous infarction, or other determined or undetermined causes. Hemorrhagic stroke may be caused by hypertension, vascular malformations, amyloid angiopathy, an infarct with secondary hemorrhage, spontaneous subarachnoid hemorrhage usually due to aneurysm rupture, anticoagulants or antiplatelet drugs, coagulation and platelet disorders, and a small proportion of them are of undetermined etiologies.

Large vessel disease encompasses pathology involving the extracranial carotid vessels and the intracranial arteries including the proximal branches of the Circle of Willis. Pathological processes affecting large arteries include atherosclerosis, arterial dissection, inflammatory arteritis, vasculitis, fibromuscular dysplasia, Moyamoya disease, and cerebral angiographic vasospasm after subarachnoid hemorrhage. Atherosclerosis is the predominant cause of large vessel disease. Thrombi can form on atherosclerotic plaques or spontaneously in the setting of an underlying

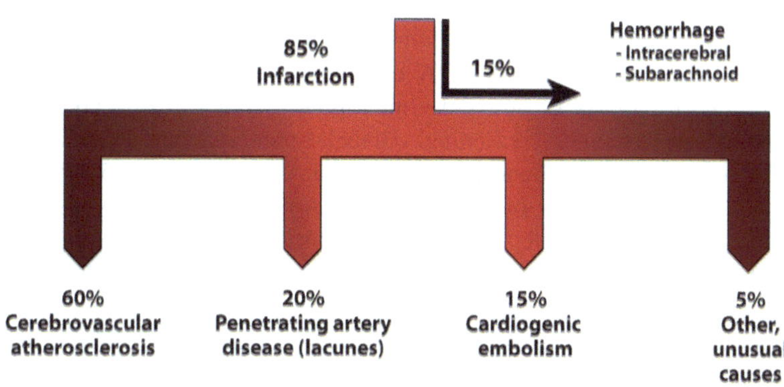

Fig. 1.2 Stroke etiologies

coagulopathy. Identification of the specific underlying vascular pathology and its extent is integral to determining the potential value of subsequent surgical and medical interventions.

Transient ischemic attacks (TIAs) are "a falsely benign form of brain attack" [14] with about 1 in 20 suffering a stroke within the subsequent 48 h, presenting a golden opportunity for prevention of further brain ischemia [15]. The 2009 American Heart Association (AHA) scientific statement on the definition and evaluation of TIA adopted an imaging-based definition [16]; a change from the prior temporal-based definition. The statement formally defines TIA as "a transient episode of neurological dysfunction caused by focal brain, spinal cord, or retinal ischemia, without acute infarction." The change in the definition was required because up to 50 % of TIAs diagnosed by the former definition have evidence of infarction on magnetic resonance imaging (MRI) (diffusion-weighted). TIA Class 1 recommendations of the AHA include imaging within 24 h, either MRI with diffusion weighting or computed tomography (CT)—though MRI is preferred. The proposed change in definition from time-based to imaging-based has implications for international clinical trials involving countries in which urgent MRI is not routinely available. Noninvasive imaging of the extracranial carotid arteries and intracranial cerebral vessels is also recommended.

Silent brain infarcts are a category of stroke in which stroke is evident on imaging studies in the absence of a history of TIA or stroke. The reported prevalence of silent brain infarcts is 8–28 % of patients in eight population-based studies [17]. Despite the lack of a history of TIA or stroke, these patients have been found to have neurological deficits including cognitive impairment, depression, visual field deficits, limb dysfunction, frailty, and physical decline. Thus, it has been suggested that they be referred to as "covert strokes." Their prevalence is up to 5 times that of stroke and increases with age [17]. Covert brain infarcts detected by imaging are thus not necessarily asymptomatic. They are also not benign. Once detected, management implications are unclear and they are an important issue for ongoing stroke prevention research. Subclinical brain ischemia is potentially preventable, though there are no proven interventions to date. Definitive randomized controlled trials of current secondary preventive treatments are required to demonstrate the benefit of treatment in this patient population.

Clinically recognized strokes therefore represent the tip of the iceberg of vascular-mediated injury to the brain. A recently published prospective cohort study of the timing of onset of cognitive decline followed 10,308 British civil servants, performing tests of memory, reasoning, vocabulary, and semantic fluency; assessed 3 times over 10 years and found cognitive decline evident at ages 45–49 [18]. The Honolulu-Asia Aging study of 3,735 Japanese-American men showed that every 10 mmHg increase in midlife systolic pressure increased cognitive impairment by 7 % [19]. Whilst blood pressure lowering has not been shown to reduce vascular cognitive decline, there may be several reasons why this may be the case, including outcome measures that lack the sensitivity to detect appreciable differences when they are present.

Cardio-embolic stroke constitutes approximately 20 % of ischemic stroke and this proportion has remained relatively stable over the past few decades. This subtype of

Table 1.2 Criteria for the clinical diagnosis of cardio-embolic stroke

Identification of potential cardio-embolic source (stratified as major-risk vs. minor-risk)
Absence of other likely causes of stroke
Neurological features
Demonstration of embolic occlusion (i.e., vanishing occlusions)

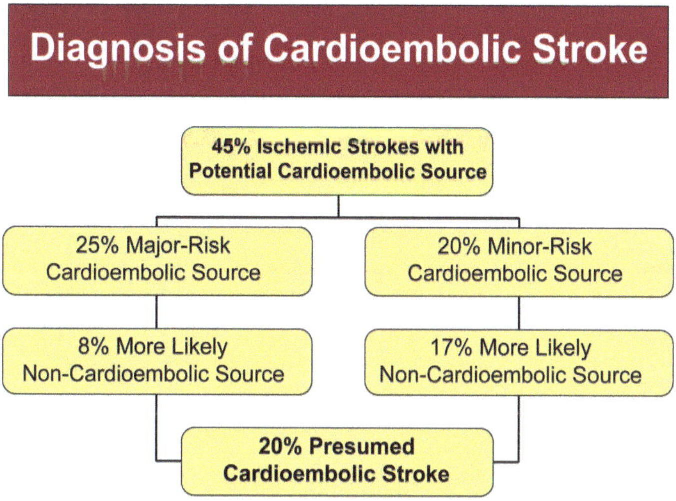

Fig. 1.3 Diagnosis of cardio-embolic stroke

stroke is diagnosed when the probable cause of the stroke is presumed to be a consequence of embolism arising from the heart. The clinical diagnosis of cardio-embolic stroke can be made using the criteria outlined in Table 1.2.

Examples of major-risk sources include atrial fibrillation, intracardiac thrombus, atrial myxomas, and infective endocarditis. Minor-risk sources include calcific aortic stenosis, patent foramen ovale, mitral valve prolapse with myxomatous valves, and left ventricular systolic dysfunction with no associated thrombus (Figs. 1.3 and 1.4; Table 1.3).

Cerebral small vessel disease can be defined as pathology of small perforating arteries and arterioles that supply subcortical structures and result in distinctive clinical, radiological, and pathological changes [20].

Radiological manifestations include subcortical deep brain infarcts, cerebral white matter lesions, and subcortical hemorrhages. Subcortical infarcts include lacunar infarcts and range in size from 3 to 20 mm on CT or MRI imaging. Broadly, it includes sporadic small vessel disease and cerebral autosomal dominant arteriopathy with subcortical infarcts and leukoencephalopathy (CADASIL). Sporadic small vessel disease is an expansive category that includes "deep brain infarcts,

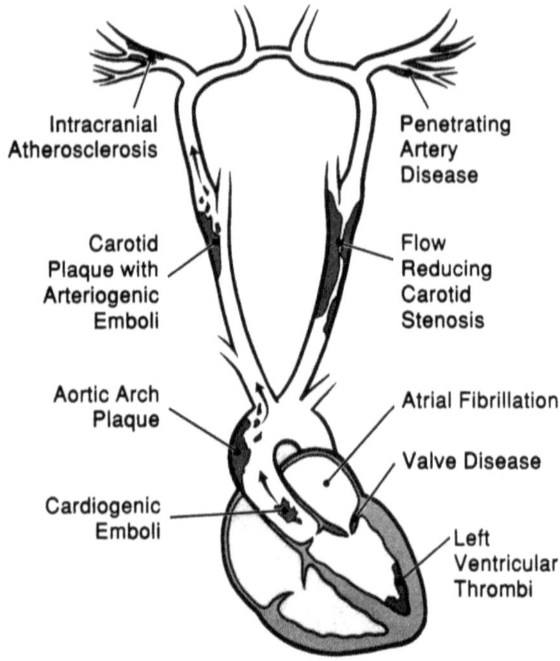

Fig. 1.4 Embolic stroke mechanisms. Reprinted with permission from Robert G. Hart, M.D.

age-related white matter lesions, deep intracerebral hemorrhages and cerebral microbleeds" [20]. Clinically, patients may be categorized as symptomatic or asymptomatic. Symptomatic subcortical/deep brain infarcts comprise 20–30 % of all strokes and are associated with a mortality of 2.4–10 % at 1 year [21, 22]. Histopathology demonstrates irregular cavities circumscribed by an area of gliosis, fragmented blood vessels, and lipid and hemosiderin laden macrophages [17]. Generally, acute deep brain infarcts have better short-term functional outcomes than cardio-embolic strokes [21, 22], though they have a higher risk of recurrence and a greater risk of neurocognitive and functional impairment in the long term [17, 23, 24].

An individual with extensive white matter lesions can manifest with profound neurocognitive disability. The prevalence of white matter lesions in Caucasian-Americans over the age of 60 years is 80 % and is found in more than 60 % of patients with dementia [25, 26].

Risk Factors

Potentially modifiable risk factors for stroke include cigarette smoking, hypertension, diabetes mellitus, dyslipidemia, atrial fibrillation, carotid artery stenosis, post-menopausal hormone replacement therapy, oral contraceptive use, physical inactivity, obesity, coronary heart disease, heart failure, and peripheral arterial

Table 1.3 Sources of cardiogenic brain embolism

Valvular diseases
Rheumatic mitral stenosis
Prosthetic valves
Calcific aortic stenosis
Mitral annulus calcification
Marantic endocarditic and other prothrombotic states (e.g., antiphospholipid antibodies, diffuse intravascular coagulation, Wegener's granulomatosis) causing valve thrombi
Infective endocarditis
Inflammatory valvulitis: Libman-Sacks endocarditis, Behcet's disease, syphilitic
Myxomatous mitral valvulopathy
Left ventricular thrombi
Due to ischemic heart disease: acute myocardial infarct, ventricular akinesis, or aneurysm
Due to nonischemic, dilating cardiomyopathies: hypertrophic, amyloid, rheumatic myocarditis, neuromuscular disorders, catecholamine induced, doxorubicin, idiopathic, viral, peripartum, contusion, hypereosinophilic, sarcoid, alcoholic, Chaga's disease, crack cocaine use, oxalosis, echinococcosis
Left atrial thrombi without mitral valve disease
Due to primary arrhythmias: atrial fibrillation, sick-sinus syndrome/atrial asystole, atrial flutter
Due to other structural disease: atrial septal aneurysm, Chiari network
Cardiac tumors
Atrial myxoma
Cardiac sarcoma
Papillary fibroelastoma
Metastatic
Paradoxical emboli (trans-cardiac embolism)
Atrial septal defects
Patent foramen ovale
Ventricular septal defects
Pulmonary arteriovenous fistulas
Miscellaneous
Postcardiac catheterization
Postvalvuloplasty
Esophageal-atrial fistula

disease. Identified stroke risk factors account for 60 % of attributable risk, compared to 90 % of attributable risk accounted for by risk factors in ischemic heart disease [27, 28].

The benefit of early initiation of secondary prevention strategies after the onset of a TIA has been demonstrated by the EXPRESS study [27] and the SOS-TIA study [28]. The EXPRESS study was a prospective observational study, nested within a larger study. It demonstrated an 80 % reduction in the 90-day risk of stroke in patients that were assessed within a day at the study clinic with early initiation of medical treatment. The SOS-TIA study, published the same year in 2007, affirmed the findings of the EXPRESS study. It found a 90-day stroke rate of 1.24 % with early clinic assessment and treatment initiation, compared to an ABCD score

Table 1.4 Key interventions and timelines that have had a significant effect on risk reduction in secondary prevention of ischemic stroke

No.	Year	Intervention
1.	1978	ASA
2.	1991	Carotid endarterectomy
3.	1993	Warfarin for atrial fibrillation
4.	1996	Clopidogrel
5.	1996	ASA and dipyridamole
6.	2001	Antihypertensive therapy with ACE inhibitors (ramipril, perindopril)
7.	2004	Lipid lowering with simvastatin
8.	2006	Lipid lowering with atorvastatin
9.	2008	Early assessment and initiation of secondary prevention in transient ischemic attacks
10.	2010	Carotid stenting/angioplasty

Note: The year is when convincing phase III randomized controlled trials validated the benefit of the intervention

(based on five parameters: age, blood pressure, clinical features, duration of TIA, and presence of diabetes) predicted stroke rate of 5.96 %.

Carotid endarterectomy has been shown to be effective as a secondary prevention strategy when performed within 2 weeks of the onset of a TIA or minor completed stroke in patients with high-grade internal carotid artery stenosis. The safety of the intervention when performed in the first 48 h is unclear [29, 30]. Carotid artery stenting produced similar results to carotid endarterectomy and is also an option for similar patients [31].

Whilst it is recognized that primary prevention will have the most profound effect on reducing the burden of disease, at present there is no effective method of identifying patients at risk with a high degree of certainty. The focus of clinical research over the past few decades has centered on secondary prevention with a number of key interventions as outlined in Table 1.4. The evidence-based mainstays of medical treatment for secondary stroke prevention are blood pressure reduction (current recommendation is systolic blood pressure <130 mmHg), antiplatelet therapy, and administration of HMG-CoA reductase inhibitors (statins). These interventions have become the standard of care in developed countries, though it is less clear whether they are consistently applied in developing countries, which likely contributes to the high stroke mortality rates in some regions (Table 1.4).

Pathophysiology

Brain tissue dies within minutes of arterial occlusion, the time depending on the severity and duration of the blood flow reduction. In the minutes to hours following cerebral vessel occlusion in an ischemic infarct, the central core of the infarct is surrounded by an area of structurally intact but functionally impaired tissue at risk of further injury known as the ischemic penumbra. A sequence of events follows including energy depletion, calcium channel disruption, glutamate and free radical release, cell membrane and ion homeostasis disruption, and inflammation [32]. These changes can result in further tissue damage and research efforts have been directed to develop neuroprotective drugs to interrupt this sequence of events and protect the tissue at risk in the zone of the ischemic penumbra, to prevent further extension of the infarct. So far the more successful approach has been to administer thrombolytic drugs such as tissue plasminogen activator (tPA) within 3–4.5 h of the ischemic stroke in order to restore blood flow.

Clinical Presentation

The clinical presentation of stroke depends on several factors including the nature of the underlying pathological process and the extent and location of the infarct or hemorrhage. Typically the onset is acute and the neurological deficits are overt. Motor deficits, speech deficits, and impaired level of consciousness are generally fairly straightforward to detect. Cognitive deficits and subtle behavioral changes are more challenging to detect and require a more detailed patient evaluation. TIAs may precede an acute ischemic stroke and efforts are being made to identify and initiate treatment early in this patient group.

Complications

Complications of acute ischemic stroke include cerebral and cerebellar edema, hemorrhagic transformation, raised intracranial pressure, delirium, recurrent stroke, central post-stroke pain, headache, and sleep disorders including sleep disordered breathing. Complications of hemorrhagic stroke include hematoma expansion, peri-hematoma edema, intraventricular extension of hemorrhage, hydrocephalus, hyperglycemia, elevated blood pressure, and seizures. Patients are also at risk of various medical complications including venous thromboembolism. Specific therapeutic interventions such as cerebrospinal fluid diversion and surgical intervention can be considered based on the clinical scenario.

Imaging Investigations

Non-contrast CT is the initial investigation of choice and allows for the early differentiation of ischemic and hemorrhagic stroke. CT perfusion and MRI can provide additional information that may be helpful in initial management, though accessibility to these investigations may be limited in some centers. MRI technology has progressed rapidly in recent times, with gradient echo and susceptibility weighted imaging sequences being quite sensitive in detecting very small hemorrhages. Diffusion-weighted imaging is helpful in acute ischemic stroke. Newer MRI modalities that measure the integrity of the white matter fiber tracts in the brain such as diffusion tensor imaging continue to be investigated. Carotid artery ultrasound and echocardiography supplement the imaging investigations, especially when an extracranial embolic source is suspected (see Chap. 14 for more detailed information on imaging techniques in stroke).

Management

In the acute presentation of an ischemic stroke the care of patients in stroke units, early thrombolysis, treatment with aspirin and decompressive craniectomy (in patients with malignant infarction) are interventions that have been proven to be beneficial [33]. Patients managed in stroke care units have lower mortality rates, improved functional and behavioral outcomes. The precise components of care in a stroke unit that account for these improvements are uncertain, however compliance with best practice and a dedicated team approach contribute significantly.

Scores of clinical trials have tested various neuroprotective agents based on efficacy in pre-clinical stroke models, but to date none have been shown to be beneficial. Administration of tPA within 3–4.5 h of ischemic stroke onset is to date the main approved intervention. In selected patients in tertiary stroke care centers, intra-arterial interventions including selective injections of tPA and clot extraction within 6 h of stroke onset appear to improve outcomes.

Beyond the acute phase of treatment the process of recovery is complex and a multidisciplinary team best facilitates stroke rehabilitation. A high level of patient and family motivation correlates with improved outcomes. Comprehensive, integrated, and coordinated stroke rehabilitation is costly and resource intensive, though it does have the ability to improve patient outcomes and reduce the burden of disease.

Summary

The recognition and management of cognitive deficits and the psychological consequences of stroke is gaining increased attention as a potential opportunity to improve patient outcomes. Depression and anxiety occur commonly following stroke. Early recognition and effective treatment can influence the long-term success of rehabilitation. More importantly, it reasserts the importance of comprehensive patient care and goes beyond simply addressing acute medical issues of a disease that can have devastating consequences.

References

1. Maclachlan D. The diseases and infirmities of advanced life. London: J. Burchill and Sons; 1863.
2. Pound P, Bury M, Ebrahim S. From apoplexy to stroke. Age Ageing. 1997;26:331–7.
3. Robinson N. A discourse upon the nature and cause of sudden deaths. London: T. Warner; 1732.
4. Murray CJ, Lopez AD. Mortality by cause for eight regions of the world: Global Burden of Disease Study. Lancet. 1997;349:1269–76.
5. Bonita R. Epidemiology of stroke. Lancet. 1992;339:342–4.
6. World Health Organization. The World Health Report, 2002.
7. Feigin VL, Lawes CM, Bennett DA, Barker-Collo SL, Parag V. Worldwide stroke incidence and early case fatality reported in 56 population-based studies: a systematic review. Lancet Neurol. 2009;8(4):355–69.
8. World Health Organization. Causes of death 2008 summary tables. May 2011
9. World Health Organization. Estimated age standardized death rate, cerebrovascular disease, both sexes, 2004. WHO Global Infobase.
10. Donnan GA, Fisher M, Macleod M, Davis SM. Stroke. Lancet. 2008;371:1612–23.
11. Medical Expenditure Panel Survey (MEPS) of the Agency for Healthcare Research and Quality (AHRQ). Household component summary data table. Table 4: total expenses and percent distribution for selected conditions by source of payment: United States, 2008.
12. Roger VL, Go AS, Lloyd-Jones DM, Benjamin EJ, Berry JD, Borden WB, American Heart Association Statistics Committee and Stroke Statistics Subcommittee, et al. Heart disease and stroke statistics–2012 update: a report from the American Heart Association. Circulation. 2012;125:e12–230.
13. Hart RG, Diener H-C, Yang S, Connolly SJ, Wallentin L, Reilly PA, et al. Intracranial hemorrhage in atrial fibrillation patients during anticoagulation with warfarin or dabigatran: the RE-LY trial. Stroke. 2012;43(6):1511–7.
14. Amarenco P, Benavente O. EXPRESS transient ischemic attack study. Speed the process! Stroke. 2008;39:2400–1.
15. Giles MF, Rothwell PM. Risk of stroke early after transient ischemic attack: a systematic review and meta-analysis. Lancet Neurol. 2007;6:1063–72.
16. Easton JD, Saver JL, Albers GW, Alberts MJ, Chaturvedi S, Feldmann E, et al. AHA/ASA scientific statement: definition and evolution of transient ischemic attack. Stroke. 2009;40:2276–93.
17. Vermeer SE, Longstreth Jr WT, Koudstaal PJ. Silent brain infarcts: a systematic review. Lancet Neurol. 2007;6:611–9.
18. Singh-Manoux A, Kivimaki M, Glymour MM, Elbaz A, Berr C, Ebmeier KP, et al. Timing of onset of cognitive decline: results from Whitehall II prospective cohort study. BMJ. 2012;344:d7622.
19. Launer LJ, Masaki K, Petrovich H, Foley D, Havlik RJ. The association between midlife blood pressure levels and late-life cognitive function: the Honolulu-Asia Aging Study. JAMA. 1995;274(23):1846–51.
20. Moran C, Phan TG, Srikanth VK. Cerebral small vessel disease: a review of clinical, radiological, and histopathological phenotypes. Int J Stroke. 2012;7:36–46.
21. Donnan G, Norrving B, Bamford J, Bogousslavsky J, editors. Subcortical stroke. 2nd ed. New York, NY: Oxford University Press; 2002.
22. Wardlaw JM. Differing risk factors and outcomes in ischemic stroke subtypes: focus on lacunar stroke. Future Neurol. 2011;6:201–22.
23. Arboix A, Marti-Vilalta JL. Lacunar stroke. Expert Rev Neurother. 2009;9:179–96.
24. Baezner H, Blahak C, Poggesi A, Pantoni L, Inzitari D, Chabriat H, LADIS Study Group, et al. Association of gait and balance disorders with age-related white matter changes: the LADIS study. Neurology. 2008;70:935–42.

25. De Leeuw FE, de Groot JC, Achten E, Oudkerk M, Ramos LM, Heijboer R, et al. Prevalence of cerebral white matter lesions in elderly people: a population based magnetic resonance imaging study. The Rotterdam Scan Study. J Neurol Neurosurg Psychiatry. 2001;70:9–14.
26. Steingart A, Hachinski VC, Lau C, Fox AJ, Fox H, Lee D, et al. Cognitive and neurologic findings in demented patients with diffuse white matter lucencies on computed tomographic scan (leuko-araiosis). Arch Neurol. 1987;44:36–9.
27. Rothwell PM, Giles MF, Chandratheva A, Marquardt L, Geraghty O, Redgrave JN, et al. Effect of urgent treatment of transient ischaemic attack and minor stroke on early recurrent stroke (EXPRESS study): a prospective population-based sequential comparison. Lancet. 2007;370: 1432–42.
28. Lavallee PC, Meseguer E, Abboud H, Cabrejo L, Olivot JM, Simon O, et al. A transient ischaemic attack clinic with round-the-clock access (SOS-TIA): feasibility and effects. Lancet Neurol. 2007;6:953–60.
29. Rothwell PM, Eliasziw M, Gutnikov SA, Warlow CP, Barnett HJ, Carotid Endarterectomy Trialists Collaboration. Endarterectomy for symptomatic carotid stenosis in relation to clinical subgroups and timing of surgery. Lancet. 2004;363:915–24.
30. Rerkasem K, Rothwell PM. Systematic review of the operative risks of carotid endarterectomy for recently symptomatic stenosis in relation to the timing of surgery. Stroke. 2009;40: e564–72.
31. Brott TG, Hobson II RW, Howard G, Roubin GS, Clark WM, Brooks W, CREST Investigators, et al. Stenting versus endarterectomy for treatment of carotid-artery stenosis. N Engl J Med. 2010;363:11–23.
32. Dirnagl U, Iadecola C, Moskowitz MA. Pathobiology of ischaemic stroke: an integrated view. Trends Neurosci. 1991;22:391–7.
33. Vahedi K, Vicaut E, Mateo J, Kurtz A, Orabi M, Guichard JP, DECIMAL Investigators, et al. Sequential-design, multicenter, randomized, controlled trial of early decompressive craniectomy in malignant middle cerebral artery infarction (DECIMAL Trial). Stroke. 2007;38: 2509–17.

Chapter 2
Treatment of Stroke

Douglas J. Cook and Michael Tymianski

Introduction

Stroke is defined by the World Health Organization (WHO) as "rapidly developing clinical signs of focal (or global) disturbance of cerebral function, with symptoms lasting 24 h or longer or leading to death, with no apparent cause other than that of vascular origin" [1]. Therefore, stroke may be the result of an ischemic or hemorrhagic etiology. In either case there is a "core" area of immediate cell death related to the primary insult surrounded by a zone of viable but at-risk tissue. In the case of ischemic stroke the core is defined as a severely hypoperfused region with blood flow estimated to be less than 8 mL/100 g tissue/min [2]. The core is surrounded by a zone of tissue that suffers from critically low blood flow estimated to be between 8 and 20 mL/100 g tissue/min known as the "penumbra." With continued ischemia the penumbra goes on to die [2]. In the case of hemorrhagic stroke, the core area comprises tissue that is destroyed immediately by shear force and mass effect relating to the hemorrhage [3]. The surrounding tissue probably does not have impaired blood flow as in ischemic stroke [4], but is at risk of further damage if the hematoma expands or re-bleeds and creates further shear force or mass effect [5].

The common aim of stroke treatment is the protection of tissue at risk. This is generally achieved by restoring blood flow in ischemic stroke or by preventing hematoma expansion from re-bleeding or swelling in hemorrhagic stroke. Therefore, the treatment employed in stroke cases varies widely between ischemic and hemorrhagic etiologies and will be discussed separately in this chapter.

D.J. Cook, M.D.
Department of Neurosurgery, Lucile Packard Children's Hospital,
Stanford Hospital and Clinics, Stanford, CA, USA

M. Tymianski, M.D., Ph.D., F.R.C.S.C. (✉)
Department of Neurosurgery, Toronto Western Hospital, Toronto, ON, Canada
e-mail: mike_t@uhnres.utoronto.ca

T.A. Schweizer and R.L. Macdonald (eds.), *The Behavioral Consequences of Stroke*,
DOI 10.1007/978-1-4614-7672-6_2, © Springer Science+Business Media New York 2014

Treatment of Ischemic Stroke

Ischemic stroke results from a lack of blood flow to brain tissue secondary to severe hypoperfusion in a vascular territory due to proximal occlusion of a cerebral artery or complete occlusion of an end artery [6]. Common causes for arterial occlusion include emboli originating from proximal in the arterial system as in carotid artery stenosis, atrial fibrillation, and valvular heart disease; cerebral arterial thrombosis due to arterial dissections, critical stenoses and atherosclerotic plaque rupture or hypercoagulable state; and from inflammatory or idiopathic processes that lead to vessel occlusion such as moyamoya disease and central nervous system vasculitides.

As previously discussed, ischemic stroke results in the formation of core and penumbral regions that represent dead and at-risk tissue, respectively. If untreated, the penumbral zone progresses to tissue death, becoming core tissue. Therefore, there is a positive correlation between time to reperfusion, tissue salvaged by reperfusion therapy, and patient outcome. Saver quantified the "Time is Brain" concept, noting that for a typical large artery stroke, 1.9 million neurons die each minute, emphasizing the urgency to achieve recanalization and reperfusion after stroke onset [7]. The time window for recanalization varies based on the methodology utilized (intravenous or intra-arterial thrombolysis, mechanical thrombectomy, or combination thereof), the vascular distribution involved, and etiology of the occlusion.

Acute Medical Management of Ischemic Stroke

The early management of stroke has been reviewed and summarized in the American Heart Association (AHA) guidelines on the early management of acute ischemic stroke [8]. In the acute phase, the guidelines focus on the evaluation and optimization of airway and ventilation, blood pressure, blood sugar, and avoidance of hyperthermia. In patients who are obtunded and are at risk of airway obstruction or in patients with lower cranial nerve deficits that elevate the risk of obstruction or aspiration, endotracheal intubation may be required to protect the airway and avoid hypoxia. The requirement for intubation must be weighed against the risk of ventilator acquired pneumonia, a complication estimated to occur in 21 % of intubated stroke patients and associated with higher mortality and worsened functional outcome [9]. The AHA guidelines recommend the avoidance of hypoxia in all stroke patients by maintaining an oxygen saturation ≥92 % with supplemental oxygen as required.

Both hypotension and hypertension during the early management of ischemic stroke are independent predictors of worsened neurological outcome [10, 11]. Hypertension is associated with an increased risk of hemorrhagic conversion that is further elevated with thrombolysis [12]. The AHA guidelines suggest a cautious approach to the treatment of hypertension in patients not proceeding to recanalization therapies, recommending that hypertension be treated in patients with other medical indications for treatment or a reduction of 15 % in patients who present with systolic pressure >220 mmHg or diastolic pressure >120 mmHg. In patients proceeding to intravenous thrombolysis or other recanalization therapies, the guidelines recommend treating hypertension to reduce systolic pressure to ≤185 mmHg or diastolic pressure ≤110 mmHg before initiating thrombolysis.

Systemic hypotension results in a decline in cerebral perfusion pressure and theoretically results in decreased collateral perfusion and increased core volume during ischemic stroke [13]. The AHA guidelines recommend investigations to identify the cause of hypotension where baseline systolic blood pressure is <100 mmHg or diastolic pressure is <70 mmHg. The correction of hypotension due to hypovolemia and cardiac arrhythmia using volume expansion and or vasopressors is recommended; however, induced hypertension or hypervolemia as a form of treatment are not suggested.

Fever is an independent predictor of worsened neurological outcome following stroke probably related to increased metabolic demands and depleted energy stores [14]. Reduction of fever should be achieved by treating the source of fever and by administering antipyretic drugs such as acetaminophen. While hypothermia has been shown to improve stroke outcome in animal models, active cooling of stroke patients has not proven effective in clinical trials and is not recommended outside of study protocols [15].

Both hypoglycemia and hyperglycemia should be avoided in the acute management of stroke. Hypoglycemia may result in neurologic findings that mimic stroke or in some cases result in medical complications such as seizures that are deleterious to outcome following stroke [16]. Prolonged hyperglycemia after the onset of stroke results in worsened functional outcome [17]. The AHA guidelines recommend treatment of hyperglycemia with insulin to achieve a blood sugar of 140–185 mg/dL.

Early imaging in acute ischemic stroke is critical to drive treatment decisions (see Chap. 14 for more detail). A plain computed tomography (CT) scan is recommended prior to initiating therapy, especially thrombolytics, to rule out hemorrhage or tumor. Additional imaging to visualize vessel patency or to estimate tissue perfusion can be obtained in the acute phase of stroke management and may have prognostic significance [18, 19]. Perfusion imaging may be obtained with dynamic contrast-enhanced CT or magnetic resonance imaging (MRI) scanning and can be used to approximate cerebral blood flow, mean transit time, and time to peak that can be used to estimate the relative size of the stroke core and penumbra within the perfusion deficit. MRI diffusion-weighted imaging (DWI) can be used to identify completed stroke and for the identification of penumbra as the zone of tissue in the perfusion–diffusion mismatch. However, additional imaging should not delay the administration of thrombolytics.

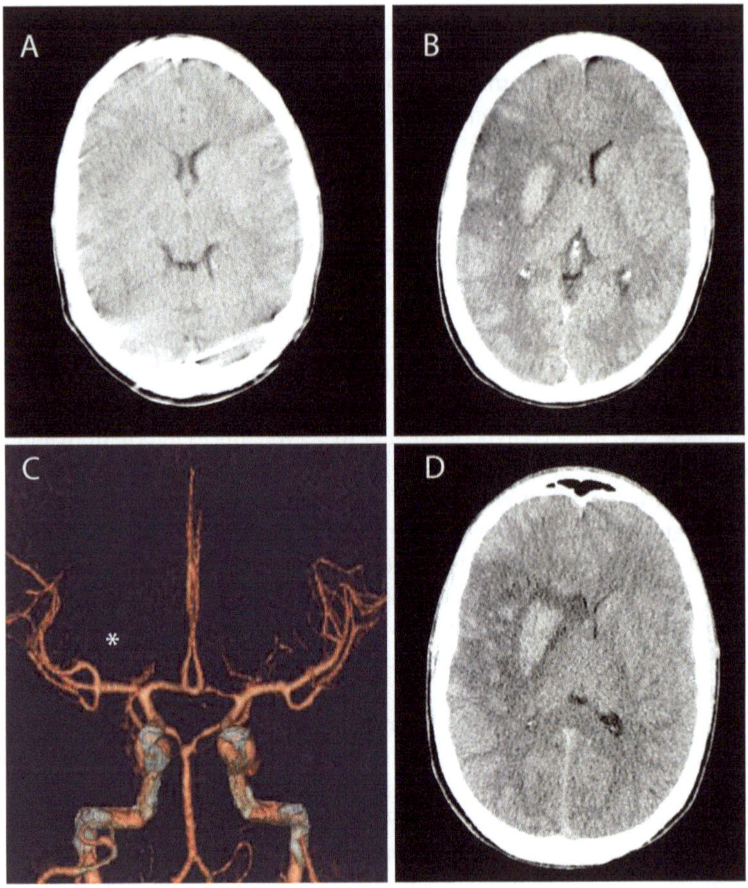

Fig. 2.1 Intravenous recombinant tissue plasminogen activator (rtPA) for acute ischemic stroke. This 58-year-old man presented to the emergency department 3.5 h after the acute onset left hemiplegia. (**a**) Initial computed tomography (CT) scan was interpreted as having no hemorrhage and evidence of early right middle cerebral artery (MCA) distribution ischemic changes. Intravenous rtPA was administered without neurologic improvement. (**b**) Eight hours after administration of rtPA the patient had a decreased level of consciousness and underwent repeat CT scan demonstrating early hemorrhage in the putamen ipsilateral to the stroke. (**c**) CT angiography obtained at that revealed a patent right M2 branch with absent flow in the lenticulostriate arteries on the right side (*asterisk*). (**d**) Twenty-four hours after presentation, a plain CT was obtained that demonstrated extension of hemorrhage and increased stroke and peri-hematoma edema resulting in midline shift

Intravenous Thrombolysis

Intravenous thrombolysis for acute ischemic stroke has become standard practice in select cases (Fig. 2.1). The National Institute of Neurological Disorders and Stroke (NINDS) trial, in which 624 patients with acute ischemic stroke presenting to hospital within 3 h of stroke onset were randomized to 0.9 mg/kg (maximum dose 90 mg) intravenous recombinant tissue plasminogen activator (rtPA) vs. placebo, was the first to show a benefit of thrombolysis for stroke [20]. This study demonstrated that rtPA administration resulted in favorable functional outcomes being reached in 31–50 % of rtPA-treated patients compared to 20–38 % of placebo-treated patients at 90 days following stroke and persisting at 1 year [20, 21]. The mortality rate between groups was not significantly different (17 % vs. 20 % at 90 days); however, the rate of symptomatic hemorrhagic conversion was greater in the rtPA group (0.6 % vs. 6.4 %), (Fig. 2.1b, d). Close monitoring of neurological conditions and vital signs is required after administration of rtPA in order to detect hemorrhagic complications, reactive angioedema, and airway compromise.

Given the short window in which rtPA therapy is initiated, there are a limited number of cases that meet the inclusion criteria for thrombolytic therapy. The number of cases that are eligible for rtPA therapy is further diminished by contraindications as rtPA cannot be used in cases with evidence of intracranial hemorrhage, clinical evidence of subarachnoid hemorrhage (SAH), active internal bleeding, known bleeding diathesis with elevated prothrombin time or international normalized ratio (INR), intracranial surgery within 3 months of presentation, uncontrolled hypertension (systolic >185 mmHg, diastolic >110 mmHg), history of intracranial hemorrhage, and/or known aneurysm or arteriovenous malformation.

The use of rtPA between 3 and 4.5 h after stroke onset was studied by the European Cooperative Acute Stroke Study using the same dosing regimen as in the NINDS trial [22]. This trial demonstrated a modest but significant improvement in functional outcomes with a significant associated increase in symptomatic intracranial hemorrhage in rtPA-treated patients; however, there was no difference in mortality between treatment groups. This evidence supports the extension of the therapeutic window to initiate intravenous thrombolytic therapy to 4.5 h following stroke onset.

Other promising pharmacological thrombolytic therapies are currently being explored for stroke thombolysis. Desmoteplase is a fibrin-specific thrombolytic agent based on the anti-hemostatic component of the saliva of the vampire bat. The Desmoteplase in Acute Stroke trial part 2 (DIAS-2) utilized this thrombolytic agent in patients between 3 and 9 h following stroke onset with a demonstrated ischemic penumbra depicted by a diffusion–perfusion-weighted imaging mismatch on MRI [23]. This study demonstrated neither an improvement in functional status nor an increase in treatment-related hemorrhage or mortality; although there was an increase in delayed and non-neurological deaths in the treatment arm. Desmoteplase will be further studied in two subsequent trials (DIAS-3 and DIAS-4) initiated in 2009. Tenecteplase is a promising thrombolytic agent based on rtPA with modifications in the rtPA amino acid sequence that increase fibrin selectivity, increase half-life, and decrease the risk of hemorrhage [24]. Tenecteplase has undergone early investigation in safety and dose finding studies and is presently being evaluated in a Phase II trial comparing it against alteplase (rtPA) for acute ischemic stroke administered within 4.5 h of stroke onset (ATTEST trial, NCT01472926) [25].

Intra-arterial Thrombolysis and Mechanical Thrombolysis/Thrombectomy

Intra-arterial thrombolysis is delivered through an endovascular catheter proximal to the site of arterial occlusion or from within the thrombus (Fig. 2.2). The main advantage of intra-arterial thrombolysis is that an elevated dose of thrombolytic agent can be delivered at the site of occlusion while reducing the systemic dose of drug compared to intravenous therapy. Intra-arterial therapy was effective when delivered up to 6 h following stroke onset, hence it may be utilized in patients who present outside of the 4.5-h window for intravenous rtPA therapy. However, this technique requires an interventional neuroradiology team to place the catheter in the occluded arterial branch prior to thrombolysis, potentially delaying therapy onset. No trial has directly compared intravenous and intra-arterial therapy to date.

The evidence supporting intra-arterial therapy includes two Phase III randomized controlled trials, the North American Prolyse in Acute Cerebral Thromboembolism study part II (PROACT-II) and the Japanese Middle Cerebral Artery Embolism Local Fibrinolytic Intervention Trial (MELT) [26, 27]. PROACT-II utilized 9 mg of recombinant pro-urokinase or heparin (control group) given as an intra-arterial infusion over 2 h in cases of middle cerebral artery (MCA) occlusion presenting less than 6 h following stroke onset [26]. In this study, 40 % of treatment patients reached a modified Rankin score (mRS) of <2 % vs. 25 % of control patients. Recanalization rates were 66 % in the treatment group vs. 18 % in controls. Symptomatic intracranial hemorrhage was higher in the treatment group at 10 % vs. 2 % for controls; however, mortality was equal in the two groups. The MELT trial utilized urokinase delivered in bolus doses every 5 min until recanalization was achieved in cases of MCA occlusion presenting within 6 h of stroke onset [27]. The control arm of the MELT trial received standard medical therapy that did not include intravenous rtPA as it was not approved in Japan during the trial. The MELT trial was halted early when intravenous rtPA was approved for use in acute ischemic stroke in Japan. The subsequent analysis of the truncated dataset was promising but did not reach significance for the primary endpoint, mRS 0–2 at 90 days, where 49 % of the treatment arm reached this endpoint vs. 39 % in the control group. However, secondary analyses were significantly improved including excellent outcome (mRS 0–1 at 90 days) achieved in 42 % of treatment patients vs. 23 % of controls and a National Institutes of Health Stroke Scale (NIHSS) score of 0–1 at 90 days was 5.3 % in the treatment arm and 3.5 % in the control arm ($P = 0.017$). Rates of intracerebral hemorrhage (ICH) were 9 % in treatment patients and 2 % in controls.

No thrombolytic therapy has received regulatory approval for intra-arterial use in the United States. Therefore, the use of intra-arterial thrombolytics is considered off label [28]. Furthermore, pro-urokinase is not clinically available in the United States and urokinase has not been approved for use in acute ischemic stroke, leaving rtPA as the thrombolytic of choice for intra-arterial thrombolysis in most centers. The American College of Chest Physicians has included intra-arterial rtPA as an option for acute ischemic stroke <6 h in an MCA distribution or in the basilar distribution in centers with expertise in endovascular therapy [28].

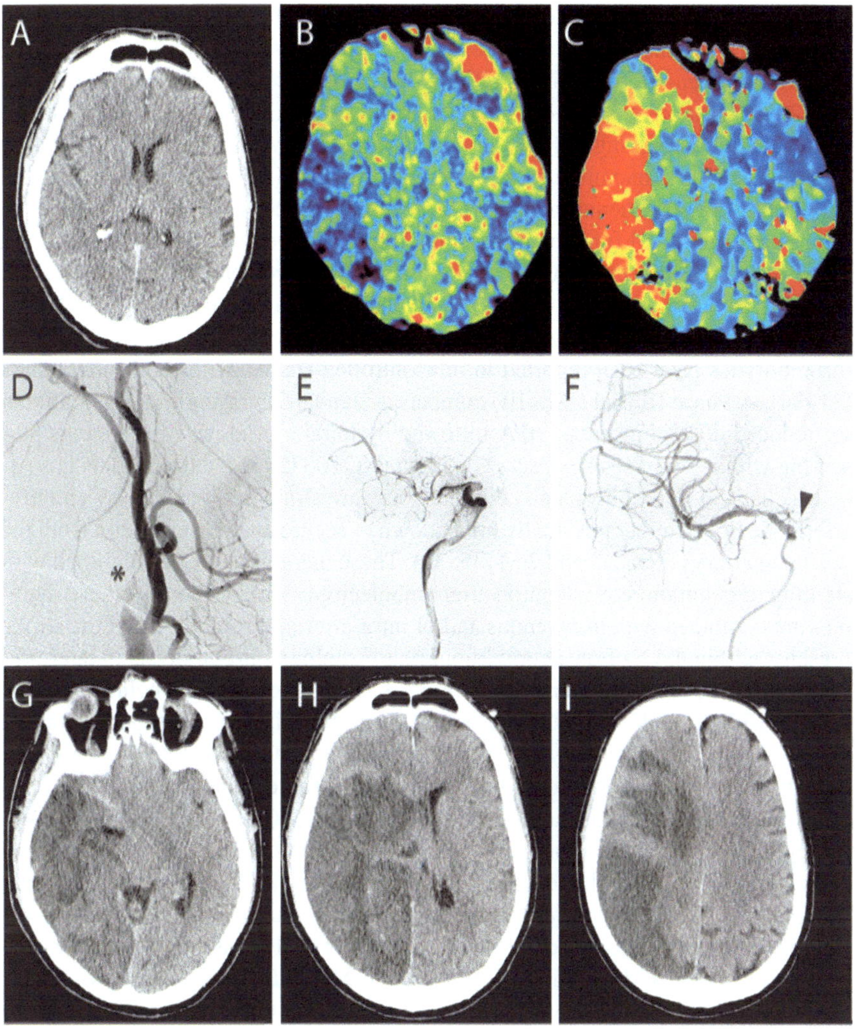

Fig. 2.2 Intra-arterial and mechanical thrombolysis for acute stroke. This 36-year-old female presented 4 h after acute onset arm and facial weakness with dysarthria. (**a**) Initial CT scan demonstrated early ischemic changes in the right MCA distribution. (**b, c**) CT perfusion demonstrated decreased cerebral blood flow (**b**) and elevated time to peak concentration (**c**) in the right MCA distribution. The patient received 0.6 mg/kg of rtPA and was taken to angiography. (**d**) Angiogram revealed an occlusion of the right internal carotid artery (*asterisk*). (**e**) A Penumbra device was used to aspirate clot from the internal carotid artery and achieve reperfusion. (**f**) Intra-arterial rtPA was infused through a microcatheter (catheter tip at *black arrow*) to achieve recanalization of the MCA. (**g–i**) The patient did not have neurological recovery following recanalization and went on to have a completed stroke in the right MCA distribution as demonstrated by hypodensity on plain CT 24 h after rtPA. Early uncal herniation (**g**) and midline shift (**h**) were evident

Combination therapy with intravenous rtPA followed by intra-arterial thrombolysis is a strategy that provides the benefit of early intravenous thrombolytic dosing followed by intra-arterial administration of thrombolytic drug directly to the thrombus [29]. This strategy has been studied in the Interventional Management of Stroke (IMS) II trial comparing reduced dose (0.6 mg/kg) rtPA intravenous followed by intra-arterial rtPA (maximum dose 22 mg) to historical controls from the NINDS study [30]. IMS II patients had a mortality of 16 % vs. 24 % in the NINDS placebo arm. Furthermore, all functional measures were improved in IMS II patients compared to NINDS placebo patients and the rate of intracerebral hematoma was not different between the IMS II cohort and the intravenous rtPA arm of NINDS. Combination strategies utilizing various dosing of intravenous and intra-arterial thrombolytics have been reported in noncontrolled studies with promising results [31–34]. A Phase III trial (IMS III) comparing standard intravenous rtPA treatment vs. reduced dose intravenous rtPA followed by intra-arterial rtPA for patients presenting within 3 h of stroke onset is ongoing (NCT00359424, Clinicaltrials.gov).

Mechanical thrombolysis and embolectomy are strategies that employ an intra-arterial device to either physically break down or secure and retrieve a clot from the occluded artery (Figs. 2.2 and 2.3) [35, 36]. These therapies may be combined with stenting to maintain recanalization after embolectomy [37]. Generally, these therapies are combined with intravenous and/or intra-arterial thrombolytics if the stroke presents within 4.5 h of onset or without pharmacologic thrombolysis in later presentations. The techniques employed in mechanical thrombolysis include clot disruption with an endovascular wire or sonothrombolysis where ultrasound pulses are delivered with thrombolytics at the tip of the catheter [38]. Embolectomy can be achieved with wire loops to capture emboli, retrievable stent strategies (Fig. 2.3e) or aspiration catheters that suction clot from the vessel (Fig. 2.2e) [38]. Devices to achieve vessel recanalization have largely been approved based on safety and recanalization results in large animal models and in small human studies, limiting functional outcome measures in support of these strategies. The Mechanical Embolectomy in Acute Ischemic Stroke (MERCI) trial is the main evidence supporting mechanical clot retrieval [39, 40]. The MERCI device is an intra-arterial coiled wire that is "corkscrewed" around or into the clot to trap it and then used to physically retrieve the clot by pulling it from the parent artery. In the MERCI trial patients presenting within 3 h of stroke onset in any vascular distribution were given intravenous rtPA at standard doses. If the occlusion did not resolve on follow-up angiography, then the MERCI device was utilized. The device resulted in a 57 % recanalization rate that increased to 70 % with additional intra-arterial rtPA. An mRS outcome of 0–2 was achieved in 36 % of patients, the mortality rate was 34 % and the risk of procedure-related morbidity was 5.5 %. In summary, mechanical means of recanalization show promise in early clinical trials and warrant further study.

In vascular distributions other than the MCA, there is less evidence to guide the use of intra-arterial thrombolytic therapy and/or mechanical thrombolysis. In multiple case series of intra-arterial therapy for basilar artery occlusion, a potentially

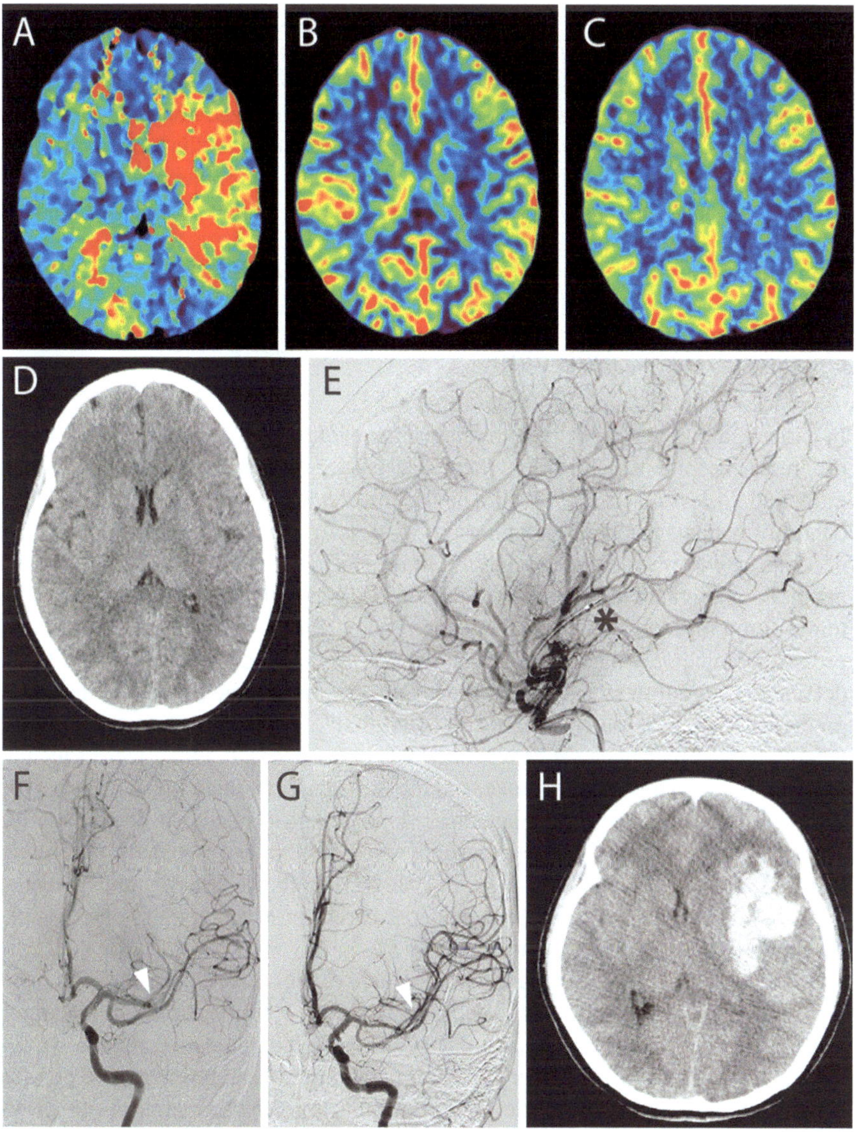

Fig. 2.3 This 42-year-old patient presented 3 h after acute onset aphasia and right arm and leg hemiparesis. (**a–c**) Early CT perfusion demonstrated an elevated time to peak concentration (**a**) with minimal deficit in cerebral blood flow (**b**) or cerebral blood volume (**c**) suggesting good collateralization in the MCA territory. (**d**) Plain CT demonstrates early stroke in left insular cortex. The patient received 0.6 mg/kg of rtPA and was taken to the angiography suite. (**e**) A retrievable stent was deployed in the occluded M2 branch of the MCA (*asterisk*) and clot was retrieved. (**f**) Angiogram prior to stent retrieval shows M2 occlusion (*arrow*). (**g**) Angiogram following clot retrieval shows recanalization of the occluded M2 branch (*arrow*). (**h**) Nine hours after the procedure the patient declined with worsening weakness and decreased level of consciousness and was found to have an intracerebral hematoma in the distribution of the stroke. The patient went on to surgery to evacuate this clot without major clinical improvements postoperatively

positive effect has been demonstrated when recanalization is achieved [41–44]. Stent-assisted recanalization of basilar occlusion has been reviewed in case series and has been shown to achieve improved recanalization rates with an acceptable risk profile up to 8 h after stroke onset [45, 46]. Randomized controlled trials for basilar occlusion might clarify the role for these treatments in this indication; however, such a trial would be difficult to execute given the poor outcomes experienced by these patients, the relatively infrequent presentation of these cases, and the lack of a standard therapy to act as a control.

Decompressive Craniectomy

In cases where reperfusion fails to reverse the evolution of stroke or where stroke progresses despite recanalization, there is a risk for malignant edema and elevated intracranial pressure (ICP) that may be life threatening (Fig. 2.4). Mortality in this cohort of patients has been as high as 78 % [47]. In severe cases there is a role for decompressive craniectomy to relieve brain shift, herniation, elevated ICP, and death. For malignant MCA stroke, the procedure generally includes a large craniectomy, durotomy, and subtemporal decompression (Fig. 2.4e, f). Results for decompressive craniectomy in case series and randomized trials demonstrate a significant decrease in mortality after this procedure [48, 49]. Patient age is the only definitive predictive factor of outcome after decompressive hemicraniectomy for stroke, where younger patients have better functional outcomes [50]. Other factors including timing of surgery, side of stroke, and the presence of herniation have not proven to be significant predictors of outcome. A potential downside of decompressive craniectomy for stroke is that the procedure saves lives but leaves patients in severely disabled or persistent vegetative states [51, 52]. Both the benefits and potential harms of this procedure must be considered carefully in the context of a specific case before treatment is undertaken.

In the case of cerebellar infarction, there is a risk of direct brainstem compression and hydrocephalus if swelling occurs (Fig. 2.5) [53]. Suboccipital craniectomy, in addition to expansion duroplasty, can be used to accommodate cerebellar swelling. In severe cases of cerebellar edema, infarcted tissue can be excised to provide more room for swelling (Fig. 2.5e, f). Early craniectomy in these cases may be justified to prevent impending herniation or brainstem compression.

Treatment of Hemorrhagic Stroke

Hemorrhagic stroke is broadly organized into ICH that involves bleeding into the parenchyma of the brain (Fig. 2.6), SAH that involves bleeding in the subarachnoid space from a ruptured vessel and intraventricular hemorrhage (IVH) where bleeding occurs into the ventricular system. All three can be caused by trauma, but

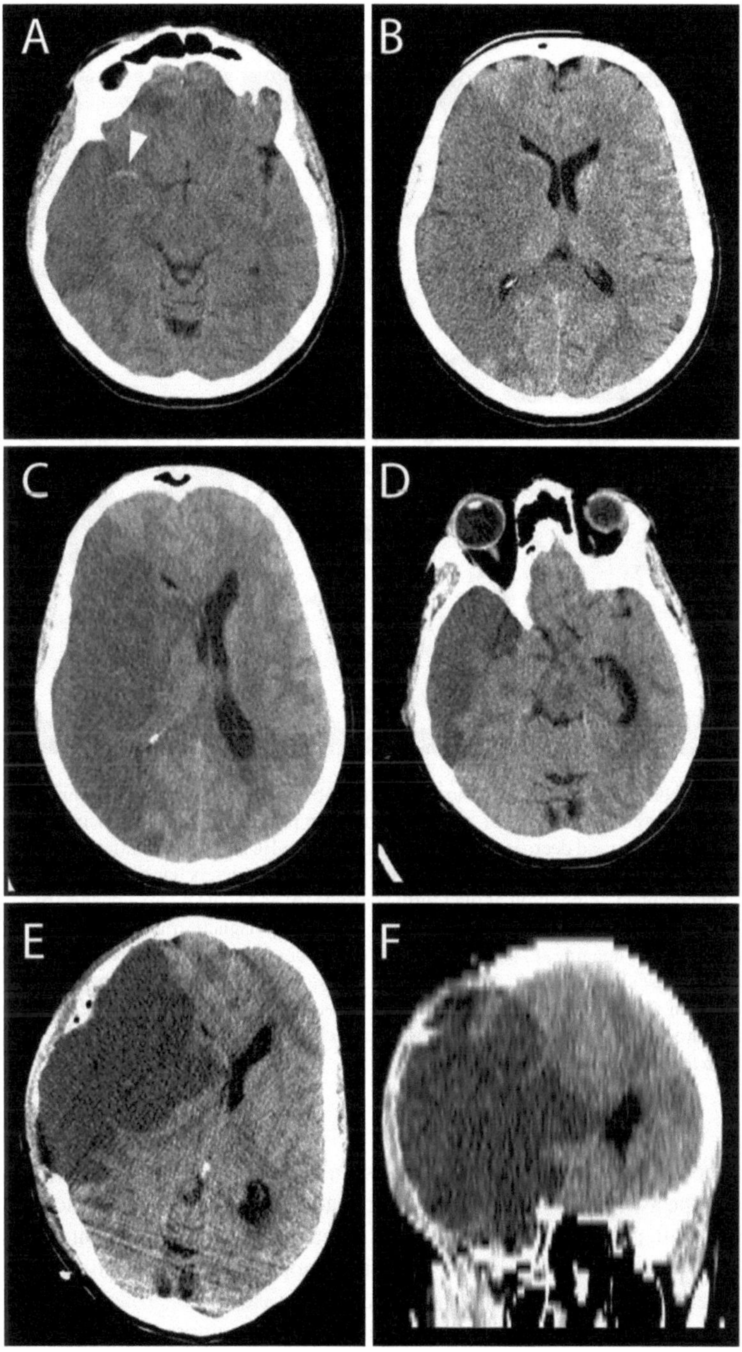

Fig. 2.4 Decompressive craniectomy for malignant MCA stroke. This 48-year-old female experienced left hemiparesis following a cardiac procedure on cardiorespiratory bypass. (**a**) A hyperdense right MCA sign was observed (*arrow*) with MCA stroke. (**b**) The entire right MCA distribution was involved with diffuse hypodensity. The next day the patient had an acute loss of consciousness with a Cushing's response and dilated right pupil. (**c**, **d**) Urgent CT scan revealed malignant edema with midline shift (**c**) and uncal herniation (**d**). (**e**, **f**) Emergent decompressive craniectomy, subtemporal decompression, and duroplasty were undertaken to allow external herniation of edematous brain

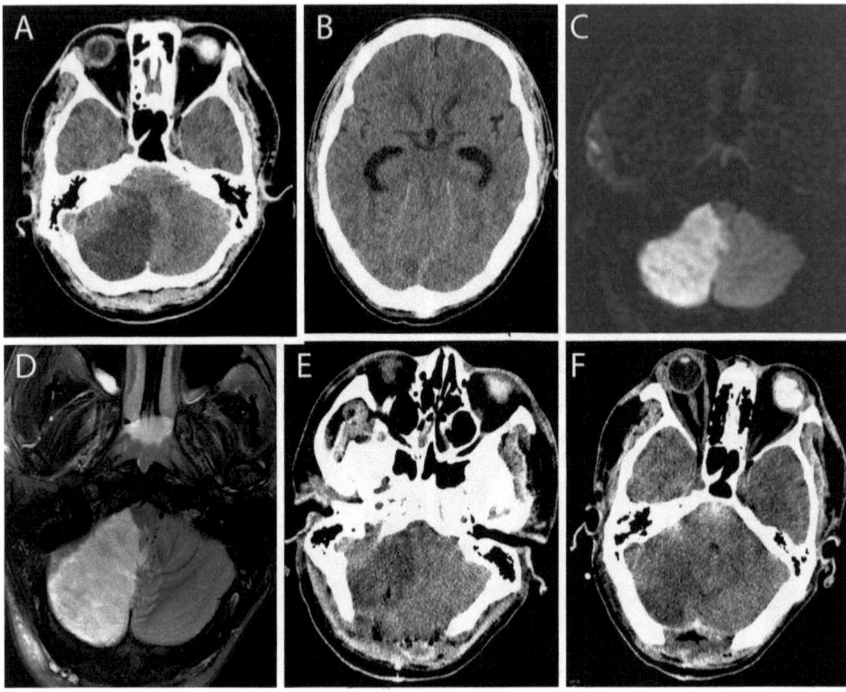

Fig. 2.5 Cerebellar stroke may lead to edema and obstructive hydrocephalus. This 27-year-old man with a history of an intracardiac arteriovenous shunt presented with a 4-h history of headache, nausea, and vomiting and decreased level of consciousness. He presented in extremis with bilateral fixed pupils and no motor response. (**a**, **b**) A posterior inferior cerebellar artery infarct was diagnosed with complete obliteration of the fourth ventricle (**a**) and acute hydrocephalus (**b**). (**c**, **d**) An emergent extraventricular drain was inserted followed by magnetic resonance imaging (MRI) diffusion-weighted imaging (**c**) and T2 imaging (**d**) that confirmed a completed stroke as hyperintensity in the vascular distribution. (**e**, **f**) The patient was taken for posterior fossa decompression and expansion duroplasty. Edematous, infracted tissue was resected from the cerebellum during the procedure to facilitate closure. The fourth ventricle was patent after the case (**f**). The patient went on to make a full neurological recovery

spontaneous cases are classified as strokes. In all three cases the acute management relates to the avoidance of secondary deterioration to preserve neurological function remaining after the primary injury. The common etiologies of spontaneous ICH include hypertensive hemorrhages, amyloid angiopathy, coagulopathy or anticoagulation therapy-related hemorrhages, bleeding into ischemic stroke, bleeding from ruptured vascular malformations, and bleeding from tumors. As discussed in the introduction there is no convincing evidence for an ischemic penumbra surrounding an ICH, although there may be a metabolic penumbra that promotes cell death in the subacute phase [54]. Generally, secondary deterioration in ICH relates to enlargement of the clot and compression of adjacent structures. Therefore the central tenet of treatment of ICH is prevention of clot expansion and treatment of the underlying

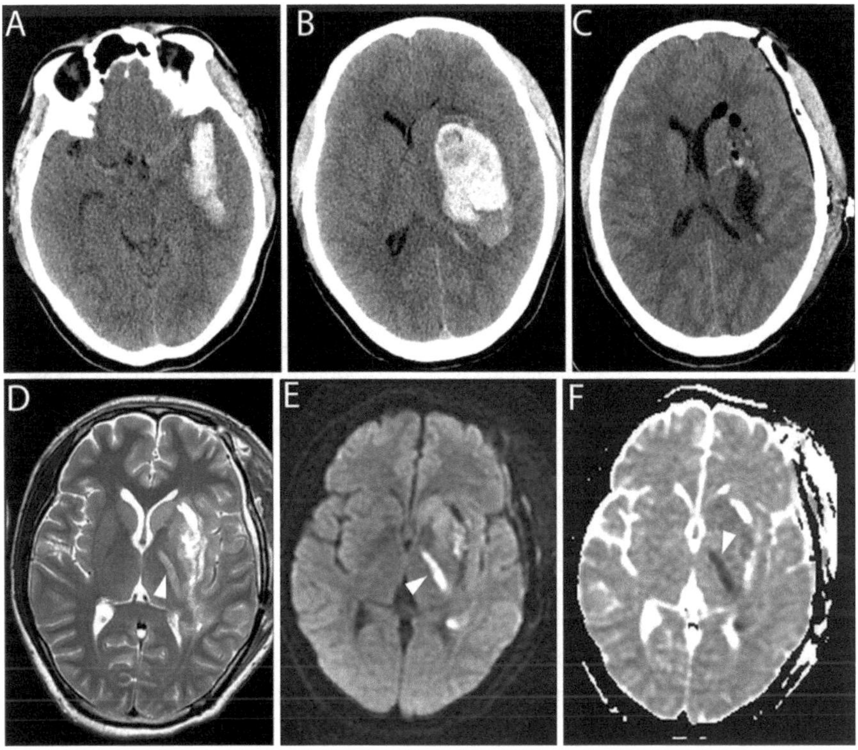

Fig. 2.6 Intracerebral hemorrhage requiring surgical evacuation. This 37-year-old male presented with acute onset headache, nausea and vomiting, aphasia, and progressive right hemiparesis. (**a**, **b**) CT revealed a left putamenal hemorrhage with heterogeneous densities of blood. The hematoma was found to come within 1 cm of the cortical surface (**a**) and was deemed a surgical lesion. (**c**) Emergent craniotomy was performed and the entire hematoma was removed. (**d**–**f**) The patient had no clinical improvement following surgery. Delayed MRI revealed persistent edema in tissue adjacent to the hematoma cavity with hyperintensity in the posterior limb of the internal capsule (*arrow*). Diffusion-weighted imaging (**e**) and apparent diffusion coefficient map (**f**) confirmed the stroke (*arrow* (**e**) and (**f**))

source of bleeding and/or decompression of the hematoma in cases where ICP is elevated.

Spontaneous SAH is generally related to ruptured brain aneurysms (about 85 % of cases), a group of idiopathic cases (about two-thirds being benign perimesencephalic SAH) and occasionally from arteriovenous or other brain vascular malformations. In aneurysmal SAH, there is a risk of re-hemorrhage with clot expansion and a risk of delayed cerebral ischemia during which time there is an increased risk of stroke [55]. Hence, treatment of SAH involves rapid diagnosis and treatment of the source of bleeding and treatment of aneurysmal SAH involves prevention and treatment of angiographic vasospasm, which is the main cause of delayed cerebral ischemia.

IVH is generally related to a preexisting ICH or SAH with extension into the ventricles and is associated with worse outcome in the case of ICH [56]. In IVH, deterioration relates to obstructive or communicating hydrocephalus and elevated ICP necessitating cerebral spinal fluid diversion [57]. The rest of the chapter will focus on the treatment of ICH.

Acute Medical Management

Principles of acute medical management for ICH are similar to those discussed for acute ischemic stroke with the exception of major differences in blood pressure management and the addition of managing anticoagulant reversal in the acute phase. Evidence-based guidelines for the management of ICH have been assembled and endorsed by the AHA and collaborating groups [58].

Acute blood pressure management is important in avoiding early re-hemorrhage in hypertensive ICH. The AHA guidelines recommend reduction of the blood pressure as follows: aggressive reduction of blood pressure if systolic >200 mmHg or mean arterial pressure >150 mmHg, reducing blood pressure to target a cerebral perfusion pressure of >60–80 mmHg in cases where systolic pressure is >180 mmHg or mean arterial pressure >130 mmHg and there is evidence of elevated ICP, and reduce blood pressure modestly to systolic ~160 mmHg and mean arterial pressure ~110 mmHg in cases where systolic pressure is >180 mmHg or mean arterial pressure is >130 mmHg and there is no evidence of raised ICP. There are several intravenous, oral, and transdermal agents that can be used for acute lowering of blood pressure including beta-blockers like labetalol or metoprolol, calcium channel blockers like nicardipine, and smooth muscle relaxants like hydralazine or nitrates.

For patients presenting with anticoagulant or fibrinolytic therapy-related hemorrhage, urgent reversal of the underlying therapy is necessary to prevent rebleeding. Patients treated with warfarin are deficient in vitamin K dependent clotting factors, therefore urgent replacement of these factors is required to reverse the elevated INR associated with warfarin. Vitamin K can be administered with fresh frozen plasma or with one of several different prothrombin complex concentrate products [59]. Patients on heparin receive protamine reversal. Antiplatelet therapy is associated with worsened outcomes following ICH [60]. Replacement of platelets does not seem to change outcome in these patients [60]. However, platelet replacement may be considered before, during, and after surgical procedures to avoid bleeding in these cases. Patients who have undergone fibrinolysis and subsequently suffer an ICH are at an elevated risk for mortality and poor neurologic outcome. Replacement of deficient coagulation factors is recommended in these cases but there is little evidence to guide management. Administration of recombinant factor VIIa is no longer recommended for the treatment of ICH due to a lack of improvement in functional outcome and an elevated risk of thromboembolic complications in the randomized controlled trial in ICH [61].

Surgical Treatment of Intracerebral Hemorrhage

Surgery for ICH involves removal of the hematoma with or without treatment of the primary source of bleeding. Several approaches to hematoma evacuation have been studied including open approaches via craniotomy (Fig. 2.6) [62], endoscopic approaches [63], and stereotactic needle aspiration with or without thrombolytic infusion into the hematoma [64–67]. Superiority of one technique compared to another has not been shown. Hematoma evacuation can be achieved with excellent results using each approach; however, re-accumulation of hematoma or re-bleeding is a common postoperative complication and is associated with further neurological decline [68].

Timing of surgery for ICH is a controversial topic as surgery has been proposed in the ultra-early timeframe (<7 h) out to 72 h following the ictus. In a randomized trial of ultra-early ICH evacuation, craniotomy performed <4 h from the onset of symptoms demonstrated no significant difference between surgery and conservative management with a trend towards higher mortality in surgery than conservative arms related to re-bleeding in the surgical arm [68]. Therefore, ultra-early evacuation must be done with the recognition that re-bleeding is a common and potentially detrimental complication. For surgery in the 8–12 h timeframe, a randomized controlled trial of 20 patients treated with surgery or conservative therapy following ICH showed no difference in mortality and a minor improvement in functional status for surgical patients in secondary analyses [69]. Surgery at later time points has not shown an advantage over conservative therapy in controlled studies [70, 71]. The International Surgical Trial in Intracerebral Haemorrhage (STICH) trial was a randomized trial comparing surgical evacuation within 24 h of randomization with initial conservative management [72]. Surgery occurred at a mean of 30 h after symptom onset in the initial surgery arm. In the conservative management arm, 26 % of patients went on to surgery at a mean time of 60 h following symptom onset. There was no overall difference in outcome between the two groups; however, in subgroup analysis, there was a benefit for patients with hemorrhage that presented less than 1 cm from the cortical surface. Further clinical trials will be required to establish definitive evidence to support this finding. Analysis of individual patient data from eight trials of surgery for ICH suggested that patients operated within 8 h of ictus, with hematoma volumes between 20 and 50 mL, Glasgow Coma Scores of 9–12 or age between 50 and 69, might benefit from surgery [73]. ICH location plays a key role in decision-making [74]. Approximately 80 % of hemorrhages occur in the supratentorial compartment and 20 % in the infratentorial compartment with site-specific incidence estimated to be 34 % in the putamen, 33 % in the thalamus, 14 % in lobar locations, 9 % in the brainstem, 7 % in the cerebellum, and 2 % in the caudate nucleus [74]. Surgery for supratentorial ICH is discussed above. Spontaneous hemorrhage in the brainstem carries a high risk of mortality and has no proven benefit from surgical evacuation in the acute phase [74, 75]. Cerebellar hemorrhages are unique because there is a high risk of brainstem compression, hydrocephalus, and hemorrhage in this location, carrying a relatively high mortality rate [76, 77]. A retrospective review of patients presenting with cerebellar hematoma, where cases were separated by ventricular compression and hydrocephalus, concluded that there may have been a benefit favoring surgery in cases where the fourth ventricle was compressed or obliterated and there was evidence of hydrocephalus [75]. Hematoma size >3 cm was a predictive factor for poor outcome in all cases of cerebellar hemorrhage regardless of presenting neurological status; whereas hematoma size <3 cm in patients with Glasgow Coma Scale (GCS) 13–15 was not a predictor of poor outcome [78]. Kobayashi and colleagues proposed a management protocol for cerebellar hemorrhage based on a retrospective review of 101 patients where patients were managed with conservative therapy when GCS was 13–15 and hematoma <40 mm, surgery for patients with GCS <13 and/or hematoma >40 mm and palliation for patients lacking brainstem reflexes [79]. This protocol was tested on 49 prospectively collected patients by the same group and emulated in another prospective case series from Italy and demonstrated acceptable outcomes [79, 80]. Data from these studies and other noncontrolled studies have been applied to the AHA guidelines that recommend surgical evacuation of cerebellar hematomas >3 cm or in cases with hydrocephalus or brainstem compression.

Summary

Treatment of ischemic stroke currently is limited to rtPA and decompressive craniectomy in selected cases and to supportive medical care. Hemorrhagic strokes are surgically evacuated in some cases, and the offending aneurysm or source of hemorrhage obliterated. Outcomes overall remain suboptimal (see Chap. 1). Stroke remains among the leading causes of disability. Reducing this likely requires addressing underlying risk factors as well as developing better acute treatments than those reviewed here and understanding and developing better rehabilitation strategies, as discussed in Chap. 16.

References

1. Thorvaldsen P, Asplund K, Kuulasmaa K, Rajakangas AM, Schroll M. Stroke incidence, case fatality, and mortality in the WHO MONICA project. World Health Organization Monitoring Trends and Determinants in Cardiovascular Disease. Stroke. 1995;26(3):361–7.
2. Baron JC. Perfusion thresholds in human cerebral ischemia: historical perspective and therapeutic implications. Cerebrovasc Dis. 2001;11 Suppl 1:2–8.
3. Mayer SA, Rincon F. Treatment of intracerebral haemorrhage. Lancet Neurol. 2005;4(10):662–72.
4. Zaleska MM, Mercado ML, Chavez J, Feuerstein GZ, Pangalos MN, Wood A. The development of stroke therapeutics: promising mechanisms and translational challenges. Neuropharmacology. 2009;56(2):329–41.

5. Davis SM, Broderick J, Hennerici M, Brun NC, Diringer MN, Mayer SA, et al. Hematoma growth is a determinant of mortality and poor outcome after intracerebral hemorrhage. Neurology. 2006;66(8):1175–81.
6. van der Worp HB, van Gijn J. Clinical practice. Acute ischemic stroke. N Engl J Med. 2007;357(6):572–9.
7. Saver JL. Time is brain–quantified. Stroke. 2006;37(1):263–6.
8. Adams Jr HP, del Zoppo G, Alberts MJ, Bhatt DL, Brass L, Furlan A, et al. Guidelines for the early management of adults with ischemic stroke: a guideline from the American Heart Association/American Stroke Association Stroke Council, Clinical Cardiology Council, Cardiovascular Radiology and Intervention Council, and the Atherosclerotic Peripheral Vascular Disease and Quality of Care Outcomes in Research Interdisciplinary Working Groups: the American Academy of Neurology affirms the value of this guideline as an educational tool for neurologists. Stroke. 2007;38(5):1655–711.
9. Hilker R, Poetter C, Findeisen N, Sobesky J, Jacobs A, Neveling M, et al. Nosocomial pneumonia after acute stroke: implications for neurological intensive care medicine. Stroke. 2003;34(4):975–81.
10. Leonardi-Bee J, Bath PM, Phillips SJ, Sandercock PA; IST Collaborative Group. Blood pressure and clinical outcomes in the International Stroke Trial. Stroke. 2002;33(5):1315–20.
11. Stead LG, Gilmore RM, Decker WW, Weaver AL, Brown Jr RD. Initial emergency department blood pressure as predictor of survival after acute ischemic stroke. Neurology. 2005;65(8):1179–83.
12. Tanne D, Kasner SE, Demchuk AM, Koren-Morag N, Hanson S, Grond M, et al. Markers of increased risk of intracerebral hemorrhage after intravenous recombinant tissue plasminogen activator therapy for acute ischemic stroke in clinical practice: the Multicenter rt-PA Stroke Survey. Circulation. 2002;105(14):1679–85.
13. Markus HS. Cerebral perfusion and stroke. J Neurol Neurosurg Psychiatry. 2004;75(3):353–61.
14. Greer DM, Funk SE, Reaven NL, Ouzounelli M, Uman GC. Impact of fever on outcome in patients with stroke and neurologic injury: a comprehensive meta-analysis. Stroke. 2008;39(11):3029–35.
15. Hemmen TM, Lyden PD. Induced hypothermia for acute stroke. Stroke. 2007;38(2 Suppl):794–9.
16. Malouf R, Brust JC. Hypoglycemia: causes, neurological manifestations, and outcome. Ann Neurol. 1985;17(5):421–30.
17. Yong M, Kaste M. Dynamic of hyperglycemia as a predictor of stroke outcome in the ECASS-II trial. Stroke. 2008;39(10):2749–55.
18. Wintermark M, Reichhart M, Cuisenaire O, Maeder P, Thiran JP, Schnyder P, et al. Comparison of admission perfusion computed tomography and qualitative diffusion- and perfusion-weighted magnetic resonance imaging in acute stroke patients. Stroke. 2002;33(8):2025–31.
19. Chalela JA, Alsop DC, Gonzalez-Atavales JB, Maldjian JA, Kasner SE, Detre JA. Magnetic resonance perfusion imaging in acute ischemic stroke using continuous arterial spin labeling. Stroke. 2000;31(3):680–7.
20. NINDS tSTSG. Tissue plasminogen activator for acute ischemic stroke. The National Institute of Neurological Disorders and Stroke rt-PA Stroke Study Group. N Engl J Med. 1995;333(24):1581–7.
21. Kwiatkowski TG, Libman RB, Frankel M, Tilley BC, Morgenstern LB, Lu M, et al. Effects of tissue plasminogen activator for acute ischemic stroke at one year. National Institute of Neurological Disorders and Stroke Recombinant Tissue Plasminogen Activator Stroke Study Group. N Engl J Med. 1999;340(23):1781–7.
22. Hacke W, Kaste M, Bluhmki E, Brozman M, Dávalos A, Guidetti D, et al.; ECASS Investigators. Thrombolysis with alteplase 3 to 4.5 hours after acute ischemic stroke. N Engl J Med. 2008;359(13):1317–29.

23. Hacke W, Albers G, Al-Rawi Y, Bogousslavsky J, Davalos A, Eliasziw M, et al.; DIAS Study Group. The Desmoteplase in Acute Ischemic Stroke Trial (DIAS): a phase II MRI-based 9-hour window acute stroke thrombolysis trial with intravenous desmoteplase. Stroke. 2005;36(1):66–73.
24. Tanswell P, Modi N, Combs D, Danays T. Pharmacokinetics and pharmacodynamics of tenecteplase in fibrinolytic therapy of acute myocardial infarction. Clin Pharmacokinet. 2002;41(15):1229–45.
25. Haley EC Jr, Lyden PD, Johnston KC, Hemmen TM; TNK in Stroke Investigators. A pilot dose-escalation safety study of tenecteplase in acute ischemic stroke. Stroke. 2005;36(3):607–12.
26. Furlan A, Higashida R, Wechsler L, Gent M, Rowley H, Kase C, et al. Intra-arterial prourokinase for acute ischemic stroke. The PROACT II study: a randomized controlled trial. Prolyse in Acute Cerebral Thromboembolism. JAMA. 1999;282(21):2003–11.
27. Ogawa A, Mori E, Minematsu K, Taki W, Takahashi A, Nemoto S, et al.; MELT Japan Study Group. Randomized trial of intraarterial infusion of urokinase within 6 hours of middle cerebral artery stroke: the middle cerebral artery embolism local fibrinolytic intervention trial (MELT) Japan. Stroke. 2007;38(10):2633–39.
28. Albers GW, Amarenco P, Easton JD, Sacco RL, Teal P; American College of Chest Physicians. Antithrombotic and thrombolytic therapy for ischemic stroke: American College of Chest Physicians Evidence-Based Clinical Practice Guidelines (8th Edition). Chest. 2008;133 (6 Suppl):630S–69S.
29. Burns TC, Rodriguez GJ, Patel S, Hussein HM, Georgiadis AL, Lakshminarayan K, et al. Endovascular interventions following intravenous thrombolysis may improve survival and recovery in patients with acute ischemic stroke: a case–control study. AJNR Am J Neuroradiol. 2008;29(10):1918–24.
30. IMS II Trial Investigators. The Interventional Management of Stroke (IMS) II Study. Stroke. 2007;38(7):2127–35.
31. Flaherty ML, Woo D, Kissela B, Jauch E, Pancioli A, Carrozzella J, et al. Combined IV and intra-arterial thrombolysis for acute ischemic stroke. Neurology. 2005;64(2):386–8.
32. Lee KY, Kim DI, Kim SH, Lee SI, Chung HW, Shim YW, et al. Sequential combination of intravenous recombinant tissue plasminogen activator and intra-arterial urokinase in acute ischemic stroke. AJNR Am J Neuroradiol. 2004;25(9):1470–5.
33. Shaltoni HM, Albright KC, Gonzales NR, Weir RU, Khaja AM, Sugg RM, et al. Is intra-arterial thrombolysis safe after full-dose intravenous recombinant tissue plasminogen activator for acute ischemic stroke? Stroke. 2007;38(1):80–4.
34. Kim DJ, Kim DI, Kim SH, Lee KY, Heo JH, Han SW. Rescue localized intra-arterial thrombolysis for hyperacute MCA ischemic stroke patients after early non-responsive intravenous tissue plasminogen activator therapy. Neuroradiology. 2005;47(8):616–21.
35. Stead LG, Gilmore RM, Bellolio MF, Rabinstein AA, Decker WW. Percutaneous clot removal devices in acute ischemic stroke: a systematic review and meta-analysis. Arch Neurol. 2008;65(8):1024–30.
36. Leary MC, Saver JL, Gobin YP, Jahan R, Duckwiler GR, Vinuela F, et al. Beyond tissue plasminogen activator: mechanical intervention in acute stroke. Ann Emerg Med. 2003;41(6):838–46.
37. Levy EI, Ecker RD, Horowitz MB, Gupta R, Hanel RA, Sauvageau E, et al. Stent-assisted intracranial recanalization for acute stroke: early results. Neurosurgery. 2006;58(3):458–63. discussion 458–463.
38. Nogueira RG, Schwamm LH, Hirsch JA. Endovascular approaches to acute stroke, part 1: drugs, devices, and data. AJNR Am J Neuroradiol. 2009;30(4):649–61.
39. Smith WS. Safety of mechanical thrombectomy and intravenous tissue plasminogen activator in acute ischemic stroke. Results of the multi Mechanical Embolus Removal in Cerebral Ischemia (MERCI) trial, part I. AJNR Am J Neuroradiol. 2006;27(6):1177–82.

40. Smith WS, Sung G, Saver J, Budzik R, Duckwiler G, Liebeskind DS, et al.; Multi MERCI Investigators. Mechanical thrombectomy for acute ischemic stroke: final results of the Multi MERCI trial. Stroke. 2008;39(4):1205–12.
41. Zeumer H, Ringelstein EB, Hacke W. [Vascular recanalization procedure in interventional neuroradiology]. Rofo. 1983;139(5):467–75.
42. Hacke W, Zeumer H, Ferbert A, Brückmann H, del Zoppo GJ. Intra-arterial thrombolytic therapy improves outcome in patients with acute vertebrobasilar occlusive disease. Stroke. 1988;19(10):1216–22.
43. Jung S, Mono ML, Fischer U, Galimanis A, Findling O, De Marchis GM, et al. Three-month and long-term outcomes and their predictors in acute basilar artery occlusion treated with intra-arterial thrombolysis. Stroke. 2011;42(7):1946–51.
44. Bergui M, Stura G, Daniele D, Cerrato P, Berardino M, Bradac GB. Mechanical thrombolysis in ischemic stroke attributable to basilar artery occlusion as first-line treatment. Stroke. 2006;37(1):145–50.
45. Xavier AR, Tiwari A, Purai N, Rayes M, Pandey P, Kansara A, et al. Safety and efficacy of intracranial stenting for acute ischemic stroke beyond 8 h of symptom onset. J Neurointerv Surg. 2012;4(2):94–100.
46. Lutsep HL, Rymer MM, Nesbit GM. Vertebrobasilar revascularization rates and outcomes in the MERCI and multi-MERCI trials. J Stroke Cerebrovasc Dis. 2008;17(2):55–7.
47. Hacke W, Schwab S, Horn M, Spranger M, De Georgia M, von Kummer R. 'Malignant' middle cerebral artery territory infarction: clinical course and prognostic signs. Arch Neurol. 1996;53(4):309–15.
48. Walz B, Zimmermann C, Böttger S, Haberl RL. Prognosis of patients after hemicraniectomy in malignant middle cerebral artery infarction. J Neurol. 2002;249(9):1183–90.
49. Vahedi K, Vicaut E, Mateo J, Kurtz A, Orabi M, Guichard JP, et al.; DECIMAL Investigators. Sequential-design, multicenter, randomized, controlled trial of early decompressive craniectomy in malignant middle cerebral artery infarction (DECIMAL Trial). Stroke. 2007;38(9):2506–17.
50. Gupta R, Connolly ES, Mayer S, Elkind MS. Hemicraniectomy for massive middle cerebral artery territory infarction: a systematic review. Stroke. 2004;35(2):539–43.
51. Carter BS, Ogilvy CS, Candia GJ, Rosas HD, Buonanno F. One-year outcome after decompressive surgery for massive nondominant hemispheric infarction. Neurosurgery. 1997;40(6):1168–75. discussion 1175–1166.
52. Leonhardt G, Wilhelm H, Doerfler A, Ehrenfeld CE, Schoch B, Rauhut F, et al. Clinical outcome and neuropsychological deficits after right decompressive hemicraniectomy in MCA infarction. J Neurol. 2002;249(10):1433–40.
53. Chen HJ, Lee TC, Wei CP. Treatment of cerebellar infarction by decompressive suboccipital craniectomy. Stroke. 1992;23(7):957–61.
54. Schellinger PD, Fiebach JB, Hoffmann K, Becker K, Orakcioglu B, Kollmar R, et al. Stroke MRI in intracerebral hemorrhage: is there a perihemorrhagic penumbra? Stroke. 2003;34(7):1674–9.
55. Macdonald RL, Pluta RM, Zhang JH. Cerebral vasospasm after subarachnoid hemorrhage: the emerging revolution. Nat Clin Pract Neurol. 2007;3(5):256–63.
56. Tuhrim S, Horowitz DR, Sacher M, Godbold JH. Volume of ventricular blood is an important determinant of outcome in supratentorial intracerebral hemorrhage. Crit Care Med. 1999;27(3):617–21.
57. Bhattathiri PS, Gregson B, Prasad KS, Mendelow AD; STICH Investigators. Intraventricular hemorrhage and hydrocephalus after spontaneous intracerebral hemorrhage: results from the STICH trial. Acta Neurochir Suppl. 2006;96:65–8.
58. Broderick J, Connolly S, Feldmann E, Hanley D, Kase C, Krieger D, et al. Guidelines for the management of spontaneous intracerebral hemorrhage in adults: 2007 update—a guideline from the American Heart Association/American Stroke Association Stroke Council, High

Blood Pressure Research Council, and the Quality of Care and Outcomes in Research Interdisciplinary Working Group. Stroke. 2007;38(6):2001–23.
59. Lubetsky A, Hoffman R, Zimlichman R, Eldor A, Zvi J, Kostenko V, et al. Efficacy and safety of a prothrombin complex concentrate (Octaplex) for rapid reversal of oral anticoagulation. Thromb Res. 2004;113(6):371–8.
60. Creutzfeldt CJ, Weinstein JR, Longstreth Jr WT, Becker KJ, McPharlin TO, Tirschwell DL. Prior antiplatelet therapy, platelet infusion therapy, and outcome after intracerebral hemorrhage. J Stroke Cerebrovasc Dis. 2009;18(3):221–8.
61. Mayer SA, Brun NC, Begtrup K, Broderick J, Davis S, Diringer MN, et al.; FAST Trial Investigators. Efficacy and safety of recombinant activated factor VII for acute intracerebral hemorrhage. N Engl J Med. 2008;358(20):2127–37.
62. Rabinstein AA, Atkinson JL, Wijdicks EF. Emergency craniotomy in patients worsening due to expanded cerebral hematoma: to what purpose? Neurology. 2002;58(9):1367–72.
63. Kuo LT, Chen CM, Li CH, Tsai JC, Chiu HC, Liu LC, et al. Early endoscope-assisted hematoma evacuation in patients with supratentorial intracerebral hemorrhage: case selection, surgical technique, and long-term results. Neurosurg Focus. 2011;30(4):E9.
64. Zhou H, Zhang Y, Liu L, Han X, Tao Y, Tang Y, et al. A prospective controlled study: minimally invasive stereotactic puncture therapy versus conventional craniotomy in the treatment of acute intracerebral hemorrhage. BMC Neurol. 2011;11:76.
65. Zhou H, Zhang Y, Liu L, Huang Y, Tang Y, Su J, et al. Minimally invasive stereotactic puncture and thrombolysis therapy improves long-term outcome after acute intracerebral hemorrhage. J Neurol. 2011;258(4):661–9.
66. Carhuapoma JR, Barrett RJ, Keyl PM, Hanley DF, Johnson RR. Stereotactic aspiration-thrombolysis of intracerebral hemorrhage and its impact on perihematoma brain edema. Neurocrit Care. 2008;8(3):322–9.
67. Umebayashi D, Mandai A, Osaka Y, Nakahara Y, Tenjin H. Effects and complications of stereotactic aspiration for spontaneous intracerebral hemorrhage. Neurol Med Chir (Tokyo). 2010;50(7):538–44.
68. Morgenstern LB, Demchuk AM, Kim DH, Frankowski RF, Grotta JC. Rebleeding leads to poor outcome in ultra-early craniotomy for intracerebral hemorrhage. Neurology. 2001;56(10):1294–9.
69. Zuccarello M, Brott T, Derex L, Kothari R, Sauerbeck L, Tew J, et al. Early surgical treatment for supratentorial intracerebral hemorrhage: a randomized feasibility study. Stroke. 1999;30(9):1833–9.
70. Tan SH, Ng PY, Yeo TT, Wong SH, Ong PL, Venketasubramanian N. Hypertensive basal ganglia hemorrhage: a prospective study comparing surgical and nonsurgical management. Surg Neurol. 2001;56(5):287–92. discussion 292–283.
71. Juvela S, Heiskanen O, Poranen A, Valtonen S, Kuurne T, Kaste M, et al. The treatment of spontaneous intracerebral hemorrhage. A prospective randomized trial of surgical and conservative treatment. J Neurosurg. 1989;70(5):755–8.
72. Mendelow AD, Gregson BA, Fernandes HM, Murray GD, Teasdale GM, Hope DT, et al. Early surgery versus initial conservative treatment in patients with spontaneous supratentorial intracerebral haematomas in the International Surgical Trial in Intracerebral Haemorrhage (STICH): a randomised trial. Lancet. 2005;365(9457):387–97.
73. Gregson BA, Broderick JP, Auer LM, Batjer H, Chen XC, Juvela S, et al. Individual patient data subgroup meta-analysis of surgery for spontaneous supratentorial intracerebral hemorrhage. Stroke. 2012;43:1496–504.
74. Inagawa T, Ohbayashi N, Takechi A, Shibukawa M, Yahara K. Primary intracerebral hemorrhage in Izumo City, Japan: incidence rates and outcome in relation to the site of hemorrhage. Neurosurgery. 2003;53(6):1283–97. discussion 1297–1288.
75. Da Pian R, Bazzan A, Pasqualin A. Surgical versus medical treatment of spontaneous posterior fossa haematomas: a cooperative study on 205 cases. Neurol Res. 1984;6(3):145–51.

76. Barbato G, Moul DE, Schwartz P, Rosenthal NE, Oren DA. Spontaneous eye blink rate in winter seasonal affective disorder. Psychiatry Res. 1993;47(1):79–85.
77. Kirollos RW, Tyagi AK, Ross SA, van Hille PT, Marks PV. Management of spontaneous cerebellar hematomas: a prospective treatment protocol. Neurosurgery. 2001;49(6):1378–86. discussion 1386–1377.
78. Cohen ZR, Ram Z, Knoller N, Peles E, Hadani M. Management and outcome of non-traumatic cerebellar haemorrhage. Cerebrovasc Dis. 2002;14(3–4):207–13.
79. Kobayashi S, Sato A, Kageyama Y, Nakamura H, Watanabe Y, Yamaura A. Treatment of hypertensive cerebellar hemorrhage–surgical or conservative management? Neurosurgery. 1994;34(2):246–50. discussion 250–241.
80. Luparello V, Canavero S. Treatment of hypertensive cerebellar hemorrhage–surgical or conservative management? Neurosurgery. 1995;37(3):552–3.

Chapter 3
Sensorimotor Control After Stroke

W. Richard Staines, David A.E. Bolton, and William E. McIlroy

Introduction

Approximately two-thirds of stroke survivors have residual neurological deficits that impair function and approximately 50 % are left with disabilities that render them dependent on others for activities of daily living. While the economic burden of stroke on the health care system is substantial (approximately $2.7 billion and 3 million hospital days annually) [1], the human cost to stroke survivors, and their families is incalculable. Despite improvements in acute stroke care, understanding of recovery processes is still relatively underdeveloped and there is a need for new innovative approaches to improve rehabilitation, promote recovery, lessen disability, and prevent subsequent stroke.

The types of sensorimotor impairments that occur after stroke are heterogeneous and vary with the particular regions of the central nervous system (CNS) that have sustained damage. Among the more common are physical impairments in upper limb use and in walking. Upper limb dysfunction remains an important problem for many stroke survivors. Only 5 % of adult stroke survivors regain full function of the upper limb and 20 % regain no functional use [2]. Advances in therapeutic approaches based on motor relearning principles (such as Constraint Induced Movement Therapy) have revealed the capacity to improve recovery in the subacute and chronic phases after stroke [3, 4]. However, further understanding of fundamental motor control mechanisms is critical to advancing the development of evidence-based, patient-specific intervention strategies.

It is clear that the integrity of the somatosensory system is important to motor recovery following stroke [5]. Previous work has revealed the importance of effective task-related sensory modulation for recovery of sensorimotor behavior following stroke-related brain injury [6]. The integration of structural and functional

W.R. Staines, Ph.D. (✉) • D.A.E. Bolton, Ph.D. • W.E. McIlroy, Ph.D.
Department of Kinesiology, University of Waterloo, Waterloo, ON, Canada
e-mail: rstaines@uwaterloo.ca

T.A. Schweizer and R.L. Macdonald (eds.), *The Behavioral Consequences of Stroke*,
DOI 10.1007/978-1-4614-7672-6_3, © Springer Science+Business Media New York 2014

neuroimaging techniques with electrophysiological and behavioral measurement is providing support for the hypothesis that it is not simply loss of sensory representation that leads to sensorimotor impairments but the much more subtle impairment in the ability to effectively regulate sensory information. The specific mechanisms and regions responsible for such modulation are not yet fully known. Traditionally, clinical determination of sensory status involves testing perceptual thresholds [7, 8] to peripheral stimuli. While such approaches reveal evidence of primary sensory loss, they typically do not reveal the capacity of the CNS to modulate afferent information. To achieve normal interaction with our environment, relevant sensory information must be identified and extracted from a vast array of concurrent inputs. Inability to do this will influence sensorimotor or perceptual ability, even if the primary sensory representation is intact. In the somatosensory system, such extraction of sensory information is achieved through highly selective mechanisms involving both inhibition of ascending afferent paths carrying task-irrelevant information—a phenomenon often termed "sensory gating"—and facilitation of task-relevant information [9–11]. Although the precise mechanisms remain unclear, interactions between the prefrontal cortex (PFC), thalamus, and primary sensory cortex appear to play an important role in this process [12–14].

There are reports in the literature describing sensory deficits following thalamic lesions [15, 16]. However, relatively few investigate alterations in cortical representation after thalamic stroke in humans. Remy and colleagues [17] reported seven patients with stroke in the ventroposterior thalamic nuclei. Positron emission tomography (PET) demonstrated resting hypoperfusion in the ipsilateral primary somatosensory cortex (SI), although SI was activated normally during vibration of the hand [17]. This characterizes primary representation but not the ability to modulate sensory inputs evident in tasks featuring multiple inputs as discussed previously.

We have shown that stroke-induced lesions involving the thalamus have a specific influence on the capacity to extract and gate sensory information [6]. Figure 3.1 shows data from an individual with a right thalamic lacunar infarction in the ventroposterolateral (VPL) nucleus that presented with mild left hemisensory loss, without evidence of clinical tactile extinction to light touch, and transient left homonymous hemianopia. Figure 3.1b illustrates two important observations regarding perceptual thresholds tested on the stroke-affected side. At 2 weeks poststroke, the patient was equally able to perceive vibratory stimuli applied to either thumb on the affected or unaffected limb. However, with the stroke-affected hand there was a significant increase in threshold detection with contralateral competing

Fig. 3.1 (continued) in the presence of contralateral vibration. (**c**) Median nerve SEPs from CP3 in the absence (*thick line*) and presence of vibration of the contralateral hand (*thin line*) at 2 weeks poststroke. The arrow indicates stimulus onset. (**d**) N20-P27 amplitudes with contralateral vibration during recovery expressed as a percentage of the resting amplitude. The *grey horizontal bar* represents the 95 % CI surrounding the mean task-related N20-P27 amplitude difference in controls ($n=15$). Reprinted with permission from Staines WR, Black SE, Graham SJ, McIlroy WE. Somatosensory gating and recovery from stroke involving the thalamus. *Stroke*, 2002, 33, 2642–2651

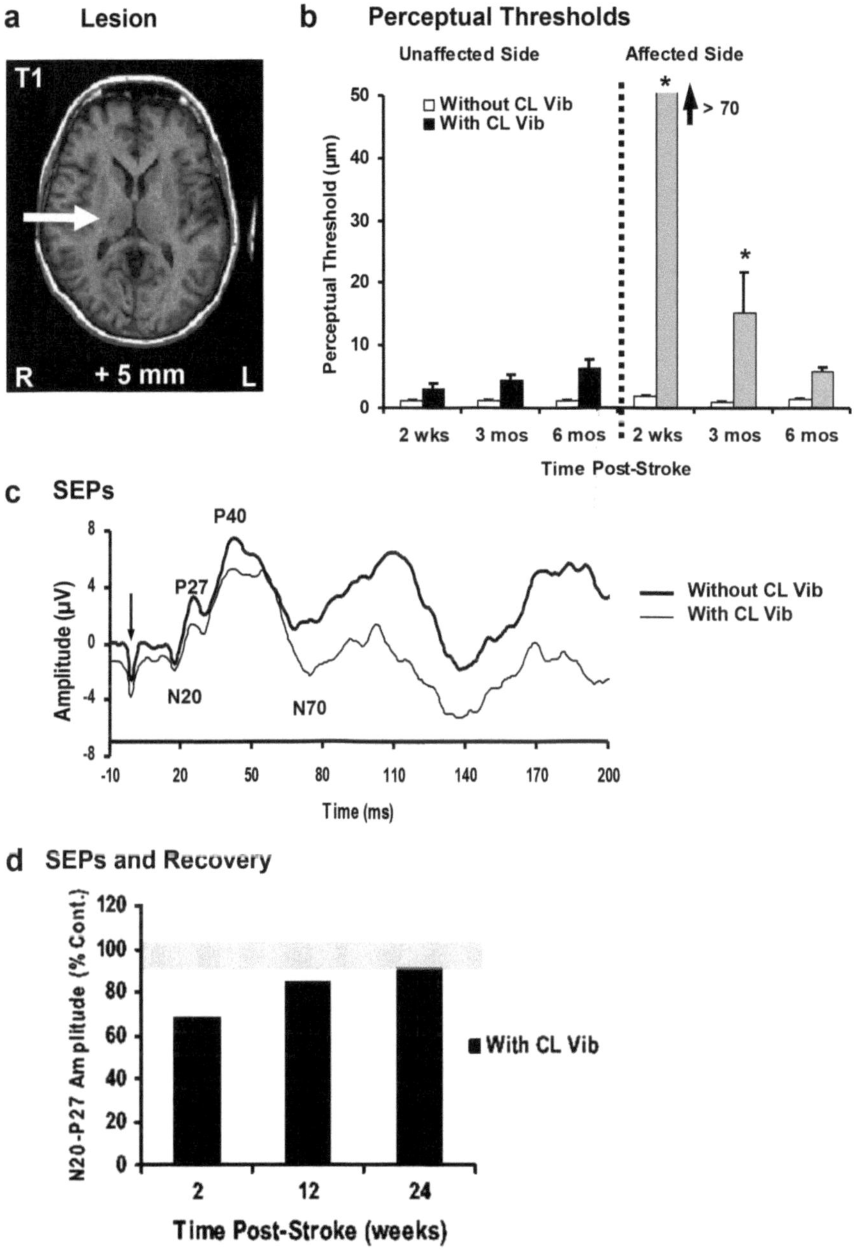

Fig. 3.1 Patient 1 (**a**) T1-weighted MRI scan illustrating the lesion location (*white arrow*) in the right ventroposterolateral (VPL) nucleus, 3 months post-stroke. Image location is 5 mm above the line adjoining the anterior and posterior commissures (AC-PC line). *R* right, *L* left. (**b**) Perceptual thresholds (PTs) to vibratory stimuli applied to the pad of the thumb in the absence (*open bars*) or presence (*filled bars*) of competing vibratory stimulation applied to the contralateral (CL) hand. Asterisks represent PTs exceeding the 95 % confidence interval (CI) for PTs of the unaffected hand

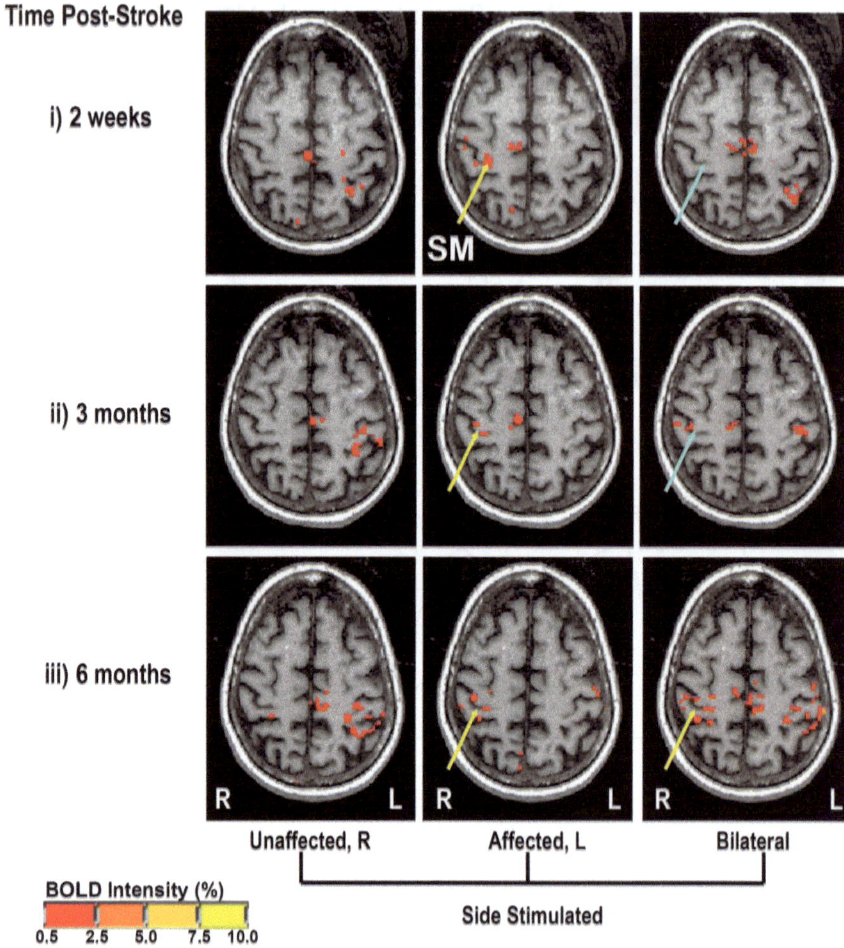

Fig. 3.2 Patient 1 fMRI scans (Talairach space, 47 mm above the AC-PC line) during tactile stimulation of the index finger on the right hand alone (Unaffected, R), the stroke-affected left hand alone (Affected, L) or both hands simultaneously (Bilateral). Each column represents repeats of the same scans at different times post-stroke: (*i*) 2 weeks, (*ii*) 3 months, and (*iii*) 6 months. *Yellow arrows* indicate the presence and *blue arrows* the absence of significant BOLD response in the area of SI. Reprinted with permission from Staines WR, Black SE, Graham SJ, McIlroy WE. Somatosensory gating and recovery from stroke involving the thalamus. *Stroke*, 2002, 33, 2642–2651

stimulation that improved markedly with recovery at 6 months post-stroke. Electrophysiological and functional neuroimaging measures paralleled these behavioral deficits. At 2 weeks post-stroke, N20-P27 somatosensory evoked potential (SEP) amplitude, representing the first reception of the somatosensory stimuli at the primary somatosensory cortex (S1) in response to median nerve stimulation, was attenuated in the presence of contralateral vibration compared to rest (Fig. 3.1c, thin vs. thick line). Similar to perceptual threshold differences under conditions of competing stimulation, task-related SEP amplitude differences resolved with recovery (Fig. 3.1).

This was also evident using functional magnetic resonance imaging (fMRI) conducted at similar time points (Fig. 3.2). At 2 weeks, unilateral stimulation of the stroke-affected hand significantly activated voxels in the area of contralateral SI (top panel). However, during bilateral stimulation there was no significant activation in the ipsilesional SI. As recovery progressed, significant activation within the same spatial regions of the ipsilesional SI was observed during both unilateral and bilateral stimulation. We showed that efficient task-related modulation of sensory inputs is an important factor contributing to recovery from thalamic stroke. Specifically, deficits in the ability to detect contralesional tactile stimuli when presented bilaterally are reflected by deficiencies in sensory gating mechanisms measured at the level of SI. The impact that disrupted modulation has on sensorimotor control and recovery after stroke is not clear, but has the potential to be substantial (Fig. 3.2).

Importance of PFC

The PFC influences distributed neural networks and is critically involved in the top-down modulation of sensory signals both at the unimodal sensory processing level, and in higher order association regions [12, 18]. This modulation has been demonstrated across different sensory modalities including visual, auditory, and somatosensory processing [12]. Numerous lines of evidence, including lesion studies in humans [19, 20] and imaging work [21] have established clear links between prefrontal activity, the suppression of distracting sensory information, and subsequent cognitive performance. In patients with frontal lesions for example, the loss of normal PFC function is associated with an impaired ability to filter out irrelevant distractor sensory information, and a loss of excitatory control over sensory pathways carrying task-relevant information even at early modality-specific cortical processing stages [12]. Such clinical findings relate to the extensive anatomical connections between prefrontal regions and distributed neural networks including sensory processing sites [22]. Indeed, PFC acts to suppress irrelevant information early in the processing stream, including at the earliest cortical stages of sensory processing and even at the thalamus prior to cortical entry [23–26]. Due to this capacity to gate out irrelevant sensory information, the PFC has a critical role in providing resistance to distraction and in sparing limited cognitive resources from becoming overwhelmed by a surplus of sensory input.

In animal models, cryogenic blockade of PFC results in larger sensory evoked responses at SI revealing tonic PFC inhibition over early somatosensory processing [27]. Additionally, studies in rats [23] and non-human primates [28] indicate that PFC modulates somatosensory signals via thalamic connections. Furthermore, evidence from humans with selective frontal lesions suggests inhibitory modulation of incoming somatosensory signals through direct fronto-parietal connections [25]. Such regulatory connections make the PFC an ideal candidate for gating irrelevant and boosting relevant information across all sensory modalities, including somatic sensation.

Temporary suppression of cortical target regions can be induced experimentally using repetitive transcranial magnetic stimulation (TMS). For instance, the effect of suppressing prefrontal activity on somatosensory or other sensory cortical processing can be addressed. In one study, TMS and electroencephalography (EEG) were combined to investigate the role of the dorsolateral prefrontal cortex (DLPFC) in regulating attention-based modulation of somatosensory signal processing [13]. Following the application of continuous theta burst stimulation over the right DLPFC, it was noted that attention-based modulation for an early modality-specific component of the EEG (P100) was significantly attenuated. This suggested that the DLPFC contributes to filtering irrelevant somatosensory information at early cortical processing stages, with the P100 most likely being generated in the secondary somatosensory cortex [29]. Moreover, the source of this attenuation resulted specifically from the loss of suppression of non-attended stimuli suggesting that the DLPFC contributed to sensory gating during this task to suppress irrelevant distractor information.

The impact of prefrontal regions on suppressing irrelevant distractor inputs has been corroborated by another TMS study. Hannula and colleagues noted that single-pulse TMS applied to the middle frontal gyrus at a specific time-point in the retention phase of a tactile working memory task resulted in significantly improved performance, particularly in the midst of tactile distractor stimuli [30]. Notably, these authors observed that this single pulse was associated with a reduction in SEP amplitude from the tactile distractor stimuli, which they proposed as support for the notion that the major mechanism for performance improvement in this paradigm was the suppression of irrelevant data. Interestingly, this group investigated the specificity of this link between the middle frontal gyrus to the SI by also having the tactile working memory task with visual interference. The TMS stimulation did not improve performance in this case suggesting that the inhibitory interference control was specific to somatosensory inputs [31]. Overall, damage to the PFC can impact sensory-related activity at modality-specific cortical areas and demonstrate a behavioral consequence.

What Factors Affect Sensorimotor Recovery?

Following a cerebral insult, such as a stroke, the CNS undergoes reorganization or plasticity during the process of functional recovery. However, the degree of recovery is variable [32] and the processes subserving this recovery are incompletely understood. Neurophysiological changes associated with recovery often begin very early after the onset of stroke (within 1–2 weeks) and may plateau between 2 and 3 months later, depending on the specific neurologic deficit. Historically, the recovery over the first 3 months is believed to be the most rapid. However, functional recovery has been shown to progress even years after the stroke event, albeit at a much slower rate. Several factors influence the rate and extent of recovery, most notably the initial severity of the stroke, which is typically quantified clinically or functionally, and the size and location of the lesion. Recovery can also be influenced by individual characteristics, particularly age, comorbidities, and activity patterns of the individual, specifically their participation in formal rehabilitation training. Questions remain about the mechanisms by which such improvement is mediated, however, rehabilitation training may be critical in shaping reorganization within the CNS.

In both the healthy and injured CNS, the development and training of motor skills leads to rapid cortical adaptation characterized by shifts in central representation to meet new task demands [5, 33–38]. During sensorimotor recovery from stroke, these normal training-related cortical adaptations can interact with and influence cortical reorganization associated with spontaneous recovery. Electrophysiological and functional imaging techniques, such as TMS, event-related cortical potentials, and fMRI allow for the sensitive quantification of CNS state and provide important experimental opportunities to reveal treatment-related short-term changes as a foundation for long-term stroke recovery.

For many deficits, the key to minimizing disability following stroke is rehabilitation to retrain patients to improve capacity. The range of these approaches is largely distinguished on the basis of the specific deficit and severity, which can be quite heterogeneous. While research continues to establish specific characteristics of the most effective rehabilitation techniques, the common thread among the more successful approaches is a requirement for patients to practice "use" of the affected body part [39, 40]. In addition, the use needs to feature increased challenges that are task-related (as opposed to simple movement repetition) [41]. In the last decade, studies have shown that such task-related motor retraining may lead to improved recovery by facilitating brain repair. Such brain repair may also be augmented by specific neuropharmacological agents, the use of factors to stimulate cell growth and/or cell replacement therapies whereby endogenous stem cells are stimulated to migrate to the injury site [37, 42, 43]. These approaches are still largely at the experimental stage in animal models. One of the challenges in translating these findings to humans is to understand why and with whom to use a specific therapeutic approach in order to maximize recovery.

Neurophysiological effects occur both in the injured and uninjured hemispheres following brain injury due to stroke, illustrating the potential importance of the interaction between the two for motor recovery. Of a variety of treatment strategies, bimanual movement training (BMT) of the upper limbs has specifically shown potential to promote behavioral and functional improvement in both the subacute and chronic phases of stroke recovery [44–49]. Although there is some evidence that bimanual movement, involving both the damaged and intact hemispheres, may enhance motor-related brain activity in the stroke-hemisphere in some patients [50, 51], the neurophysiological mechanisms that contribute to the benefits of BMT are not yet clear. More specifically, few studies have reported on the cortical changes that underlie observed behavioral improvements following BMT in stroke patients [52]. The findings, however, have been consistent with reports in healthy individuals, and have shown a facilitation of primary motor cortical (M1) excitability during movements that involve activation of the contralateral homologous muscle representations [52–56].

Changes in the balance between local intracortical inhibition and excitation occur in both the affected and unaffected hemispheres following stroke [57–59]. In the subacute stage, complex changes in the function of local GABAergic interneurons (i.e., inhibitory interneurons) occur in both hemispheres. M1 lesions in animal models induce a decrease in $GABA_A$ receptor expression both in the areas immediately adjacent to the lesion and in structurally intact but connected regions, including contralesional M1 [59]. Using TMS in human stroke patients, there is evidence for both increases and decreases of cortical inhibition *in the same patients* [57], suggesting the presence of distinct, independent inhibitory interneuronal circuits. Importantly, the amount of cortical disinhibition appears to be correlated with the degree of hemiparesis, being more pronounced with less severe hemiparesis.

Cortical activity is enhanced in both hemispheres in damaged and healthy M1 when homologous muscles are activated together [50, 51]. It is thought that transcallosal neural activity of homologous representations in M1 act to excite and/or release inhibition from the contralateral hemisphere [60], which may contribute to M1 plasticity. Intracortical inhibition is decreased in M1 when the two upper limbs co-activate the homologous muscles simultaneously, but inhibition remains when they are not activated simultaneously [60]. Premotor cortical areas, such as the dorsal premotor cortex (PMd), have extensive reciprocal neuronal projections with M1 [61, 62] and have the potential to influence motor cortical excitability. In-phase bimanual training, in which homologous muscles are activated simultaneously in the two limbs, increases fMRI activity in the lateral premotor cortical areas, including PMd [63]. Also, short-term in-phase bimanual training, particularly involving visual cues, has been associated with increases in lateral premotor cortical activity [33, 34]. Contralesional PMd activity has been shown to influence ipsilesional sensorimotor regions, which could provide a mechanism to contribute to recovery of motor function post-stroke [64, 65]. Therefore, the liberation of inhibition due to homologous M1 extensor carpi radialis representations activated together, along with this activity engaging areas in PMd that could further facilitate muscle representations in M1, may have implications for enhancing motor function following stroke.

There now remains no debate about the importance of aerobic training after stroke. It has been shown to positively influence both stroke-related outcomes and risk factors for recurrent stroke. There exists convincing evidence of the benefits of aerobic training as reflected by improved cardiovascular function, reduced neurological impairment, enhanced sensorimotor function and endurance, and as a stimulus to promote neuroplasticity [66–72]. The evidence is so compelling that physical exercise, specifically aerobic training, is an important element of stroke management and promotes post-stroke recovery; it is part of the best-practice recommendations [73].

It is important to recognize that focused, patient-specific therapy is an important element in minimizing post-stroke disability. Determinants of the relative success of such therapy, beyond the specific techniques applied, include the intensity of therapy, patient motivation/participation in the program, and the timing of the therapy. It is becoming clear that more time directed at retraining following stroke, including patient-appropriate practice beyond the clinical setting, can enhance recovery. Within the first month after stroke, the standard discipline-specific therapy occurs on average 3 times a week for 30–60 min per session. While intensifying the routine at this stage of recovery may improve outcomes, there is growing evidence that a lack of long-term opportunities for rehabilitation services will, in many patients, result in progressive worsening of outcomes. For example, persisting disability in some patients leads to further disuse and musculoskeletal disorders that can further limit the ability of the individual to participate in activities of daily life, decrease their aerobic capacity, and compromise their physical activity. Objectively tracking functional status over time may help to document changes and screen for and establish the need for continuing therapy to offset the consequences of persisting disability. It should also be reinforced that prescribed physical activity training, both aerobic and strength training, is essential to the rehabilitation process. The difficulty in training for stroke patients can be the suitability and/or safety of the exercise activity selected. While a walking program may seem right for some people, they must be able to walk safely and rapidly enough to achieve target heart rates. For those unable to walk quickly enough or unable to walk at all, a recumbent cycle ergometer or stepper is an effective modality of training; it has been successfully used in the subacute and chronic stages after stroke and can help counter the deconditioning that occurs after a stroke [72, 74–76]. The concern for compliance can be evaluated and even motivated by monitoring activity logs using wearable activity monitors [77–79]. Cardiac rehabilitation, an established model of care for individuals with cardiac disease, may also be adapted to serve as an effective approach for those who have had a stroke [80, 81]. The next step in treatment is to couple the benefits of aerobic training with task-specific retraining in order to maximize sensorimotor recovery after stroke.

Summary

Here we have briefly discussed two classes of functional impairments that can occur following stroke: hemiparesis in the upper limbs and changes in the modulation of somatosensory information—more specifically, the inability to efficiently filter non-task relevant information. We have discussed what effect this can have on sensorimotor function, as well as some advances in knowledge on recovery and treatment following stroke. Although the etiology of these may differ, the neurophysiological substrates for recovery are similar. For example, Kautz and colleagues have shown limited benefits to the control of walking associated with an aerobic pedaling program, despite improvements in walking velocity [82]. They concluded that to achieve gains in the control of walking, "task-specific intervention may be required to improve coordination, consistent with principles of use-dependent plasticity." In this way the substrates for recovery between the differing levels of impairment converge. The underpinning of functional recovery may be training-related adaptation in the CNS that is maximized with use of the paretic limb in challenging, skilled tasks [83]. Optimally, this could be combined with physical activity or aerobic training to achieve the cardiovascular benefits associated with decreased risk factors for recurrent stroke.

References

1. Heart and Stroke Foundation of Ontario. Facts you should know about heart disease and stroke. Toronto, ON: Ontario Heart and Stroke Foundation; 1999.
2. Basmajian JV, Gowland C, Brandstater ME, Swanson L, Trotter J. EMG feedback treatment of upper limb in hemiplegic stroke patients: a pilot study. Arch Phys Med Rehabil. 1982;63(12): 613–6.
3. Barreca S, Wolf SL, Fasoli S, Bohannon R. Treatment interventions for the paretic upper limb of stroke survivors: a critical review. Neurorehabil Neural Repair. 2003;17:220–6.
4. Wolf SL, Winstein CJ, Miller JP, Taub E, Uswatte G, Morris D, Giuliani C, Light KE, Nichols-Larsen D, EXCITE Investigators. Effect of constraint-induced movement therapy on upper extremity function 3 to 9 months after stroke: the EXCITE randomized clinical trial. JAMA. 2006;296:2095–104.
5. Nudo RJ, Friel KM, Delia SW. Role of sensory deficits in motor impairments after injury to primary motor cortex. Neuropharmacology. 2000;39:733–42.
6. Staines WR, Black SE, Graham SJ, McIlroy WE. Somatosensory gating and recovery from stroke involving the thalamus. Stroke. 2002;33:2642–51.
7. Bell-Krotoski J, Weinstein S, Weinstein C. Testing sensibility, including touch-pressure, two-point discrimination, point localization, and vibration. J Hand Ther. 1993;6:114–23.
8. Bell-Krotoski JA, Fess EE, Figarola JH, Hiltz D. Threshold detection and Semmes-Weinstein monofilaments. J Hand Ther. 1995;8:155–62.
9. Eguibar JR, Quevedo J, Jimenez I, Rudomin P. Selective cortical control of information flow through different intraspinal collaterals of the same muscle afferent fiber. Brain Res. 1994;643:328–33.
10. Knecht S, Kunesch E, Buchner H, Freund H-J. Facilitation of somatosensory evoked potentials by exploratory finger movements. Exp Brain Res. 1993;95:330–8.
11. Staines WR, Brooke JD, McIlroy WE. Task-relevant selective modulation of somatosensory afferent paths from the lower limb. Neuroreport. 2000;11:1713–9.
12. Knight RT, Staines WR, Swick D, Chao LL. Prefrontal cortex regulates inhibition and excitation in distributed neural networks. Acta Psychol. 1999;101:159–78.
13. Bolton DA, Staines WR. Transient inhibition of the dorsolateral prefrontal cortex disrupts attention-based modulation of tactile stimuli at early stages of somatosensory processing. Neuropsychologia. 2011;49:1928–37.
14. Bolton DA, Brown KE, McIlroy WE, Staines WR. Transient inhibition of the dorsolateral prefrontal cortex disrupts somatosensory modulation during standing balance as measured by electroencephalography. Neuroreport. 2012;23:369–72.
15. Kim JS. Pure sensory stroke. Clinical-radiological correlates of 21 cases. Stroke. 1992;23:983–7.
16. Kim JS. Lenticulocapsular hemorrhages presenting as pure sensory stroke. Eur Neurol. 1999;42:128–31.
17. Remy P, Zilbovicius M, Cesaro P, Amarenco P, Degos JD, Samson Y. Primary somatosensory cortex activation is not altered in patients with ventroposterior thalamic lesions: a PET study. Stroke. 1999;30:2651–8.
18. Fuster JM. The prefrontal cortex. 4th ed. London: Academic; 2008.
19. Chao LL, Knight RT. Human prefrontal lesions increase distractibility to irrelevant sensory inputs. Neuroreport. 1995;6:1605–10.
20. Chao LL, Knight RT. Contribution of human prefrontal cortex to delay performance. J Cogn Neurosci. 1998;10:167–77.
21. Dolcos F, Miller B, Kragel P, Jha A, McCarthy G. Regional brain differences in the effect of distraction during the delay interval of a working memory task. Brain Res. 2007;1152:171–81.
22. Croxson PL, Johansen-Berg H, Behrens TE, Robson MD, Pinsk MA, Gross CG, et al. Quantitative investigation of connections of the prefrontal cortex in the human and macaque using probabilistic diffusion tractography. J Neurosci. 2005;25:8854–66.

23. Cao XH, Wang DH, Bai J, Zhou SC, Zhou YD. Prefrontal modulation of tactile responses in the ventrobasal thalamus of rats. Neurosci Lett. 2008;435:152–7.
24. Zikopoulos B, Barbas H. Prefrontal projections to the thalamic reticular nucleus form a unique circuit for attentional mechanisms. J Neurosci. 2006;26:7348–61.
25. Yamaguchi S, Knight RT. Gating of somatosensory input by human prefrontal cortex. Brain Res. 1990;521:281–8.
26. Yingling CD, Skinner JE. Selective regulation of thalamic sensory relay nuclei by nucleus reticularis thalami. Electroencephalogr Clin Neurophysiol. 1976;41:476–82.
27. Skinner JE, Yingling CD. Central gating mechanisms that regulate event-related potentials and behavior. Prog Clin Neurophysiol. 1977;1:70–96.
28. Pandya DN, Barnes CL. Architecture and connections of the frontal lobe. In: Perecman E, editor. The frontal lobes revisited. New York, NY: IRBN; 1987. p. 41–72.
29. Hamalainen H, Kekoni J, Sams M, Reinikainen K, Näätänen R. Human somatosensory evoked potentials to mechanical pulses and vibration: contributions of SI and SII somatosensory cortices to P50 and P100 components. Electroencephalogr Clin Neurophysiol. 1990;75:13–21.
30. Hannula H, Neuvonen T, Savolainen P, Hiltunen J, Ma YY, Antila H, et al. Increasing top-down suppression from prefrontal cortex facilitates tactile working memory. Neuroimage. 2010;49:1091–8.
31. Savolainen P, Carlson S, Boldt R, Neuvonen T, Hannula H, Hiltunen J, et al. Facilitation of tactile working memory by top-down suppression from prefrontal to primary somatosensory cortex during sensory interference. Behav Brain Res. 2011;219:387–90.
32. Duncan PW, Goldstein LB, Matchar D, Divine GW, Feussner J. Measurement of motor recovery after stroke. Outcome assessment and sample size requirements. Stroke. 1992;23:1084–9.
33. Smith AL, Staines WR. Cortical adaptations and motor performance improvements associated with short-term bimanual training. Brain Res. 2006;1071:165–74.
34. Smith AL, Staines WR. Cortical and behavioural adaptations in response to short-term inphase versus antiphase bimanual movement training. Exp Brain Res. 2010;205:465–77.
35. Smith AL, Staines WR. Externally cued inphase bimanual training enhances preparatory premotor activity. Clin Neurophysiol. 2012;123(9):1846–57.
36. Neva JL, Legon W, Staines WR. Primary motor cortex excitability is modulated with bimanual training. Neurosci Lett. 2012;514:147–51.
37. Cramer SC. Repairing the human brain after stroke. I. Mechanisms of spontaneous recovery. Ann Neurol. 2008;63:272–87.
38. Butler AJ, Wolf SL. Putting the brain on the map: use of transcranial magnetic stimulation to assess and induce cortical plasticity of upper-extremity movement. Phys Ther. 2007;87:719–36.
39. Taub E, Miller NE, Novak TA, Cook EW, Fleming WC, Nepomuceno CS, et al. Technique to improve chronic motor deficit after stroke. Arch Phys Med Rehabil. 1993;74:347–54.
40. Wolf SL, Lecraw DE, Barton LA, Jann BB. Forced use of hemiplegic upper extremities to reverse the effect of learned nonuse among chronic stroke and head injured patients. Exp Neurol. 1989;104:104–32.
41. Kleim JA, Jones TA. Principles of experience-dependent neural plasticity: implications for rehabilitation after brain damage. J Speech Lang Hear Res. 2008;5:S225–39.
42. Hallett M. Recent advances in stroke rehabilitation. Neurorehabil Neural Repair. 2002;16:211–7.
43. Cramer SC. Repairing the human brain after stroke. II. Restorative therapies. Ann Neurol. 2008;63:549–60.
44. Cuadrado ML, Arias JA. Bilateral movement enhances ipsilesional cortical activity in acute stroke: a pilot functional MRI study. Neurology. 2001;57:1740–1.
45. Luft AR, McCombe Waller S, Whitall J, Forrester LW, Macko R, Sorkin JD, et al. Repetitive bilateral arm training and motor cortex activation in chronic stroke: a randomized controlled trial. JAMA. 2004;292:1853–61.
46. Whitall J, McCombe Waller S, Silver KH, Macko RF. Repetitive bilateral arm training with rhythmic auditory cueing improves motor function in chronic hemiparetic stroke. Stroke. 2000;31:2390–5.

47. Mudie MH, Matyas TA. Can simultaneous bilateral movement involve the undamaged hemisphere in reconstruction of neural networks damaged by stroke? Disabil Rehabil. 2000;22:23–37.
48. Stewart KC, Cauraugh JH, Summers JJ. Bilateral movement training and stroke rehabilitation: a systematic review and meta-analysis. J Neurol Sci. 2006;244:89–95.
49. McCombe Waller S, Whitall J. Bilateral arm training: why and who benefits? NeuroRehabilitation. 2008;23:29–41.
50. Silvestrini M, Cupini LM, Placidi F, Diomedi M, Bernardi G. Bilateral hemispheric activation in the early recovery of motor function after stroke. Stroke. 1998;29:1305–10.
51. Staines WR, McIlroy WE, Graham SJ, Black SE. Bilateral movement enhances ipsilesional cortical activity in acute stroke: a pilot functional MRI study. Neurology. 2001;56:401–4.
52. Stinear CM, Barber PA, Coxon JP, Fleming MK, Byblow WD. Priming the motor system enhances the effects of upper limb therapy in chronic stroke. Brain. 2008;131:1381–90.
53. Carson RG, Riek S, Mackey DC, Meichenbaum DP, Willms K, Forner M, Byblow WD. Excitability changes in human forearm corticospinal projections and spinal reflex pathways during rhythmic voluntary movement of the opposite limb. J Physiol. 2004;560:929–40.
54. Carson RG, Smethurst CJ, Oytam Y, de Rugy A. Postural context alters the stability of bimanual coordination by modulating the crossed excitability of corticospinal pathways. J Neurophysiol. 2007;97:2016–23.
55. Perez MA, Cohen LG. Mechanisms underlying functional changes in the primary motor cortex ipsilateral to an active hand. J Neurosci. 2008;28:5631–40.
56. Kennedy NC, Carson RG. The effect of simultaneous contractions of ipsilateral muscles on changes in corticospinal excitability induced by paired associative stimulation (PAS). Neurosci Lett. 2008;445:7–11.
57. Liepert J, Storch P, Fritsch A, Weiller C. Motor cortex disinhibition in acute stroke. Clin Neurophysiol. 2000;111:671–6.
58. Bütefisch CM, Netz J, Wessling M, Seitz RJ, Hömberg V. Remote changes in cortical excitability after stroke. Brain. 2003;126:470–81.
59. Schiene K, Bruehl C, Zilles K, Qü M, Hagemann G, Kraemer M, Witte OW. Neuronal hyperexcitability and reduction of GABAA-receptor expression in the surround of cerebral photothrombosis. J Cereb Blood Flow Metab. 1996;16:906–14.
60. Stinear JW, Byblow WD. Disinhibition in the human motor cortex is enhanced by synchronous upper limb movements. J Physiol. 2002;543:307–16.
61. Weinrich M, Wise SP. The premotor cortex of the monkey. J Neurosci. 1982;2:1329–45.
62. Wise SP. The primate premotor cortex: past, present, and preparatory. Annu Rev Neurosci. 1985;8:1–19.
63. Seitz RJ, Kleiser R, Bütefisch CM, Jörgens S, Neuhaus O, Hartung HP, et al. Bimanual recoupling by visual cueing in callosal disconnection. Neurocase. 2004;10:316–25.
64. Bestmann S, Swayne O, Blankenburg F, Ruff CC, Teo J, Weiskopf N, et al. The role of contralesional dorsal premotor cortex after stroke as studied with concurrent TMS-fMRI. J Neurosci. 2010;30:11926–37.
65. Kantak SS, Stinear JW, Buch ER, Cohen LG. Rewiring the brain: potential role of the premotor cortex in motor control, learning, and recovery of function following brain injury. Neurorehabil Neural Repair. 2012;26:282–92.
66. Potempa K, Braun LT, Tinknell T, Popovich J. Benefits of aerobic exercise after stroke. Sports Med. 1996;21:337–46.
67. Eich HJ, Mach H, Werner C, Hesse S. Aerobic treadmill plus Bobath walking training improves walking in subacute stroke: a randomized controlled trial. Clin Rehabil. 2004;18:640–51.
68. Kluding P, Billinger SA. Exercise-induced changes of the upper extremity in chronic stroke survivors. Top Stroke Rehabil. 2005;12:58–68.
69. MacKay-Lyons MJ, Howlett J. Exercise capacity and cardiovascular adaptations to aerobic training early after stroke. Top Stroke Rehabil. 2005;12:31–44.
70. Ploughman M, Attwood Z, White N, Doré JJ, Corbett D. Endurance exercise facilitates relearning of forelimb motor skill after focal ischemia. Eur J Neurosci. 2007;25:3453–60.

71. Ploughman M, McCarthy J, Bossé M, Sullivan HJ, Corbett D. Does treadmill exercise improve performance of cognitive or upper-extremity tasks in people with chronic stroke? A randomized cross-over trial. Arch Phys Med Rehabil. 2008;89:2041–7.
72. Tang A, Sibley KM, Thomas SG, Bayley MT, Richardson D, McIlroy WE, et al. Effects of an aerobic exercise program on aerobic capacity, spatiotemporal gait parameters, and functional capacity in subacute stroke. Neurorehabil Neural Repair. 2009;23:398–406.
73. Lindsay MP, Gubitz G, Bayley M, Hill MD, Davies-Schinkel C, Singh S, Phillips S. Canadian best practice recommendations for stroke care (update 2010). On behalf of the Canadian Stroke Strategy Best Practices and Standards Writing Group. Ottawa, ON: Canadian Stroke Network; 2010. http://www.strokebestpractices.ca/wp-content/uploads/2011/04/2010BPR_ENG.pdf
74. Billinger SA, Tseng BY, Kluding PM. Modified total-body recumbent stepper exercise test for assessing peak oxygen consumption in people with chronic stroke. Phys Ther. 2008;88:1188–95.
75. Katz-Leurer M, Sender I, Keren O, Dvir Z. The influence of early cycling training on balance in stroke patients at the subacute stage. Results of a preliminary trial. Clin Rehabil. 2006;20:398–405.
76. Sibley KM, Tang A, Brooks D, Brown DA, McIlroy WE. Feasibility of adapted aerobic cycle ergometry tasks to encourage paretic limb use after stroke: a case series. J Neurol Phys Ther. 2008;32:80–7.
77. Gebruers N, Vanroy C, Truijen S, Engelborghs S, De Deyn PP. Monitoring of physical activity after stroke: a systematic review of accelerometry-based measures. Arch Phys Med Rehabil. 2010;91:288–97.
78. Prajapati SK, Gage WH, Brooks D, Black SE, McIlroy WE. A novel approach to ambulatory monitoring: investigation into the quantity and control of everyday walking in patients with subacute stroke. Neurorehabil Neural Repair. 2011;2:6–14.
79. Dobkin BH, Xu X, Batalin M, Thomas S, Kaiser W. Reliability and validity of bilateral ankle accelerometer algorithms for activity recognition and walking speed after stroke. Stroke. 2011;42:2246–50.
80. Tang A, Marzolini S, Oh P, McIlroy WE, Brooks D. Feasibility and effects of adapted cardiac rehabilitation after stroke: a prospective trial. BMC Neurol. 2010;10:40.
81. Tang A, Closson V, Marzolini S, Oh P, McIlroy W, Brooks D. Cardiac rehabilitation after stroke-need and opportunity. J Cardiopulm Rehabil Prev. 2009;29:97–104.
82. Kautz SA, Duncan PW, Perera S, Neptune RR, Studenski SA. Coordination of hemiparetic locomotion after stroke rehabilitation. Neurorehabil Neural Repair. 2005;19:250–8.
83. Taub E, Uswatte G, Elbert T. New treatments in neurorehabilitation founded on basic research. Nat Rev Neurosci. 2002;3:228–36.

Chapter 4
Limb Apraxia: Types, Neural Correlates, and Implications for Clinical Assessment and Function in Daily Living

Eric A. Roy, Sandra E. Black, Vessela Stamenova, Deborah Hebert, and David Gonzalez

Introduction

Apraxia is a neurological disorder of learned movements that cannot be explained on the basis of deficits in basic sensory or motor functions, verbal comprehension, or recognition of tools or objects [1]. While apraxia frequently arises in concert with other impairments such as ataxia, aphasia, and dementia in conditions such as stroke and Alzheimer's disease, apraxia can be dissociated from these deficits. Apraxia most often arises from left hemisphere damage (LHD) but can also occur with right hemisphere damage (RHD), with incidences ranging between 28–57 % for LHD and 0–34 % for RHD [2].

This chapter begins with an examination of classifications of limb apraxia, including a characterization of the types of errors. Discussion will then turn to a consideration of the neural correlates of apraxia both from work with stroke patients as well as studies of functional neuroimaging in healthy adults. Finally, we will

E.A. Roy, Ph.D., C.Psych (✉)
Department of Kinesiology and Psychology, University of Waterloo, Waterloo, ON, Canada N2L 3G1
e-mail: eroy@uwaterloo.ca

S.E. Black, M.D., F.R.C.P.(C)
Brill Chair in Neurology, University of Toronto and Cognitive Neurology, Sunnybrook Health Sciences Centre, Toronto, ON, Canada M4N 3M5

V. Stamenova, Ph.D.(C)
The Rotman Research Institute, Baycrest Centre for Geriatric Care, Toronto, ON, Canada M6A 2E1

D. Hebert, B.Sc.(OT), M.Sc.(Kin), Ph.D.(C)
Toronto Rehabilitation Institute, University Health Network, 550 University Ave, Toronto, ON, Canada M5G 2A2

D. Gonzalez, Ph.D.
Department of Kinesiology, University of Waterloo, Waterloo, ON, Canada N2L 3G1

T.A. Schweizer and R.L. Macdonald (eds.), *The Behavioral Consequences of Stroke*,
DOI 10.1007/978-1-4614-7672-6_4,

briefly examine models of apraxia focusing on our model and its implications for the clinical assessment of apraxia, as well as for how apraxia may impact on function in daily living.

Classification of Apraxia

The term apraxia was originally coined by Steinthal in 1871 [3]. The traditional classification of apraxia owes its roots to Liepmann in the early 1900s [3, 4]. Three traditional types of apraxia have been described. These apraxias have been defined on the basis of the task and input modality. With regard to task, the distinction has been made between the performance of a single gesture and a sequence of gestures. In the context of producing single gestures, several tasks have been of interest. In one task, termed *pantomime*, the patient is asked to produce the gesture from memory (e.g., showing how to pound a nail with a hammer). Here neither the tool (hammer) nor the object (nail) is present. The other task examined is termed *object use* where the patient performs the gesture from memory, as in pantomime, but here the tool is held and manipulated when demonstrating the gesture. The third single-gesture task examined is *imitation*. Here the examiner demonstrates or models a gesture and the patient is required to imitate or copy it. The second major task in the assessment for apraxia is one that involves a sequence of gestures often involving multiple tools/objects at the various steps in the sequence. Here the patient is required to perform the sequence in the correct order using the appropriate tools/objects correctly at each step.

The traditional types of apraxia were originally defined by Liepmann [4]. These are ideational, ideomotor, and limb-kinetic apraxia. Ideational apraxia has been defined in two ways. The first is with respect to single-gesture performance. In this context ideational apraxia is one in which the patient is impaired in both pantomime and object use conditions but improves when imitating gestures [3, 5]. Often patients with ideational apraxia are not aware of what the to-be-performed gestures should look like, suggesting some impairment in the idea or concept of the gesture. The other definition of ideational apraxia pertains to movement sequences in which the patient is impaired in performing a sequence of movements/gestures. The patient may perform the constituent gestures incorrectly or may omit constituent gestures or perform them in the wrong order (e.g., [6]).

Ideomotor apraxia (IMA) is defined with respect to single-gesture performance and involves impairment in both pantomime and imitation, with some improvement when actually holding and using the tool (tool-use condition). In contrast to ideational apraxics these patients often know what the gesture should look like, suggesting no impairment in the conceptualization of the gesture [5]. Heilman et al. [7], however, did report two forms of IMA, one involving a more posterior lesion in the left hemisphere that was associated with an impairment in gesture recognition and another that involved a more anterior lesion and which did not have an accompanying deficit in gesture recognition.

Limb-kinetic apraxia is defined as a loss of control of fine finger movements, such as in flipping a coin [8–10]. There is considerable controversy as to whether this is a true apraxia. First, unlike ideational and IMA, it is seen in only one hand as opposed to both hands as is the case with the former apraxias. Secondly, this apraxia is seen in the limb contralateral to the damaged hemisphere in the context of stroke and so it may be difficult to distinguish from weakness and loss of motor control seen in hemiparesis. Finally, limb-kinetic apraxia is independent of the task or modality such as pantomime or imitation [10].

Another type of apraxia, which like limb-kinetic apraxia affects only one hand, is termed *callosal apraxia* and was described by Liepmann [3] and others (e.g., [11]). Unlike limb-kinetic apraxia, this apraxia was affected by the input modality such that it primarily affected the left hand and was seen in pantomime to verbal command. The apraxia here was thought to arise from damage to the corpus callosum, which disconnected the language processing areas in the left hemisphere from the areas in the right hemisphere controlling movement of the left hand.

While one factor in understanding apraxia relates to the tasks or modalities assessed (pantomime, imitation, or object use), another pertains to the type of gesture. One dissociation is whether the gestures are representational or nonrepresentational. Representational gestures hold some meaning and have representations in semantics. There are two types of representational gestures: transitive or tool-use gestures, such as demonstrating how to use a spoon to eat soup, and intransitive or communicative gestures, such as waving goodbye, which do not involve tools. Given their representation in semantics, these gestures can be elicited through pantomime. This task typically involves responding to a verbal command. In the case of transitive gestures, visual input in the form of a picture of the tool or the actual tool itself has also been used. Some research has shown that these input modalities are dissociable. For example, Humphreys and colleagues reported a stroke patient who was unable to pantomime a gesture to verbal command but was able to do so based on the picture of the tool [12]. This dissociation led Humphreys to suggest two routes to action: one accessing semantics through the verbal command and the other involving a more direct link between the visual features of the tool and the motor program involved in producing the gesture.

This notion of a link between the visual features of tools and their associated actions reflects something of the affordances of the tools; that is, what functions and actions the tools may subserve [13, 14]. One type of apraxia, which appears to represent a disruption in this link, is conceptual apraxia (e.g., [15]). Here the patient is required to complete a task such as eating soup. Several tools are placed on the table and the patient is asked to select and use the appropriate tool for the task. Patients with this apraxia will often select the wrong tool to complete the task yet perform the action flawlessly [16]. They will also be unable to show which action is appropriate for the tool, despite often knowing what the tool is used for. The fact that the action itself is performed well indicates that this disorder is dissociable from IMA. Some though have likened it to ideational apraxia given the apparent conceptual deficit, but this position is not widely accepted [5].

Turning now to nonrepresentational gestures, these are meaningless or novel gestures that have no representation in semantics. Meaningless gestures were initially included in the assessment of apraxia in order to avoid the influence of familiarity and semantics on gesture production. These studies necessarily involved imitation and many showed that patients who were impaired at imitation of meaningless gestures were also impaired at pantomime and imitation of meaningful gestures. These findings suggested that apraxia is a motor deficit that cuts across tasks or modalities and does not necessarily involve a disruption to semantics. Studies by Rumiati and colleagues [17–19], however, show that imitation of meaningful gestures is more accurate than that for meaningless gestures, suggesting that access to semantics facilitates performance. As was seen for the effect of input modality on transitive gestures, so here this dissociation suggests that there are two routes to imitation, direct and indirect, and that the latter enjoys an advantage.

This idea of dissociable routes to action in the context of imitation is also seen in cases where the patient shows a selective impairment to imitation. This disorder is seen as akin to conduction aphasia where the patient is unable to repeat words yet is not impaired in speech production or comprehension of language. As such this disorder has been termed *conduction apraxia*. These patients know what the gesture should look like and are able to pantomime the gesture accurately but are not able to imitate gestures modeled by the examiner [20, 21].

In order to identify apraxia, there are a number of errors in performance that are characteristic of apraxia. Generally these errors can be categorized as content and production errors. Content errors involve using the wrong tool to perform a gesture (e.g., using a knife rather than a spoon to eat soup) while performing the action (e.g., eating soup from a bowl) correctly [16, 22]. Production errors involve impairments in the performance of an action and may be seen in single actions or in a sequence. Errors in performing single actions may involve an inability to perform the gesture at all. Here the patient may look quizzically at their hand and make some movement in an apparent attempt to perform the gesture, but the movement bears little resemblance to any identifiable gesture. Single-gesture production errors may involve the patient performing a gesture but one that involves aberrations in hand posture, in spatial characteristics of the gesture (e.g., plane of movement, location of hand, orientation of the hand) or action features of the gesture (e.g., performing a sawing action in the sagittal plane rather than a hammering action in the coronal plane when asked to demonstrate how to use a hammer to drive a nail). Other errors include temporal errors, augmentation errors, fragmentation errors, substitutive errors, and associative errors where the correct gesture is replaced with one that shares one feature. One of the most well-known hand posture errors is termed a body-part-as-tool error where the fingers are configured in a way to take on the features of the tool; for example, when asked to demonstrate using a toothbrush to clean the teeth, the patient extends the index finger in a fist-like posture and uses the finger as the toothbrush. Since these errors can arise in healthy adults when pantomiming tool use, one key feature in identifying this error relates to re-instruction. When healthy adults are re-instructed to pretend holding the tool, they immediately correct the posture error. Apraxic patients, however, do not correct their hand posture on re-instruction.

A number of studies have attempted to identify the neural correlates of these errors in apraxia. Spatial errors, for example, have been associated only with more posterior lesions, while errors in producing internal hand positions were seen with both anterior and posterior lesions [23–25]. One interesting recent study by Manuel et al. has examined the link between errors in tool-use pantomime and lesion localization in a large group of left and right hemisphere patients ($N=150$) using voxel-based lesion-symptom mapping [26]. They compared configural-spatial errors and body-part-as-tool errors (see Fig. 4.1). Their findings confirmed the importance of the left hemisphere in apraxia in that both errors were seen with damage to the left inferior frontal area. Interestingly, they found that configural-spatial errors were linked to damage in the left temporal parietal area, while body-part-as-tool errors were associated with lesions in the left inferior frontal area. They concluded that "...pantomime is subserved by a distributed, left-lateralized, frontoparietal network and that lesions to subparts of this network induce distinct error types" [26].

Errors in the context of a movement sequence such as demonstrating how to make a cup of instant coffee may involve omission of actions in the sequence or performance of the actions in the wrong order. Individual actions in the sequence may be performed incorrectly with errors similar to those described for single gestures. In some cases these aberrations in the individual actions may arise from perseverations, that is, repetition of the preceding action or some feature of it. Errors in a sequence have been described in the Naturalistic Action Test [27].

In summary, several types of apraxia have been defined on the basis of the input modality, pantomime, imitation, or tool/object use, whether the task involves a single gesture or a sequence of gestures, and in the case of single gestures whether the gesture is meaningful (transitive versus intransitive) or meaningless, and finally whether the apraxia affects both hands or not.

Neural Correlates of Apraxia

Apraxia occurs more often in patients with left as opposed to RHD and damage to the left hemisphere also results in a more severe apraxia [10, 28]. Nevertheless apraxia has been found after stroke to the right hemisphere, suggesting that the right hemisphere may also play a role in praxis [29, 30]. This observation may reveal the basis for sparing of some praxis tasks, such as intransitive gestures and recognition of gestures in cases of LHD. Interestingly, support for right hemisphere involvement in praxis was seen in an electroencephalogram (EEG) study of patients with apraxia associated with LHD that examined activation associated with practice of a tool-use gesture. The analysis showed a greater activation of a right frontoparietal network relative to controls, suggesting that training on a tool-use gesture in the face of LHD led to recruitment of the right hemisphere [10, 31].

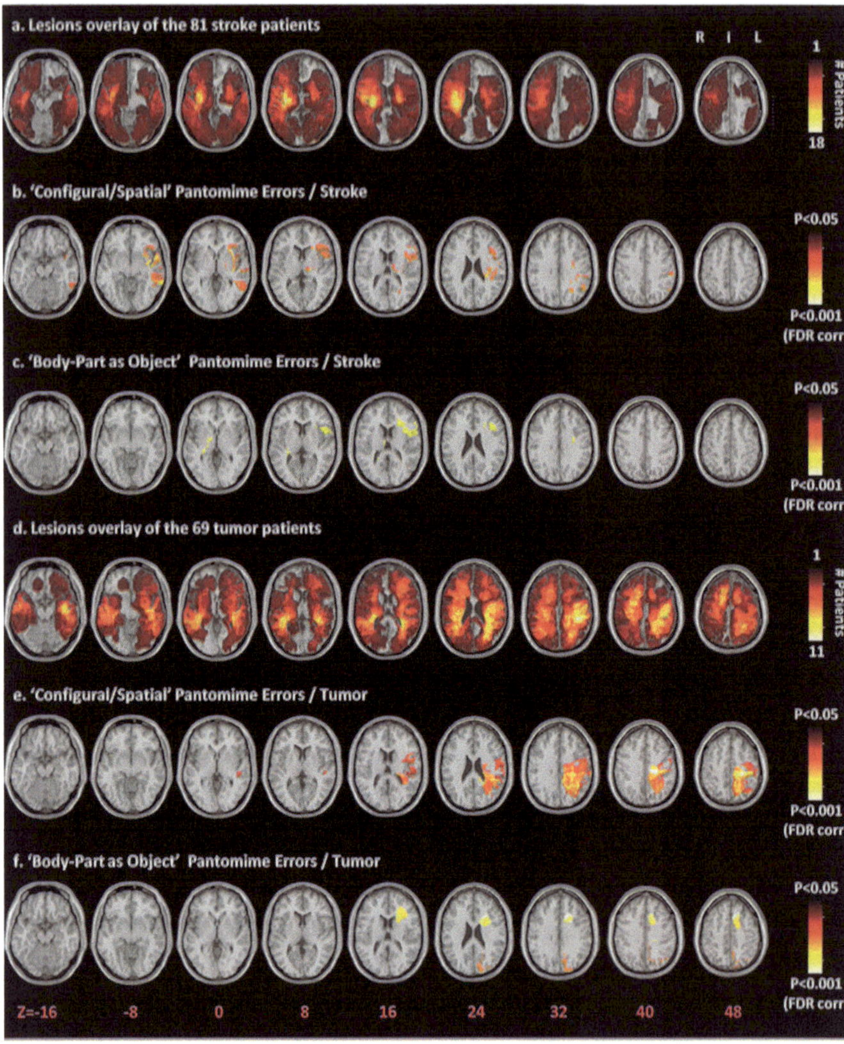

Fig. 4.1 Focus of interest is on the voxel-based lesion—symptom mapping on the stroke patients showing the relationship between brain lesions and pantomime performance. Configural-spatial errors were linked to lesions in a network involving left inferior and temporal lobe areas (**b**), while body-part-as-object errors were associated with lesions to left middle and inferior gyri of the frontal lobe and the underlying white matter particularly the superior longitudinal fasciculus (**c**) (Reprinted with permission Manuel AL, Radman N, Mesot D, Chouiter L, Clarke S, Annoni JM, et al. Inter- and intrahemispheric dissociations in ideomotor apraxia: a large-scale lesion-symptom mapping study in subacute brain-damaged patients. Cereb Cortex. 2012 Sep 17)

While much work on the neural correlates of apraxia has focused on hemispheric asymmetries, considerable research has examined for evidence of localization within the left hemisphere. A number of studies have not been able to show a reliable association between lesion locus and apraxia severity (e.g., [32, 33]). Many others though have associated IMA with damage to the parietal (left inferior parietal) and frontal (left dorsolateral frontal) cortices [23, 34] as well as the white matter tracts connecting these areas. A number of studies have linked damage in the angular and supramarginal gyri of the inferior parietal lobe to IMA [35, 36] and impairments in imitation [37]. Others have observed an association between damage in the superior parietal lobe and the ability to integrate visual and somatosensory information in gesture production [35, 38]. Lesions in the frontal and premotor areas have also been linked to apraxia [24, 35, 39]. Interestingly, it has been difficult to dissociate between lesions in the parietal and frontal lobes since lesions in these areas have been linked to similar deficits [23, 35]. Spatial errors, for example, have been associated only with more posterior lesions [23, 24, 27]. Errors in producing internal hand positions were seen with both anterior and posterior lesions, while deficits in gesture recognition and discrimination have been found with posterior but not anterior lesions [24, 40]. This connection between structure and function has largely been reported based on studies of patients with focal lesions such as those seen in stroke or in neurodegenerative conditions, such as corticobasal syndrome, which affect frontoparietal areas [41–43]. Apraxia may also be seen arising from damage in brain regions other than frontal and parietal areas such as temporal, occipital, and subcortical areas [44–46]. Anterior callosal lesions have also been associated with apraxia. In this case the apraxia would be more unilateral, affecting just the left hand and is attributed to a disconnection of the premotor and motor cortices in the right hemisphere from praxis control exerted by the left hemisphere. Damage to the basal ganglia has also been implicated in the expression of apraxia. Apraxia has been seen in various movement disorders such as corticobasal syndrome (CBS) [41–43], Parkinson's disease [47], and Huntington's disease [48]. The basal ganglia are thought to contribute to praxis through the control of sequencing, selection of appropriate motor programs, and performance of overlearned movements. There is some question as to whether damage isolated to the basal ganglia causes marked apraxia in that the majority of the cases involve surrounding white matter tracts and frontoparietal areas.

In 2012, Hong et al. reported a case of a 67-year-old man with a stroke affecting the left frontal anterior part of the precentral gyrus and prefrontal cortex, as well as the centrum semiovale [49]. This case is interesting because of its use of multiple brain imaging techniques to investigate the neural correlates of apraxia. In this case, structural magnetic resonance imaging (sMRI) and functional magnetic resonance imaging (fMRI) along with diffusion tensor imaging (DTI) were used. On apraxia assessment the patient was determined to have both ideomotor and limb-kinetic apraxia. The IMA was linked with damage to the superior portion of the superior longitudinal fasciculus in the left hemisphere while the limb-kinetic apraxia was linked with a thinning of the fiber tracts from the premotor and supplementary areas in the left hemisphere. Interestingly, the fibers through the corticospinal tract were unaffected, indicating that the corticospinal pathways subserving movement were intact as one might expect in cases of apraxia (see Fig. 4.2 for details of imaging).

Evidence for left hemisphere dominance for praxis has also come from neuroimaging studies of healthy adults. Left hemisphere activation has been found when

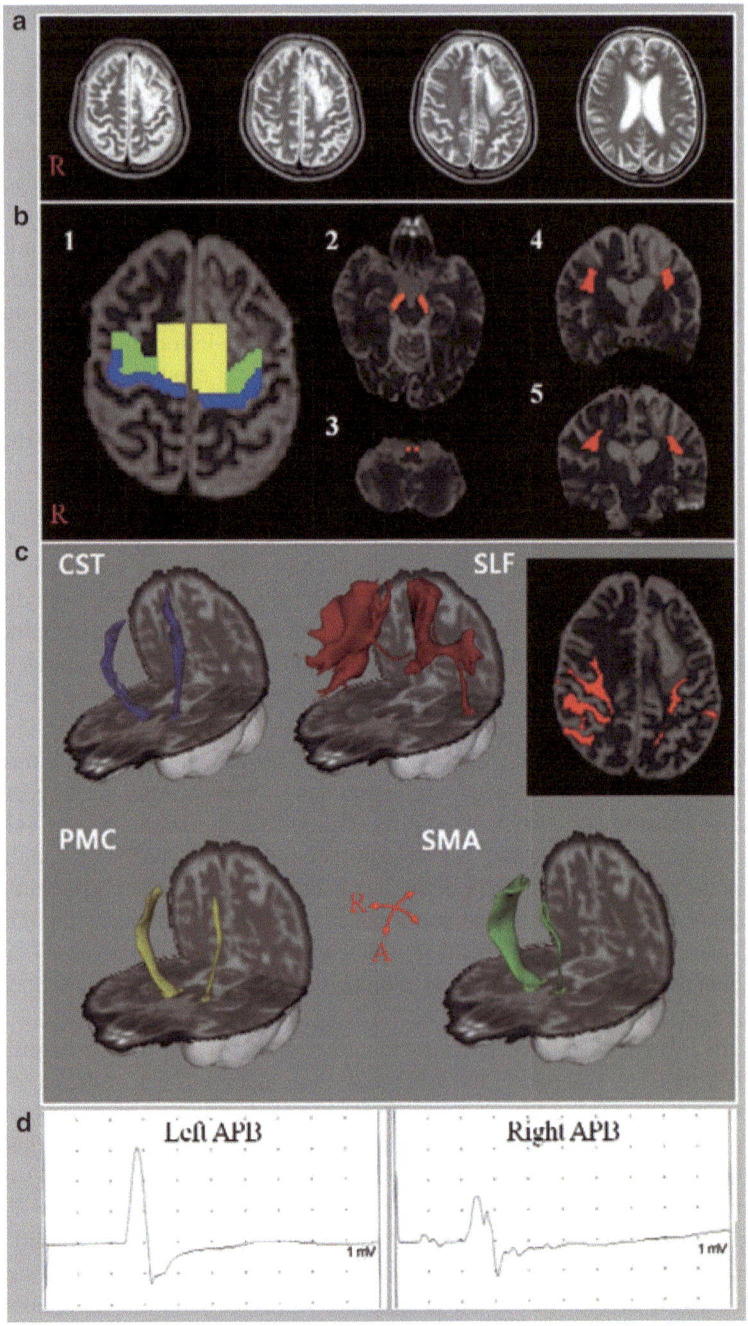

Fig. 4.2 Shows T2-weighted MR images (**a**) and diffusion tensor tractography (**c**), which shows disruption to the superior longitudinal fasciculus and fibers from the premotor cortex and the supplementary motor area but intact corticospinal tract on the affected left hemisphere. Note as well the reduced amplitude of the motor-evoked potential from the abductor pollicis brevis on the right hand (left hemisphere) (Reprinted from Hong JH, Lee J, Cho YW, Byun WM, Cho HK, Son SM, et al. Limb apraxia in a patient with cerebral infarct: Diffusion tensor tractography study. NeuroRehabilitation. 2012;30:255–259, with permission from IOS Press)

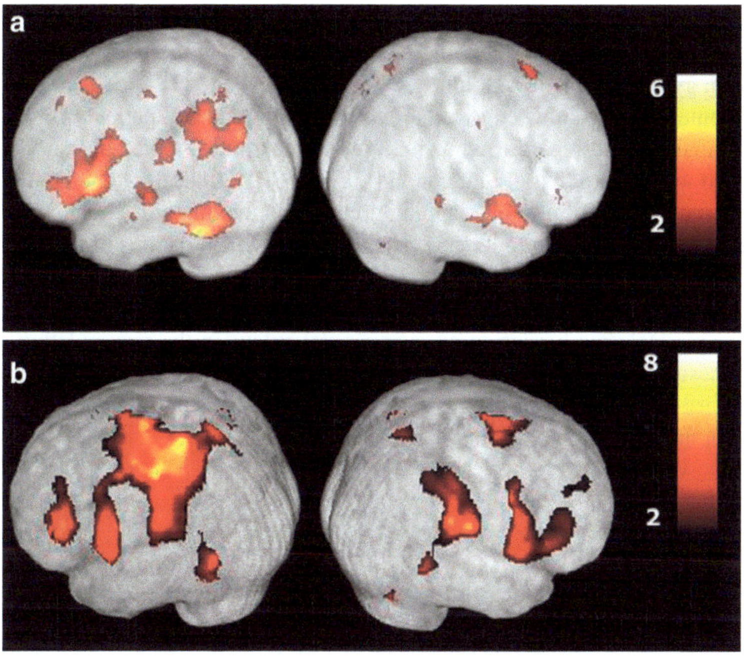

Fig. 4.3 Functional magnetic resonance imaging (fMRI) scans showing activations associated with planning and executing hand gestures. (**a**) Shows activation associated with planning tool-use gestures compared to random hand movements. Note major left hemisphere focused activations in the posterior parietal cortex, posterior temporal cortex, and dorsolateral prefrontal cortex. Gesture execution (**b**) recruits similar areas in addition to other regions associated with sensory and motor processes in both hemispheres (Reprinted with permission from Johnson-Frey SH, Newman-Norlund R, Grafton ST. A distributed left hemisphere network during planning of everyday tool use skills. Cereb Cortex. 2005;15 (6):681–695)

performing gestures with either hand [50, 51]. These neuroimaging studies have also revealed a network of brain areas subserving praxis. EEG coherence analysis revealed two networks: one involving the parietal, supplementary motor and motor areas, and the other parietal, lateral premotor and motor areas [31]. Using fMRI, Frey also identified a network associated with tool-use and communicative gestures (see Fig. 4.3) [50, 52]. Retrieving and planning transitive gestures for production using either hand was associated with increased activation in the left parietal, dorsal premotor, posterior temporal, and middle frontal areas. Interestingly, the same areas were activated when planning intransitive, communicative gestures. Frey argued that the network recruited might be task and context-dependent [53]. The parietal cortex is seen as playing an important role in selecting and producing actions based on the integration of sources of information from brain areas distributed throughout the cortex. The particular computations, and so the distributed brain areas involved, are dependent on the task demands. Pantomime of tool-use gestures involves accessing conceptual knowledge of tools in terms of their function and manipulation. The posterior temporal area appears to subserve the representation of this information.

Pantomime to verbal cues may engage the dorsal premotor and middle frontal areas, while cues to pantomime involving visual tool information appear to engage the fusiform gyrus in the temporal lobe.

In contemplating knowledge of tools, Buxbaum found that patients with apraxia were impaired with regard to knowing how to manipulate tools but they were not impaired with regard to tool function [54]. Using fMRI in healthy adults, Buxbaum demonstrated activation in overlapping areas when making either manipulation or function judgments about tools [55]. These areas included regions of the left dorsolateral and ventrolateral frontal and prefrontal areas as well as posterior temporal and inferior parietal areas. There was, however, much greater activation in the left inferior parietal area bordering on intraparietal sulcus in the manipulation condition. These findings reinforce the context- and task-dependent nature of activation in tool-use gestures in the healthy brain as well as in the expression of apraxic deficits seen in neuropathologic conditions such as stroke.

Model of Apraxia

Several models have been proposed to understand the disruptions at various stages in gesture production thought to underlay apraxia [1, 56–58]. One model developed by Roy and colleagues proposes that there are three processing systems involved in praxis, i.e., the sensory-perceptual, conceptual, and production systems [42, 58–61]. The sensory-perceptual system receives and analyzes input information from the environment. This information serves as the first stage in identifying tools and actions based on their structural features. In addition, this system is involved in the analysis of visual-gestural information of a modeled gesture that is to be imitated. The conceptual system is a storage area for action-based knowledge. Within this system, three aspects of knowledge are incorporated: the knowledge of objects and tools and their functions, knowledge of actions independent of tools, and knowledge of action sequences. Selection of appropriate tools and actions in response to environmental demands is believed to be dependent on the integrity of this knowledge base.

The production system is thought to involve four stages. The first stage, response selection, involves the selection of the appropriate movement from secondary memory based on the information received from the sensory-perceptual and conceptual systems. The second (image generation) stage involves generating an image of the action to be performed, which encompasses both visual and kinesthetic information. This stage is thought to play a role in organizing movement and in controlling movement by pre-tuning the motor system to perform the selected action and to provide information on the expected visual and kinesthetic feedback. This image also incorporates information about the tool and object to enable the patient to configure the hand and fingers to accommodate to the features (e.g., size and shape) of the tool. This image is then encoded in working memory, the third stage. Working memory is critical for the maintenance of the temporarily stored movement and tool and object information. The final stage involves response organization and control.

This model is thought to encompass three routes to action: the pantomime route and two imitation routes—concurrent and delayed imitation. The pantomime route involves producing the gesture from memory and may involve two sources of input through the sensory-perceptual system: speech-language input if performing to a verbal command or visual tool information if performing in response to a picture of a tool or the actual tool. This information is fed through the conceptual system where contact is made with the components of knowledge related to the gesture and tool. This information is then fed into the production system where the appropriate response is selected and the image of the action is generated. This image is then stored in working memory and then fed into the final stage of the production system to execute the action.

Imitation of gestures is seen to involve two routes. Both depend on analysis of the visual-gestural information in the modeled gesture. In concurrent imitation the examiner models the gesture while at the same time the patient attempts to imitate it. In this condition the product of the visual-gestural analysis is fed directly to the late stage of the production system for execution of the gesture. In delayed imitation the examiner models the gesture and then stops. Once the examiner has completed the demonstration the patient attempts imitation. In this condition the product of the visual-gestural analysis is encoded into and stored in working memory. From here this information is fed into the production system for execution of the imitated gesture.

These three routes to action are seen to give rise to eight patterns of apraxia, which reflect breakdowns at various stages in gesture production (see [58] for details). Our recent work with a large group of stroke patients has provided strong support for our model using a comprehensive apraxia battery [58]. These findings have implications for the clinical assessment of apraxia as well as for examining how apraxia may affect performance of activities involved in daily living.

Implications for Clinical Assessment of Apraxia

The model-based approach to understanding apraxia has important implications for the clinical assessment of apraxia. This approach identifies disruptions at different stages of gesture production and so demands the use of a comprehensive battery, which examines both the conceptual and motor components of gesture production.

The clinical assessment for apraxia, then, should include a comprehensive battery of tests, which tap into sensory-perceptual, conceptual, and production aspects of gesture. Some of these aspects of praxis are assessed in order to rule out several exclusion criteria. Assessment of some of the sensory-perceptual and production aspects of praxis involves looking for evidence of verbal comprehension deficits and visual object agnosia. The presence of these deficits from the input side would tend to rule out a diagnosis of apraxia. By the same token from the output side one needs to look for evidence of a basic motor deficit such as a hemiparesis. In such cases the presence of a hemiparesis would not rule out apraxia per se. Rather, the expression of apraxia in the affected limb would be difficult to dissociate from the weakness and sensory loss associated with the hemiparesis, and the patient would demonstrate apraxia in the limb not affected by the hemiparesis.

The Relationship of Apraxia to Functional Performance

The model-based approach to apraxia also has implications for how apraxia is related to the performance of activities of daily living (ADL). This relationship has been addressed through observation of function in daily living as well as examining the correlation between performance on tests of apraxia and function.

From an observational perspective, the problem of apraxia is sometimes quite evident in daily living when a person presents with unusual handling of objects or with improper sequencing of task steps. For example, obvious apraxic errors are seen in a woman with ideational apraxia (IA) performing her morning routine by combing her hair with a toothbrush rather than a hairbrush or attempting to use her hairbrush in an awkward spatial position as in IMA. There are many observable conceptual, sequencing, and production errors that have been presented earlier that might possibly emerge in functional performance. What is challenging in identifying this problem through observation alone is that the presentation of apraxic errors can be quite variable [62–64]. That is, they can occur in some situations and not in others depending on the attentional state of the person, familiarity of the environment and tools, and the competing demands on the person. Also complicating the identification of apraxic errors is that people with apraxia often achieve the goal of the task, but if carefully observed they take longer and are also sidetracked by unnecessary objects in the task environment [65]. Just as observation alone is unreliable in identification of apraxia in daily function, identification of apraxia through clinical testing does not necessarily indicate that functional performance will be compromised.

The studies examining the relationship to function in ADL differ in their methods of testing apraxia and how functional dependency is measured. For example, the earliest study by Bjorneby and Reinvang [66] examined 120 stroke patients on their level of function at admission by rating functions identified in the occupational therapy notes as independent or dependent and at 6 months after discharge based on a 25-item questionnaire. IMA was assessed through pantomime to command and imitation of meaningful and nonmeaningful gestures and ideational apraxia through object use and picture sequencing. Limb-kinetic and constructional apraxias were also assessed. The relationship between ADL function and apraxia was found to be stronger in the 6-month post-discharge period than on admission. In 1988, Sundet et al. [67] repeated this study with 145 stroke patients at 6-months post discharge using a questionnaire similar to that used in the Bjorneby and Reinvang study. Age, sex, hemianopia, and hemiplegia explained 22 % of the variance in dependence at

6 months post discharge. IMA as measured by imitative finger positioning and hand movement accounted for an additional 12 % of the variance in ADL function. Nonverbal memory, constructional apraxia, aphasia, and nonverbal intelligence explained the remaining 8 % of the total variance. Similarly, in 1999, Giaquinto et al. [68] found that IMA was also part of a predictive model for the discharge Functional Independence Measure (FIM™) [69–72] of 248 subjects with stroke. IMA, measured by De Renzi's test of gesture imitation [73], along with the cognitive and sphincter sub-items of the FIM admission scores and neglect measures of these patients, explained 72 % of the variance on the FIM™. Variables not contributing to the model included sex, lesion side, depression, comorbidities, body mass index, FIM™ motor sub-items, and aphasia [73].

Hanna-Pladdy et al. [74] found a strong relationship between apraxia severity as measured by the Florida Assessment Battery [75] and performance on the Physical Self-Maintenance Scale [76]. Apraxia severity and independence in ADL were correlated significantly ($r=0.664$, $p=0.0001$). Interestingly while functional performance was strongly correlated with gesture production, impairments in the conceptual system as reflected in gesture comprehension were not correlated with functional performance. This study was, however, quite small and involved only ten stroke patients and ten controls.

In one of the largest studies examining apraxia and functioning in daily living, Pedersen et al. [77] examined 618 persons with stroke to determine the relationship between the Barthel Index (BI) [78] and a test of manual apraxia. While apraxia had a weak association with the BI at discharge, a relationship was suggested that persons without manual apraxia were more likely to be discharged to an independent living environment. This study tested apraxia in a very superficial fashion—a three-item transitive pantomime test that was done either by pantomime for nonaphasic or imitation for aphasic participants. This raises the question as to how best to select measures to most clearly determine the extent of this relationship.

Indeed current literature and theory would suggest that the selection of tests to measure apraxia and functional performance needs to be done carefully. For example, mixing transitive and intransitive assessments as measures of apraxia or using them exclusively, as in the Pederson et al. [77] study, affects the ability to detect severity of apraxia, as it is generally known that intransitive tests are performed better than transitive tests [79–83]. Such a difference in severity would affect the potential for finding a relationship between apraxia and function in daily living.

This factor of severity is seen in a 2002 study by Butler who examined the performance of 17 patients with stroke on four test batteries of apraxia [64]. Three of the batteries were purported to measure IMA and the fourth ideational apraxia. Functional independence was also evaluated on the BI. The severity of apraxia varied for each patient based on which battery was the focus of the assessment. Further, while performance on these tests of apraxia was not significantly correlated with the measure of function in daily living, the BI, the correlations varied from $r=0.25$ to $r=0.46$ indicating that the different batteries present a very different picture of the relationship between apraxia and function in daily living.

Studies examining functional performance after intervention for apraxia have also revealed that how performance is measured is critical in examining this relationship. Indices of performance such as the BI are not well predicted by apraxia tests whereas observational measures such as the ADL Observation test do demonstrate a relation to apraxia performance [84, 85].

We examined the apraxia and function relationship in an attempt to address the limitations of the aforementioned studies. In this study, praxis function was considered broadly by utilizing measures from the Waterloo-Sunnybrook Apraxia Battery [86], which involve tests of the conceptual system and a battery of transitive tests, which have been shown to reflect disruptions at different stages in gesture production. These tests included pantomime to command, pantomime to picture, concurrent imitation, delayed imitation, and object use. Functional performance was measured in three ways. First, the FIM™ [69–72] was used as a measure of a broad range of functional skills. The second test selected was the Assessment of Motor and Process Skills (AMPS) [87], a well-established and psychometrically sound observational measure that considers both motor and process (e.g., planning strategy) skills on a number of ADL tasks. Finally a behavioral checklist (Apraxia Behavioral Checklist—or ABCL) was devised to capture any behavior characteristic of apraxia that the participants had demonstrated the preceding week [88]. It included items that were sensory-motor in nature (e.g., appears to have problems paying attention to objects presented visually), conceptual (e.g., sometimes unable to decide what to do with an object or tool), sequencing (e.g., sometimes mixes up the order in which she/he does a task), or production (e.g., sometimes uses an object on the wrong part of the body) behaviors. This checklist was filled out by the occupational therapists that were treating the patients. Thirty people with left brain damage due to stroke were given the battery of tests noted above. Time since stroke, motor function, mental status, aphasia severity, sensation, visual function, and auditory function were examined to determine if these factors contributed to performance of items on the tests.

The conceptual system was tested using several tool and action naming and identification tests, as well as tests of gesture matching and error recognition. Correlation analysis showed that the ABCL was significantly related to four tests: the tool name by action test (i.e., name the tool associated with pantomimed action demonstrated by the examiner), the action identification by tool test (i.e., point to a target pantomimed action based on a tool presented from four videotaped actions), the alternative tool test (i.e., point to the tool that you might be able to use instead of the normal tool for a function), and the gesture recognition test (i.e., point to the action matching one demonstrated by the examiner). The FIM was related only to the tool identification by function test (i.e., point to the tool used to do a particular function). The AMPS process score was related to the tool name by action test. The AMPS motor score was not significantly correlated with any of the conceptual tests. These analyses reveal that functioning in daily living is related to disruptions in conceptual functions in apraxia, particularly those involving action knowledge or tool-action relationships. This relationship appears most strongly in the observational measure of daily function, the ABCL.

Turning now to the measures of gesture production, the ABCL was correlated to all of the tests of apraxia (pantomime, pantomime by picture, tool/object use, delayed imitation, and concurrent imitation). Neither the FIM™ scores nor the AMPS scores were found to be related to any of the tests of apraxia. This finding was surprising, as one would have expected some correlation of gesture performance with the AMPS process score, given that this process score was correlated with action knowledge. When, however, an analysis of variance was performed comparing mild, moderate, and severe apraxic patients on their AMPS process and motor scores on each test of apraxia, a significant effect between groups on the AMPS process scores for the pantomime by picture and a trend for the tool-/object-use test emerged. This analysis did not reveal any significant differences for the FIM™.

Taken together these findings suggest that apraxia is related to deficits in function in daily living. The nature and strength of this relationship, however, appears to depend on the measures of apraxia and functional performance. Deficits in the conceptual system particularly reflected in knowledge of action and tool-action relationships are predictive of impairments in some measures of function (the ABCL, AMPS process scores, and the FIM) but not others (the AMPS motor scores). Similarly, deficits in gesture performance are associated with impairments on some tests of function (the ABCL and AMPS process score) but not others (AMPS motor score and the FIM™). The ABCL was the most highly correlated with apraxic functioning suggesting that observing the impact of an impairment on function might be best done by designing an observation measure with items that capture expected behavioral impairments. The variability in the expression of different apraxic impairments in functional daily living speaks to the importance of breaking down the problem of apraxia into component parts for the purpose of identifying potential areas to consider in intervention.

Summary

In summary, apraxia is a neurological disorder affecting the performance of learned purposive movements that cannot be explained on the basis of deficits of elemental motor or sensory systems, comprehension, or object or tool recognition. There are various types of apraxia that are defined based on the input modality (e.g., pantomime from memory or imitation), whether the memory representation of the gesture is impaired and whether one or both hands are affected. The nature of apraxia is also defined on the basis of the types of errors, which involve aberrations of various features (e.g., spatial, postural, or action) of the to-be-performed gesture. While apraxia most commonly occurs with LHD, there are studies examining the neural correlates of apraxia within the left hemisphere both in stroke patients and in the healthy brain using functional neuroimaging. Many of these studies have associated apraxia with damage to the parietal (left inferior parietal) and frontal (left dorsolateral frontal) cortices as well as the white matter tracts connecting these areas.

A number of other studies have linked damage in the angular and supramarginal gyri of the inferior parietal lobe to apraxia. In order to gain a better understanding of the underlying nature of apraxia, a number of groups have developed models that describe the disruptions at different stages of gestures production. This model-based approach to apraxia has implications for how to assess for the disorder but also what effect apraxia may have on function in daily living. With regard to clinical assessment, this approach argues for the use of a comprehensive battery examining for the integrity of both the conceptual and cognitive-motor stages of gesture production. Turning to the implications for function in daily living, this approach provides a framework within which to examine how apraxia may affect function in daily living. That is, does apraxia affect the conceptual aspects (e.g., knowing what to do with a tool) as well as the motor aspects (e.g., selecting and correctly executing the appropriate response) of daily living functions? Evidence suggests that both of these may be affected by apraxia.

Acknowledgement The section on Apraxia and Functional Performance is derived from work by one of the coauthors, Deborah Hebert, to be submitted as part of her doctoral dissertation at the Ontario Institute for Studies in Education, University of Toronto.

References

1. Rothi JL, Heilman KM. Apraxia: the neuropsychology of action. East Sussex: Psychology; 1997.
2. Donkervoort M, Dekker J, van den Ende E, Stehmann-Saris JC, Deelman BG. Prevalence of apraxia among patients with a first left hemisphere stroke in rehabilitation centres and nursing homes. Clin Rehabil. 2000;14(2):130–6.
3. Goldenberg G. Apraxia and beyond: life and work of Hugo Liepmann. Cortex. 2003;39:509–24.
4. Liepmann H. Apraxie. Ergebn Ges Med. 1920;1:516–43.
5. Vanbellingen T, Bohlhalter S. Apraxia in neurorehabilitation: classification, assessment and treatment. NeuroRehabilitation. 2011;28(2):91–8.
6. Poeck K. Ideational apraxia. J Neurol. 1983;230(1):1–5.
7. Heilman KM, Rothi LJ, Valenstein E. Two forms of ideomotor apraxia. Neurology. 1982;32:342–6.
8. Denes GD, Montovan GC, Gallana A, Cappelletti JY. Limb kinetic apraxia. Mov Disord. 1998;13:468–76.
9. Heilman KM, Meador KJ, Loring DW. Hemispheric asymmetries in limb kinetic apraxia: a loss of deftness. Neurology. 2000;55:523–6.
10. Goldmann Gross R, Grossman M. Update on apraxia. Curr Neurol Neurosci Rep. 2008;8:490–6.
11. Geschwind N. The apraxias: neural mechanisms of disorders of learned movements. Am Sci. 1975;63:188–95.
12. Riddoch MJ, Humphreys GW, Price CJ. Routes to action—evidence from apraxia. Cogn Neuropsychol. 1989;6:437–54.
13. Gibson JJ. The concept of affordances. In: Shaw RE, Bransford J, editors. Perceiving, acting, and knowing. Hillsdale, NJ: Lawrence Erlbaum; 1977.
14. Tucker M, Ellis R. On the relations between seen objects and components of potential actions. J Exp Psychol Hum Percept Perform. 1998;24:830–46.

15. Heilman KM, Maher LM, Greenwald ML, Rothi LJ. Conceptual apraxia from lateralized lesions. Neurology. 1997;49:457–64.
16. Ochipa C, Rothi LJG, Heilman K. Ideational apraxia: a deficit in tool selection. Ann Neurol. 1989;25:190–3.
17. Rumiati RI, Tessari A. Imitation of novel and well known actions: the role of short term memory. Exp Brain Res. 2002;142:425–33.
18. Tessari A, Rumiati RI. The strategic control of multiple routes in imitation of actions. J Exp Psychol Hum Percept Perform. 2004;30:1107–16.
19. Tessari A, Bosanac D, Rumiati RI. Effect of learning on imitation of new actions: implications for a model of memory. Exp Brain Res. 2006;173:507–13.
20. Montavani-Nagoako J, Ortiz KZ. Reviewing the limb apraxia concept from definition to cognitive neuropsychological models. Dement Neuropsychol. 2010;4:165–72.
21. Ochipa C, Rothi LJ, Heilman KM. Conduction apraxia. J Neurol Neurosurg Psychiatry. 1994;57:1241–4.
22. Ochipa C, Rothi LJ, Heilman KM. Conceptual apraxia in Alzheimer's disease. Brain. 1992;115:1061–71.
23. Haaland KY, Harrington DL, Knight RT. Neural representations of skilled movement. Brain. 2000;123:2306–13.
24. Halsband U, Schmitt J, Weyers M, Binkofski F, Grützner G, Freund HJ. Recognition and imitation of pantomimed motor acts after unilateral parietal and premotor lesions: a perspective on apraxia. Neuropsychologia. 2001;39:200–16.
25. Goldenberg G, Karnath HO. The neural basis of imitation is body part specific. J Neurosci. 2006;26(23):6282–7.
26. Manuel AL, Radman N, Mesot D, Chouiter L, Clarke S, Annoni JM, et al. Inter- and intrahemispheric dissociations in ideomotor apraxia: a large-scale lesion-symptom mapping study in subacute brain-damaged patients. Cereb Cortex. 2012;22:2451–62.
27. Schwartz MF, Segal M, Veramonti T, Ferraro M, Buxbaum LJ. The Naturalistic Action Test: a standardised assessment for everyday action impairment. Neuropsychol Rehabil. 2002;12:311–39.
28. Schnider A, Hanlon RE, Alexander DN, Benson DF. Ideomotor apraxia: behavioural dimensions and neuroanatomical basis. Brain Lang. 1997;58:125–36.
29. Roy EA, Black SE, Blair N, Dimeck PT. Analyses of deficits in gestural pantomime. J Clin Exp Neuropsychol. 1998;20(5):628–43.
30. Roy EA, Heath M, Westwood D, Schweizer TA, Dixon MJ, Black SE, et al. Task demands and limb apraxia in stroke. Brain Cogn. 2000;44(2):253–79.
31. Wheaton LA, Bohlhalter S, Nolte G, Shibasaki H, Hattori N, Fridman E, et al. Cortico-cortical networks in patients with ideomotor apraxia as revealed by EEG coherence analysis. Neurosci Lett. 2008;433:87–92.
32. Alexander MP, Baker E, Naeser MA, Kaplan E, Palumbo C. Neuropsychological and neuroanatomical dimensions of ideomotor apraxia. Brian. 1992;115:87–107.
33. Petreska B, Adriani M, Blanke O, Billard AG. Apraxia: a review. In: Von Hosten C, Rolander K, editors. Progress in brain research, vol. 164. Elsevier: Amsterdam; 2007. p. 61–83.
34. Buxbaum LJ, Johnson-Frey SH, Bartlett-Williams M. Deficient internal models for planning hand-object interactions in apraxia. Neuropsychologia. 2005;43:917–29.
35. Haaland KY, Harrington DL, Knight RT. Spatial deficits in ideomotor limb apraxia. A kinematic analysis of aiming movements. Brain. 1999;122:1169–82.
36. Buxbaum LJ, Sirigu A, Schwartz MF, Klatzky R. Cognitive representations of hand posture in ideomotor apraxia. Neuropsychologia. 2003;41:1091–113.
37. Goldenberg G. Imitation and matching of hand and finger postures. Neuroimage. 2001;14:S132–6.
38. Heilman KM, Rothi LG, Mack L, Feinberg T, Watson RT. Apraxia after a superior parietal lesion. Cortex. 1986;22(1):141–50.
39. Watson RT, Fleet WS, Rothi LJG, Heilman KM. Apraxia and the supplementary motor area. Arch Neurol. 1986;43(8):787–92.

40. Buxbaum LJ, Kyle KM, Menon R. On beyond mirror neurons: internal representations subserving imitation and recognition of skilled object-related actions in humans. Brain Res Cogn Brain Res. 2005;25:226–39.
41. Peigneux P, Salmon E, Garraux G, Laureys S, Willems S, Dujardin K, et al. Neural and cognitive bases of upper limb apraxia in corticobasal degeneration. Neurology. 2001;57:1259–68.
42. Stamenova V, Roy EA, Black SE. A model-based approach to understanding apraxia in corticobasal syndrome. Neuropsychol Rev. 2009;19(1):47–63.
43. Stamenova V, Roy EA, Black SE. Associations and dissociations of transitive and intransitive gestures in left and right hemisphere stroke patients. Brain Cogn. 2010;72:483–90.
44. De Renzi E, Lucchelli F. Ideational apraxia. Brain. 1988;111:1173–85.
45. Goldenberg G. Imitating gestures and manipulating a manikin: the representation of the human body in ideomotor apraxia. Neuropsychologia. 1995;33:63–72.
46. Bizzozero I, Costato D, Sala SD, Papagno C, Spinnler H, Venneri A. Upper and lower face apraxia: role of the right hemisphere. Brain. 2000;123:2213–30.
47. Wheaton LA, Hallett M. Ideomotor apraxia: a review. J Neurol Sci. 2007;260:1–10.
48. Pramstaller PP, Marsden CD. The basal ganglia and apraxia. Brain. 1996;119:319–40.
49. Hong JH, Lee J, Cho YW, Byun WM, Cho HK, Son SM, et al. Limb apraxia in a patient with cerebral infarct: diffusion tensor tractography study. NeuroRehabilitation. 2012;30:255–9.
50. Johnson-Frey SH, Newman-Norlund R, Grafton ST. A distributed left hemisphere network during planning of everyday tool use skills. Cereb Cortex. 2005;16:681–95.
51. Moll J, de Oliviera-Sousa R, Passman LJ. Functional MRI correlates of real and imagined tool-use pantomimes. Neurology. 2000;54:1331–6.
52. Kroliczak G, Frey SH. A common network in the left cerebral hemisphere represents planning of tool use pantomimes and familiar intransitive gestures at the hand-independent level. Cereb Cortex. 2009;19:2396–410.
53. Povinelli DJ, Reaux JE, Frey SH. Chimpanzees context-dependent tool use provides evidence for separable representations of hand and tool even during active use within peripersonal space. Neuropsychologia. 2009;48(1):243–7.
54. Buxbaum LJ, Saffran EM. Knowledge of object manipulation and object function: dissociations in apraxic and nonapraxic patients. Brain Lang. 2002;82:179–99.
55. Boronat CB, Buxbaum LJ, Coslett HB, Tang K, Saffran EM, Kimberg DY, et al. Distinctions between manipulation and function knowledge of objects: evidence from functional magnetic resonance imaging. Brain Res Cogn Brain Res. 2005;23:361–73.
56. Buxbaum LJ. Ideomotor apraxia: a call to action. Neurocase. 2001;7:445–58.
57. Cubelli R, Marchetti C, Boscolo C, Della SS. Cognition in action: testing a model of limb apraxia. Brain Cogn. 2000;44:144–65.
58. Stamenova V, Black SE, Roy EA. An update on the conceptual-production systems model of apraxia: evidence from stroke. Brain Cogn. 2012;80(1):53–63.
59. Roy EA. Hand preference, manual asymmetries, and limb apraxia. In: Elliot D, editor. Manual asymmetries in motor control. Boca Raton, FL: CRC; 1996. p. 215.
60. Roy EA, Hall C. Limb apraxia: a process approach. In: Elliott D, Proteau L, editors. Vision and motor control. Amsterdam: Elsevier; 1992. p. 261–82.
61. Roy EA, Square PA. Common considerations in the study of limb, verbal, and oral apraxia. In: Roy EA, editor. Neuropsychological studies of apraxia and related disorder. Amsterdam: North-Holland; 1985. p. 111–6.
62. Butler J. Intervention effectiveness: evidence from a case study of ideomotor and ideational apraxia. Br J Occup Ther. 1997;60(11):491–7.
63. Butler J. Rehabilitation in severe ideomotor apraxia using sensory stimulation strategies: a single-case experimental design study. Br J Occup Ther. 2000;63(7):319–28.
64. Butler J. How comparable are tests of apraxia? Clin Rehabil. 2002;16:389–98.
65. Foundas AL, Macauley BL, Raymer AM, Maher LM, Heilman KM, Gonzalez Rothi LJ. Ecological implications of limb apraxia: evidence from mealtime behaviour. J Int Neuropsychol Soc. 1995;1:62–6.

66. Bjorneby ER, Reinvang IR. Acquiring and maintaining self-care skills after stroke. Scand J Rehabil Med. 1985;17:75–80.
67. Sundet K, Finset A, Reinvang I. Neuropsychological predictors in stroke rehabilitation. J Clin Exp Neuropsychol. 1988;10:363–79.
68. Giaquinto S, Buzzelli S, Di Francesco L, Lottarini A, Montenero P, Tonin P, et al. On the prognosis of outcome after stroke. Acta Neurol Scand. 1999;100(3):202–8.
69. Guide for the Uniform Data Set for Medical Rehabilitation (including the FIM(TM) instrument), Version 5.1. Buffalo, NY 14214-3007: State University of New York at Buffalo; 1997.
70. Deutsch A, Braun S, Granger C. The functional independence measure and the functional independence measure for children: ten years of development. Crit Rev Phys Rehabil Med. 1996;8(4):267–81.
71. Granger CV, Hamilton BB, Keith RA, Zeilezny M, Sherwin FS. Advances in functional assessment for medical rehabilitation. Top Geriatr Rehabil. 1986;1:59–74.
72. Beninato M, Gill-Body KM, Salles S, Stark PC, Black-Schaffer RM, Stein J. Determination of the minimal clinically important difference in the FIM instrument in patients with stroke. Arch Phys Med Rehabil. 2006;87(1):32–9.
73. De Renzi E. Apraxia. In: Boller F, Grafman J, editors. Handbook of neuropsychology. Amsterdam: Elsevier; 1989. p. 245–63.
74. Hanna-Pladdy B, Heilman KM, Foundas AL. Ecological implications of ideomotor apraxia—evidence from physical activities of daily living. Neurology. 2003;60(3):487.
75. Gonzalez Rothi LJ, Raymer AM, Heilman KM. Limb praxis assessment. In: Gonzalez Rothi LJ, Heilman KM, editors. Apraxia: the neuropsychology of action. East Sussex: Psychology; 1997. p. 61–74.
76. Lawton MP, Brody EM. Assessment of older people: self-maintaining and instrumental activities of daily living. Gerontologist. 1969;9:179–86.
77. Pedersen PM, Jørgensen HS, Kammersgaard LP, Nakayama H, Raaschou HO, Olsen TS. Manual and oral apraxia in acute stroke, frequency and influence on functional outcome: The Copenhagen Stroke Study. Am J Phys Med Rehabil. 2001;80(9):685–92.
78. Mahoney FI, Barthel DW. Functional evaluation: The Barthel Index. Md State Med J. 1965;14:61–5.
79. Foundas AL, Macauley BL, Raymer AM, Maher LM, Rothi LJ, Heilman KM. Ideomotor apraxia in Alzheimer disease and left hemisphere stroke: limb transitive and intransitive movements. Neuropsychiatry Neuropsychol Behav Neurol. 1999;12(3):161–6.
80. Haaland KY, Flaherty D. The different types of limb apraxia errors made by patients with right and left hemisphere damage. Brain Cogn. 1984;3:370–84.
81. Mozaz M, Rothi LJ, Anderson JM, Crucian GP, Heilman KM. Postural knowledge of transitive pantomimes and intransitive gestures. J Int Neuropsychol Soc. 2002;8:958–62.
82. Rothi LJ, Mack L, Verfaellie M, Brown P, Heilman KM. Ideomotor apraxia: error pattern analysis. Aphasiology. 1988;2:381–7.
83. Roy E, Square-Storer P, Hogg S, Adams S. Analysis of task demands in apraxia. Int J Neurosci. 1991;56:177–86.
84. Donkervoort M, Dekker J, Deelman BG. Sensitivity of different ADL measures to apraxia and motor impairment. Clin Rehabil. 2002;16:299–305.
85. Donkervoort M, Dekker J, Deelman BG. A course of apraxia and ADL functioning in left hemisphere stroke patients treated in rehabilitation centres and nursing homes. Clin Rehabil. 2006;20:1085–93.
86. Stamenova V, Black SE, Roy EA. An update on the conceptual-production systems model of apraxia: evidence from stroke. Brain Cogn. 2012;80:53–63.
87. Fisher AG. AMPS: assessment of motor and process skills: development, standardization and administration manual. Colorado: Three Star; 2003.
88. Hebert D, Roy E. Limb apraxia: a clinical perspective. Geriatr Aging. 2002;5:15–21.

Chapter 5
Spatial Neglect: Not Simply Disordered Attention

James Danckert

Introduction

A common consequence of right-sided brain damage is the syndrome of spatial neglect, which is seen in 40–90 % of patients in the acute phase after stroke [1–4]. The disorder is a debilitating one in which patients fail to attend to stimuli in contralesional, left space [5–8]. Patients with neglect typically require assistance with most activities of daily living and often fail to dress or groom the left side of their body, leave food uneaten on the left side of a plate, and bump into obstacles on their left. In this chapter I first describe the "classic" case of neglect before describing recent work painting a more complicated picture of the constellation of deficits making up the syndrome. I will then review the recent debate regarding the critical lesion site for neglect before discussing recent rehabilitation attempts. The past few decades of research has shown that neglect can no longer be characterized simply as a disorder of attention.

The "Classic" Case of Spatial Neglect

Damage to right inferior parietal or superior temporal cortex often leads to the neglect syndrome, which can best be described as a disorder of consciousness in which the patient's mental model of the world is constricted to include only the right side of space. In other words, the patient behaves as if left space has simply vanished [6–8]. In most of these cases, and in examples provided in this chapter, the etiology of the brain damage is ischemic stroke, which is one of the most common

J. Danckert, Ph.D. (✉)
Department of Psychology, University of Waterloo, 200 University Avenue West, Waterloo, ON, Canada N2L 3G1
e-mail: jdancker@uwaterloo.ca

T.A. Schweizer and R.L. Macdonald (eds.), *The Behavioral Consequences of Stroke*,
DOI 10.1007/978-1-4614-7672-6_5, © Springer Science+Business Media New York 2014

causes of neglect. Neglect can be observed for right space following left parietal damage; however, this occurs less frequently and symptoms are both less severe than those of left spatial neglect patients and tend to resolve more fully [1, 2, 9, 10]. Neglect is demonstrated via a range of bedside pencil-and-paper tests [11], the three most common of which are line bisection, figure copying, and cancellation (Fig. 5.1). For line bisection the patient is shown a (set of) horizontal line(s) at their body midline and must place a mark to indicate where the center is. Neglect patients typically place their mark to the right of true center. For figure copying, patients must produce a copy of line drawings of common objects with left-sided details distorted or omitted by neglect patients. For cancellation tasks patients are shown an array of targets (e.g., stars) and must place a mark (i.e., to "cancel") through all the targets, with neglect patients failing to cancel left-sided targets. These tasks represent only the most common tests of neglect with a wide range of tasks commonly used including free drawing, clock drawing, tactile search, and arranging objects on a surface (i.e., the baking tray task) (Fig. 5.1).

A related disorder known as extinction is also assessed at the bedside [12, 13] by having patients fixate on the examiner's nose while attending to the examiner's left and right index fingers held out in the periphery. The patient must detect movement in the fingers with the examiner moving either the left or right finger alone (i.e., single stimulation) or both fingers together (i.e., double simultaneous stimulation). Patients with extinction can detect single events in left or right space, but "extinguish" the contralesional event in simultaneous stimulation trials. Neglect and extinction do co-occur; however, extinction can be evident in the absence of neglect and is equally common following left or right brain damage [12, 13]. Extinction and neglect also have distinct lesion foci [12, 14].

Although neglect is most prevalent for vision, it is commonly observed for auditory and tactile stimuli [5, 7, 8]. In addition, neglect is independent of low-level perceptual deficits and is best characterized not as a primary motor or perceptual deficit, but as a representational impairment. Perhaps the most famous demonstration of this comes from two patients tested by Bisiach and Luzzatti [15]. The patients were asked to imagine standing in a square in Milan and to report details they "saw" in their mind's eye. When the patients were asked to imagine standing in the south, facing north, they reported details of the square from the right (east) and neglected details on the left (west). Conversely, when asked to imagine standing at the north, facing south, the patients now reported details from the previously neglected west and neglected details that they had initially reported from the east [15]. Similar demonstrations of what some have called "imaginal" neglect have been shown in other patients, including two with right hemisphere ischemic strokes [16, 17]. What this shows is that neglect patients have no deficit in recalling the details of mental representations. Instead, their mental representations are impoverished, containing only the right half of egocentric space.

As any clinician familiar with neglect can attest to, few patients fit the classic case. Neglect varies in terms of the reference frame within which symptoms predominate (e.g., personal versus extra-personal space) [18–23]. While most patients show neglect for stimuli defined strictly by spatial location, some show neglect for the left half of objects regardless of location [24, 25]. Finally, a plethora of

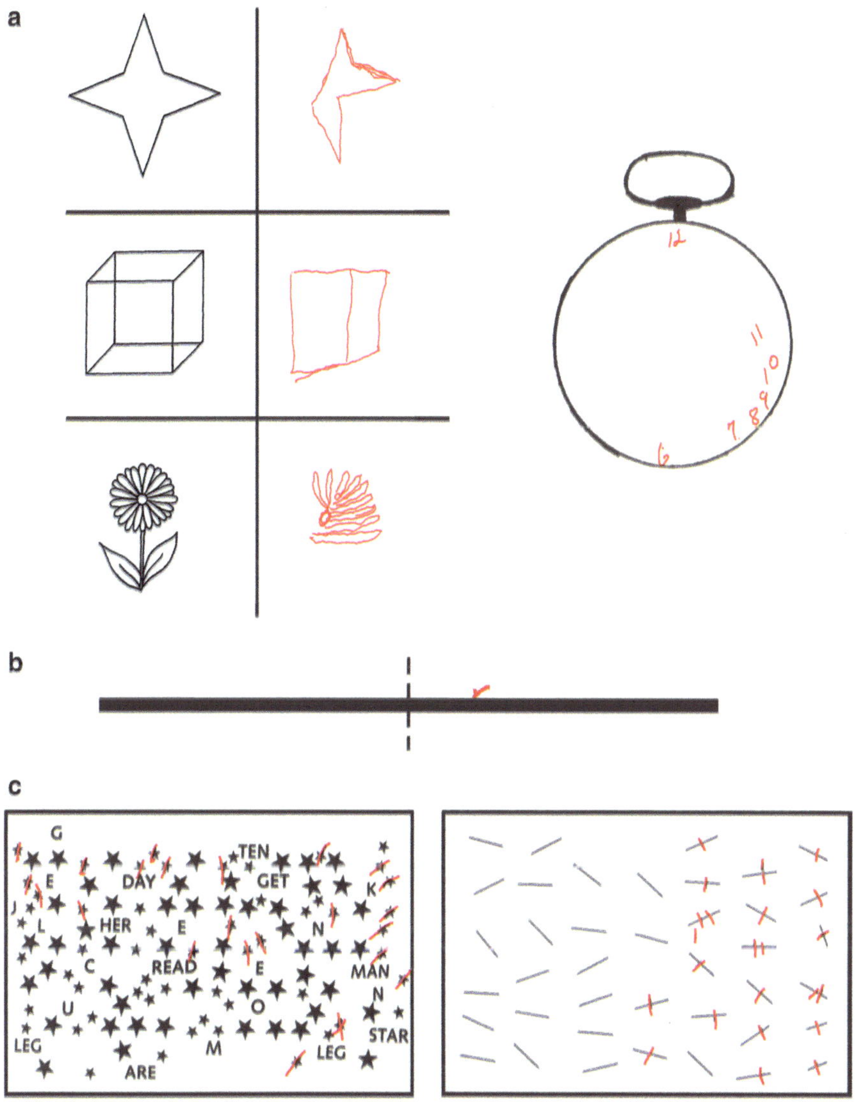

Fig. 5.1 Examples of performance on clinical tests of neglect (patient's performance is in *red* throughout). (**a**) Figure copying from the Behavioral Inattention Test [11] and clock drawing. (**b**) Line bisection performance with true center marked by the *dashed line* (*note*, this detail is absent in the actual test). (**c**) Two forms of cancellation task; star cancellation to the *left* and Albert's lines to the *right*

"sub-syndromes" of neglect symptoms abound. For example, neglect dyslexia, in which patients fail to read the left half of words, is present in a handful of patients [26–28]. Although neglect was initially described some 60 years ago [29–31], the development of a conclusive theoretical account of the syndrome has proven elusive, in part due to the heterogeneity of symptoms.

Neglect as a Disorder of Spatial Attention

Traditional models of neglect stress the spatial nature of deficits [5–8, 32, 33]. In addition, given that neglect is more common after right-sided stroke or other brain damage, some suggest that the right parietal cortex is specialized for spatial attention [8, 34–37]. There are two component deficits to attention-based models of neglect: first, attention is captured strongly by right-sided stimuli. Second, patients have difficulty redirecting attention towards left space once attention is captured by right-sided stimuli. In other words, patients have difficulty disengaging attention from the right (i.e., a disengage deficit) [38, 39] (Fig. 5.2).

Attentional capture for right-sided events can be seen in many perceptual biases in neglect. One such bias is evident on the chimeric faces task in which two vertically aligned chimerics (faces smiling on one side and neutral on the other) are shown with the patient indicating which appears to be happier. Controls choose the face shown as smiling on the left, reflective of right hemisphere dominance for emotional processing [40–42]. Neglect patients show a strong bias for choosing the face shown as smiling on the right [43]. Such biases are not unique to faces. When shown two rectangles that change in intensity from light to dark, patients choose the rectangle with the darkest end on the right as appearing darker—the opposite bias to that of controls [43]. Similar biases in intensity judgments can be seen for a broad range of stimuli, including numbers [44–47]. What these biases suggest is that for neglect patients, their attention is robustly captured by stimulus properties on the right [24]. Under some circumstances these biases actually confer a benefit to neglect patients, such that manual or saccadic reaction times are faster (relative to controls) for targets presented to restricted regions of right-sided space [48–51].

The second attentional impairment invoked in neglect models is the so-called disengage deficit. Best characterized as a difficulty in *reorienting* attention, Posner and colleagues [38, 39] showed that when patients with parietal lesions were cued to attend to ipsilesional space, their ability to detect contralesional targets was slowed (Fig. 5.2). Although evident after both left and right parietal damage and in patients with and without neglect, the impairment is most severe in neglect [52, 53]. In traditional models of neglect then, it is the combination of rightward attentional capture and a reorienting impairment (i.e., the disengage deficit) that determines the failure to respond to events in left space.

Deficits of Spatial Attention Fail to Capture the Full Syndrome of Neglect

It has become increasingly evident that attentional models of neglect do not capture the full gamut of symptoms evident in the syndrome. This has led to the suggestion that neglect is due to a constellation of symptoms that make it difficult for the patient to create and make use of a full and accurate representation of their surroundings [5]. Several of these components are discussed as follows.

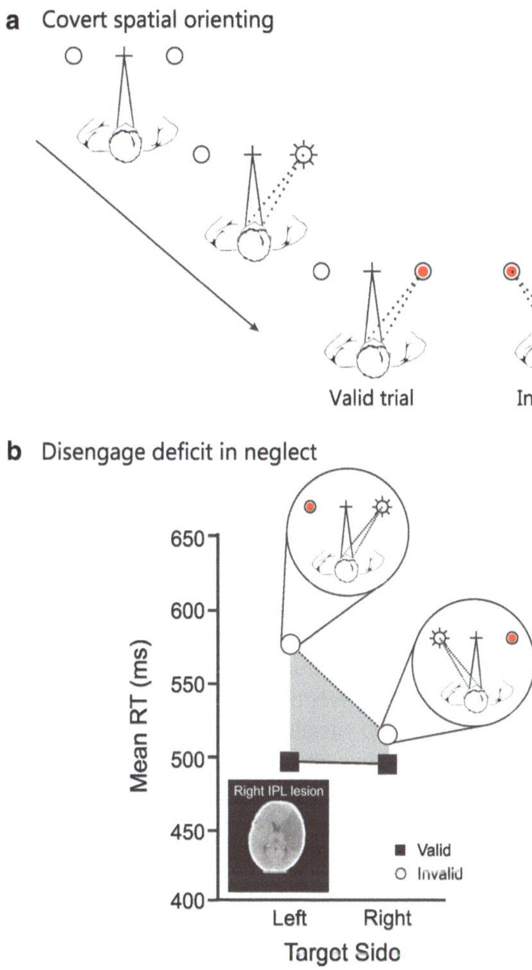

Fig. 5.2 (**a**) Schematic representation of trial sequences in the covert orienting of visual attention task [38, 39]. Patients fixate a central cross throughout while covertly attending to landmarks to the *left* and *right*. One landmark is cued (*sun symbol* in figure) drawing covert attention (*dotted line*) to that location. Targets can then appear at the cued (valid trial) or uncued (invalid trial) location with reaction times (RTs) faster on valid trials. (**b**) Performance of a neglect patient with a right inferior parietal lobule (IPL) lesion arising from a middle cerebral artery stroke. RTs are slowest to contralesional (*left*) targets appearing after an ipsilesional (*right*) cue (i.e., an invalid trial with the target appearing on the *left*)

Deficits in Nonspatial Sustained Attention

Robertson and colleagues [54, 55] first demonstrated sustained attention deficits in neglect using a simple task—the “Elevator Counting Task” [56, 57]. Patients were given between 3 and 14 tones separated by 3–5 s and had to report the number of tones presented. Neglect patients performed poorly on this task when compared with right brain-damaged patients without neglect [54]. Moreover, poor sustained attention correlated with neglect severity [54, 55]. The authors [54] suggest that poor sustained attention represents a critical marker of neglect and acts to exacerbate the more obvious spatial symptoms of the disorder [58–60]. Support for this notion comes from attempts to rehabilitate patients with neglect, which relies on modulating arousal levels [54, 55, 60, 61]. That is, when neglect patients are given brief loud tones intended to induce a *phasic* change to arousal, there is a concomitant improvement in some of the spatial symptoms of the disorder [61].

The ability to allocate attention over time is also impaired in neglect [62]. In the attentional blink task used to index this, patients attend to a stream of alpha-numeric stimuli and detect two targets embedded in the stream, with different temporal distances between the first and second targets [63, 64]. Controls show a diminished ability for discriminating target two when it is presented in close temporal proximity to target one (although there is spared capacity at the briefest lag) [65]. That is, once resources are allocated to target one, there is a refractory period—the attentional blink—during which those resources cannot be fully marshaled to discriminate target two. Neglect patients show an attentional blink almost 3 times longer than that observed in controls [62]. It should be noted that research shows an increased attentional blink with lesions of the left or right superior temporal gyrus (STG) [66], frontal cortex [67], and even cerebellum [68]. Given that impaired temporal allocation of attention is not unique to neglect, many suggest that this, and related deficits, reflect a disruption to *tonic* arousal levels that arise as a consequence of any neural insult. As such, impaired nonspatial sustained attention and poor temporal allocation of attention are thought to represent exacerbating symptoms that worsen the cardinal feature of neglect—an inability to consciously represent left space [59].

Spatial Working Memory

Two aspects of cancellation performance hint at problems with spatial working memory (SWM) in neglect that are not limited to left space. First, patients often fail to cancel targets in right space, suggesting they have a faulty representation of the spatial layout of the environment even for stimuli appearing in "good" or non-neglected space [5, 69]. Second, patients often place multiple cancellation marks on a single target, treating already processed or "old" items as if they are "new" (see Fig. 5.3). The tendency to treat old items as if they were new is especially evident when the patient's cancellation marks are hidden from view [70, 71]. Conversely, revisiting behavior can be *reduced* when the salience of targets is reduced (or the target is removed) once cancelled by the patient [72].

Cancellation tasks are essentially a test of visual search performance, involve complex stimuli, and are reliant on multiple factors (e.g., perceptual discrimination, reorienting attention) to achieve good performance. Failure on such tasks could be due to impairments on any one or a combination of those factors. To address this, we (and others) have explored SWM in neglect [69, 73]. We asked patients to keep in mind over a short delay three vertically aligned targets. After the delay a probe appeared, with the patient indicating whether the probe appeared in one of the previous target locations (Fig. 5.3). A group of four neglect patients showed a severe deficit on this task even when the information to be remembered was presented to right, non-neglected, space [69]. This was in contrast to their normal ability to maintain alphanumeric stimuli over an identical delay [69]. This highlights that neglect patients do not have a generalized working memory deficit, but have a specific difficulty in maintaining *spatial* information over short periods of time. The fact that these deficits are not restricted to left, neglected space, but are instead evident for both central and right space [69, 73], makes it difficult to explain in the context of impaired allocation of spatial attention.

Impaired Perception of Time

Impoverished perception of time is another key deficit evident in neglect that defies explanation in terms of disrupted spatial or sustained attention [74–77]. Basso and colleagues [74] showed in one neglect patient that stimuli presented in the leftmost positions of an array (all items were presented in right, non-neglected space) were overestimated, whereas rightmost stimuli were underestimated. Harrington and colleagues [78] showed that patients with right frontoparietal lesions were unable to discriminate sub-second temporal durations. Finally, we showed that neglect patients dramatically underestimated multi-second durations [75]. Patients attended to an illusory motion stimulus for between 5 and 60 s and reported aloud the appearance of single digits presented at random temporal intervals (to prevent internal counting of durations). Interestingly, many patients failed to report several of these digits, mirroring the results of the aforementioned Elevator Counting and Attentional Blink tasks (although on different time scales). Each patient massively underestimated the elapsed time, never making estimates of greater than 10 s for what was actually a 60-s interval [75]. In another patient, we showed that this deficit was evident for visual, verbal, and nonverbal auditory stimuli, suggesting that impaired time perception is multimodal in neglect [76]. Finally, we tested two right medial temporal lobectomy patients on the visual time estimation task. Despite demonstrable memory deficits, these patients produced temporal estimates that were well within the range of healthy controls (Locklin and Danckert, unpublished data). This indicates that accurate time perception does not rely heavily on working memory processes, which were deficient in these patients (Fig. 5.4).

An inability to accurately represent time in and of itself cannot explain, nor be explained by the *spatial* nature of neglect. What this kind of deficit represents is a difficulty in accounting for a key component of an ever-changing environment—the temporal dynamics of those changes. For a neglect patient who is already biased to attend only to right space, an inability to time stamp changes to incoming information will compound this deficit, making it more difficult to represent changes in the environment, particularly when they occur in left space.

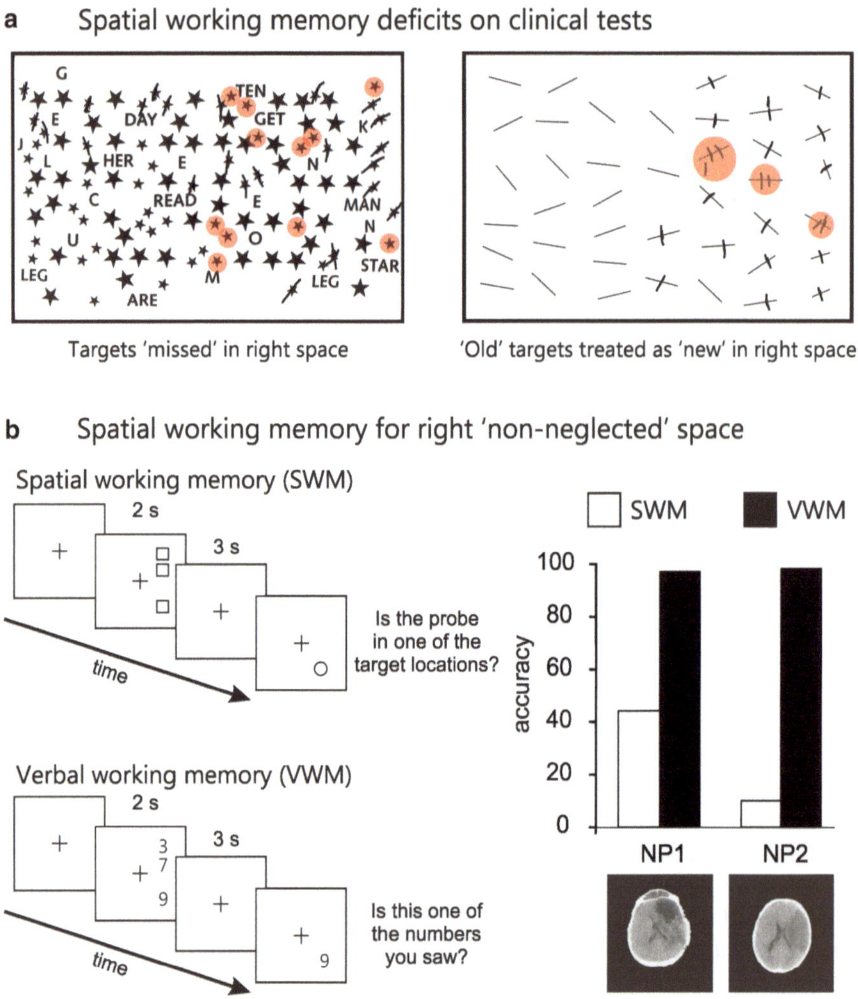

Fig. 5.3 (**a**) Spatial working memory (SWM) deficits on clinical tests of neglect—cancellation tasks. To the *left* is performance on the star cancellation task with targets missed on the *right* side of the array highlighted by *red circles*. To the *right* is performance on Albert's lines task with targets that were revisited (i.e., "old" targets treated as if they were "new") highlighted by *red circles*. (**b**) Performance of two neglect patients on a SWM and verbal working memory (VWM) task. For the SWM task, patients were presented with three vertically aligned targets and kept these locations in mind over a delay. A circle probe then appeared with the patient indicating whether the probe was presented in one of the previous target locations. For the VWM task patients were presented with three numbers to keep in mind over a delay and had to indicate whether a probe number presented after the delay was among one of the target group. To the right is accuracy (hits—false alarms) for both the SWM (*white bars*) and VWM (*black bars*) for two neglect patients. Adapted from [69]

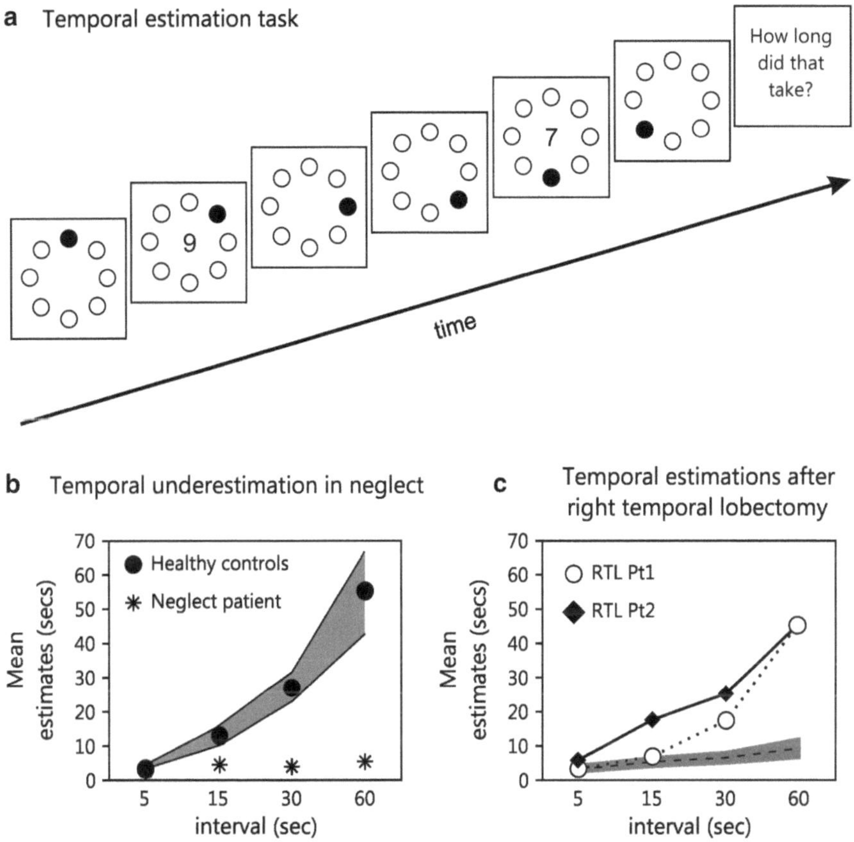

Fig. 5.4 (**a**) Schematic representation of the visual time estimation task. Patients saw an illusory motion stimulus for between 5 and 60 s and reported aloud numerals presented at random temporal intervals (to prevent internal counting of durations). (**b**) Mean estimates for healthy controls (*black circles*; SE represented by the *gray bar*) and one neglect patient (*asterisk*; see [75] for data from eight neglect patients and right brain-damaged patients without neglect). (**c**) Data from two patients who had undergone right temporal lobectomies (RTL) as treatment for medication resistant epilepsy. Both patients show adequate temporal estimations despite substantial working memory deficiencies indicating that for neglect patients, disordered time estimation is unlikely to be related to memory capacity

Representational Updating in Neglect

As discussed earlier, neglect has been considered a disorder of mental representations [15]. In other words, neglect represents an inability to create or appropriately make use of mental models of the environment. Many of the spatial and nonspatial deficits (or at the very least non-spatially lateralized deficits—cf. SWM) in neglect can be parsimoniously characterized as deficits in mental model building and updating. Mental models have been shown to subserve a wide variety of functions including inferring intention (i.e., theory of mind) [79], predicting the sensory consequences of actions (a capacity impaired after parietal damage) [80–83], and learning new skills based on prior experiences [84].

Recent theories have suggested that neglect is best characterized as a combination of deficits in spatial attention and difficulties in remapping space [85]. The task used to test spatial remapping is known as the double-step saccade task, in which two targets for successive eye movements are presented in rapid succession [86]. Programming both saccades based on retinal signals leads to an erroneous saccade to target two. Instead, we remap our model of space based on the anticipated outcome of the saccade to target one. Spatial remapping is clearly deficient after parietal lobe injury, due in the cited cases to aneurysmal subarachnoid hemorrhage, ischemic stroke, and brain tumor surgery [86, 87]. Nevertheless, deficits in spatial remapping still fail to account for many of the symptoms discussed previously including poor SWM, which is not restricted to any particular region of space and impaired perception of time [69, 75, 76]. In addition, recent work employing prisms to rehabilitate neglect (see section "Recovery from Neglect") has shown that while some spatial symptoms can be improved, many other aspects, including perceptual biases, SWM, and time perception deficits, remain unchanged [88, 89]. This suggests there are dissociable components to the neglect syndrome, only some of which are ameliorated via attentional training.

One of the challenges in theorizing about neglect arises due to the heterogeneity of symptom profiles [5, 32]. What is suggested here is that neglect likely arises as a consequence of a *combination* of key impairments including poor spatial orienting and sustained attention and what could be characterized as a generalized impairment in updating mental models. With respect to the latter, deficits in saccadic remapping, SWM, motor imagery, time perception, and to some degree even spatial attention [90] are recast as impairments to the ability to generate and make use of mental models of the environment [91, 92].

Mental models represent learned rules and expectations concerning the way in which the world operates and enable us to simplify the processing needed to control behavior in flexible and optimal ways. Necessarily, these models require frequent updating as new information indicates a change in environmental state. Such updating requires some form of comparison to determine whether new information matches the rules and expectations of our model. When mismatches arise, the degree or type of mismatch determines whether the model needs updating. This kind of comparator process has been invoked to describe the role of parietal cortex in motor control and motor imagery [83, 93–95]. What is being proposed here is that a more generic kind of mental model updating process could be ascribed to the inferior parietal lobule that, when damaged, would explain many of the key symptoms of neglect—symptoms that defy a simple attentional explanation. If the inferior parietal lobule is found to support the generation and updating of mental models, one would expect to see updating deficits evident not only in spatial (e.g., impaired saccadic remapping, SWM) domains, but also in nonspatial domains.

We examined whether neglect patients would show a deficit in updating mental models in a nonspatial domain by having them play the children's game "rock, paper, scissors" (Fig. 5.5) [96].

Their computer opponent initially played a uniform strategy choosing one option on 33 % of trials (no strategy the patient could adopt will lead to more than 33 % wins). Later, the computer adopted a biased strategy choosing one option ("paper") on 80 % of trials. The logic here is that an accurate model of your opponent's play will allow you to adjust your own strategy to exploit that bias and maximize wins (e.g., choose "scissors" most often). Five of seven neglect patients failed to take advantage of the biased play of the computer opponent (Fig. 5.5) [96]. The two who did alter their play took more than 3 times as long as controls to do so and never reached the optimal levels of controls. We also observed impaired performance in some right brain-damaged patients without neglect, indicating that an updating deficit is not unique to neglect. This is an all too common refrain in the neglect literature with even what is sometimes considered to be a cardinal deficit (i.e., impaired orienting to left space) showing up in right brain-damaged patients without neglect (note: the same is true for saccadic remapping and temporal deficits, although impairments are consistently worst for neglect patients) [53, 66, 75, 87]. What this indicates is that no single deficit alone is sufficient to produce the full neglect syndrome. Instead, neglect arises as a consequence of a constellation of key impairments, one of which is the inability to generate or update mental models of the environment across multiple domains and behaviors.

What the preceding sections emphasize is that a neglect patient already captured by events in right space will experience great difficulty exploring the world beyond that region, given that his/her internal model of the environment either is poorly constructed in the first instance or does not undergo appropriate updating relevant to changes in the environment in the second instance.

Anatomy of Neglect

As already mentioned, neglect is a heterogeneous syndrome. The precise array of deficits evident in any patient will vary with lesion location and the extent to which white matter pathways are involved. This section outlines some of the recent work exploring the critical lesion site for neglect.

What Is the Critical Lesion for Neglect?

The earliest descriptions of neglect implicated damage to the supramarginal and angular gyri [30, 31]. Many early studies did not have the benefit of imaging (although Paterson and Zangwill [30] did make use of X-rays) and often relied on autopsy, where available, to determine structure-function relationships. The first

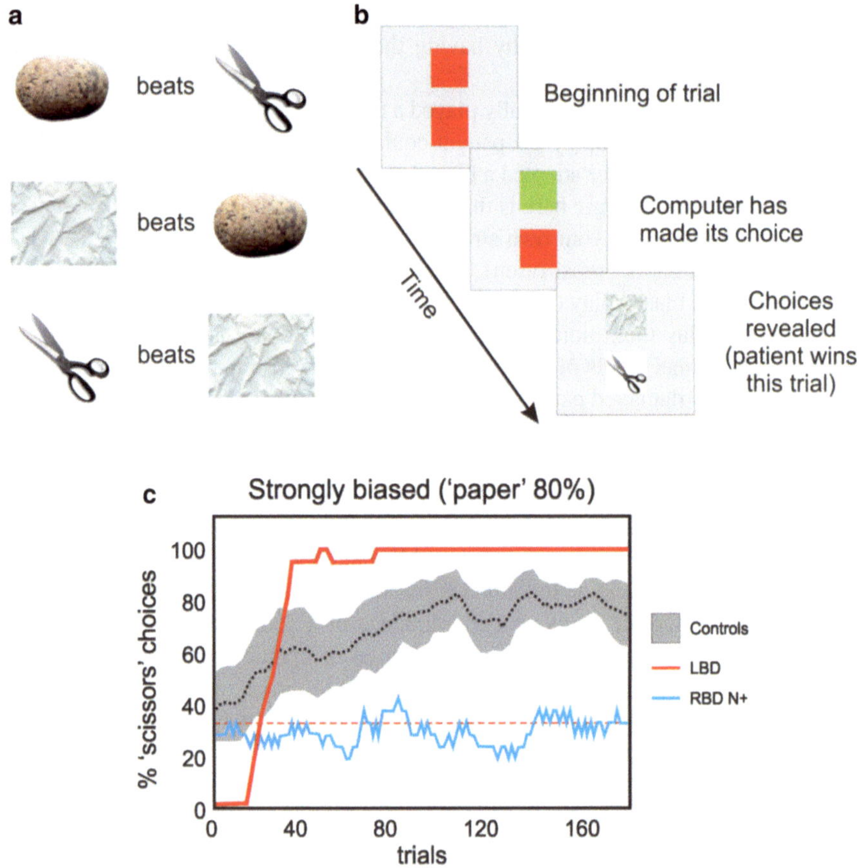

Fig. 5.5 (**a**) Representation of the rules governing the zero-sum children's game "rock, paper, scissors." (**b**) Schematic representation of the version of "rock, paper, scissors" played with neglect patients in [96]. The *top square* represents the computer's play and this changed from *red* to *green* when the computer had "locked in" its choice. Patients then made their choices and both were revealed to indicate the outcome for the patient (win, loss, tie; in the instance shown here, the patient wins). (**c**) Performance for the condition in which the computer adopted a strongly biased play strategy (i.e., choosing "paper" on 80 % of all trials). Participant choices are represented as a moving average ($n=20$ trials) of the optimal choice ("scissors"). Controls (*dotted line*, *gray bar* represents standard error) quickly discern the computer's strategy and adopt the optimal choice around 75 % of the time (i.e., matching the computer's play). A typical left brain-damaged (LBD) patient is shown in red. This patient quickly chose to exploit the biased play of the computer. Unlike controls, the patient maximized their response choices (i.e., choosing "scissors" 100 % of the time). This was a common strategy seen in 9 of 10 LBD patients. A typical right brain-damaged patient with neglect (RBD N+) is shown in blue. Clearly, this patient's play did not deviate from random, a pattern that was observed in 5 of 7 neglect patients [96]

study to attempt to determine the *common* region involved in neglect used computed tomography (CT) scans and suggested the inferior parietal cortices were most commonly involved [97]. More recently some controversy has arisen concerning the

Table 5.1 Summary of studies exploring the anatomical basis of neglect

References	# Patients	Method	Critical region associated with neglect
Buxbaum et al. [101]	166 (80 neglect, 42 acute, 38 chronic)	CT/MRI	Acute = BG, inf/mesial temp Chronic = cingulate, OFC, IPL, IT, inf/mesial temp, occ, sup/mid temp
Chechlacz et al. [102]	41 (21 RH patients; 19 left neglect, 4 right neglect)	DTI, VLSM	IPS, TPJ, SLF, ILF, SFOF, IFOF, thalamic radiation, corona radiata Allocentric neglect = pSTS, AG, mid temp, mid occ Egocentric neglect = mid front, PCG, STG, SMG
Karnath et al. [98]	33 neglect patients (25 w/o field cuts)	MRI, VLSM	STG, ventral PCG, parietal operculum
Karnath et al. [132, 133]	140 RH (78 neglect)	MRI, VLSM	STG, insula, putamen, caudate, SLF, IFOF, SFOF,
Karnath et al. [134]	54 (24 neglect; 8 chronic, 16 recovered)	MRI, VLSM	STG, MTG, BG, IFOF, extreme capsule + uncinate fasciculus in chronic patients
Mort et al. [99]	35 (24 MCA/14 neglect; 11 PCA/5 neglect)	MRI	MCA = AG, STG in 50 % PCA = Parahippocampal
Ptak and Schnider [135]	30 RBD (20 neglect)	VLSM	TPJ, pIPS, MFG
Urbanski et al. [111]	4 (2 neglect)	DTI	IPL, IFOF
Urbanski et al. [110]	12 (6 neglect)	DTI	Perisylvian w.m., ant. limb of IC, w.m. underlying IFG (ant. segment of arcuate fasciculus)
Vallar and Perani [97]	110 (47 neglect, 29 severe neglect)	CT	IPL, thalamus, BG
Verdon et al. [104]	80 (55 neglect, 16 severe neglect)	MRI, VLSM	Visuospatial neglect = IPL Motor neglect = DLPFC Allocentric neglect = parahippocampal gyrus

CT computerized tomography, *MRI* magnetic resonance imaging, *DTI* diffusion tensor imaging, *VLSM* voxel-wise lesion symptom mapping, *BG* basal ganglia, *OFC* orbitofrontal cortex, *IPL* inferior parietal lobule, *IT* inferotemporal, *IPS* intraparietal sulcus, *TPJ* temporoparietal junction, *SLF* superior longitudinal fasciculus, *ILF* inferior longitudinal fasciculus, *SFOF* superior frontal-occipital fasciculus, *IFOF* inferior frontal-occipital fasciculus, *STS* superior temporal sulcus, *AG* angular gyrus, *PCG* precentral gyrus, *STG* superior temporal gyrus, *SMG* supramarginal gyrus, *MCA* middle cerebral artery territory, *PCA* posterior cerebral artery territory, *IFG* inferior frontal gyrus, *MFG* middle frontal gyrus, *IC* internal capsule, *DLPFC* dorsolateral prefrontal cortex, *inf* inferior, *sup* superior, *occ* occipital, *temp* temporal, *front* frontal, *par* parietal, *mid* middle, *ant* anterior, *w.m.* white matter, *RBD* right brain damaged

critical lesion for neglect (Table 5.1) [98, 99]. These studies variously claim that the STG or the angular gyrus of the inferior parietal lobule (IPL) represent the critical lesion site for neglect (Table 5.1 and Fig. 5.6) [98, 99].

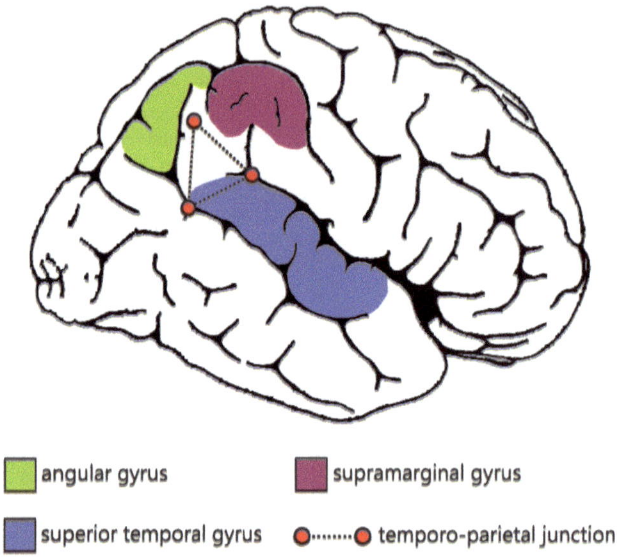

Fig. 5.6 Right hemisphere of the brain showing the critical cortical sites that produce neglect (see Table 5.1 for more details). The angular gyrus is shown in *green*, the supramarginal gyrus in *magenta*, and the superior temporal gyrus in *purple*. The temporoparietal junction (TPJ) is indicated by the *red circles* and *dotted line*

The different cortical substrates put forth as a critical lesion site for neglect may, in part, reflect different selection criteria or the presence or absence of hemianopia—a nontrivial factor given that opposing biases are expected for neglect and hemianopic patients on some tests [100]. Furthermore, some studies have shown distinct regions of cortical damage are associated with specific symptom profiles (Table 5.1) [101–104]. Despite this controversy and the heterogeneity of findings, the regions commonly associated with neglect [i.e., the IPL, the temporoparietal junction (TPJ), the STG] all represent multimodal association cortices that maintain reciprocal connections with primary sensory cortices, subcortical structures, and the frontal lobes.

Neglect as a Disconnection Syndrome

More controversial than the debate concerning the cortical lesion site for neglect is the notion that neglect be considered a disconnection syndrome [105]. Many of the previously discussed studies highlight commonly affected subcortical regions including the thalamus, caudate, putamen, and deep cortical structures such as the insula (Table 5.1). However, it is the consistent involvement of white matter pathways that has led some to suggest that the disorder should be characterized as a disconnection syndrome [105, 106]. This may go part of the way in explaining the heterogeneity of symptom profiles in neglect if one assumes that the range of deficits evident in any given patient depends on a combination of the cortical regions involved and the connections disrupted between these regions and their normal projection sites. In particular, what may be most critical, both in terms of whether a right brain-damaged patient does or does not display neglect, as well as the particular symptom profile evident, is the involvement of long-range white matter fiber pathways connecting the frontal and parietal cortices [106–111]. The challenge this work poses for theories of neglect is to develop a comprehensive model of the syndrome that reliably associates specific symptoms profiles with specific brain regions/pathways.

Recovery from Neglect

Given the prevalence of neglect and the fact that neglect is an indicator of poor prognosis, there have been many attempts to rehabilitate the disorder [112–116]. Techniques such as neck muscle vibration and caloric stimulation lead to a temporary optokinetic nystagmus and temporarily improve some symptoms of neglect [113–116]. Similarly, limb restriction, in which the patient's "good" arm is restricted to force use of their contralesional limb, also leads to temporary improvements but is untenable in the long term for obvious reasons [117]. More recently, the use of prisms has led to some broad, longer-term benefits to neglect patients [118, 119] (see [120, 121] for clinical trials with null results).

Prism Adaptation and Applications for Rehabilitation

Prism adaptation first requires the patient to point to where they think straight ahead of their body midline is while blindfolded (i.e., a proprioceptive judgment of straight ahead) (Fig. 5.7).

Patients often indicate a location to the right of true center. They then point to targets to the left and right of their midline while wearing prisms that shift visual input further to the right. Initial pointing movements show rightward errors due to perturbed input from the prisms. Patients then correct for these errors shifting their movements leftward to accommodate for the prism shift. After a brief exposure period (~5 min), patients repeat their judgments of straight ahead and show aftereffects such that movements are now to the left of pre-prism judgments (Fig. 5.7) [122, 123]. This shift in straight-ahead judgments shows that the patient successfully adapted to the prisms. More striking are the effects seen in a wide range of tasks including improved performance on clinical tasks, shifts in postural balance, visual imagery, tactile search, ocular explorations, and spatial reorienting [16, 17, 124–126]. Many of the improvements seen after prisms are maximal at around 2-h postexposure [119], with multiple adaptation sessions leading to even longer-term changes in neglect symptoms [127].

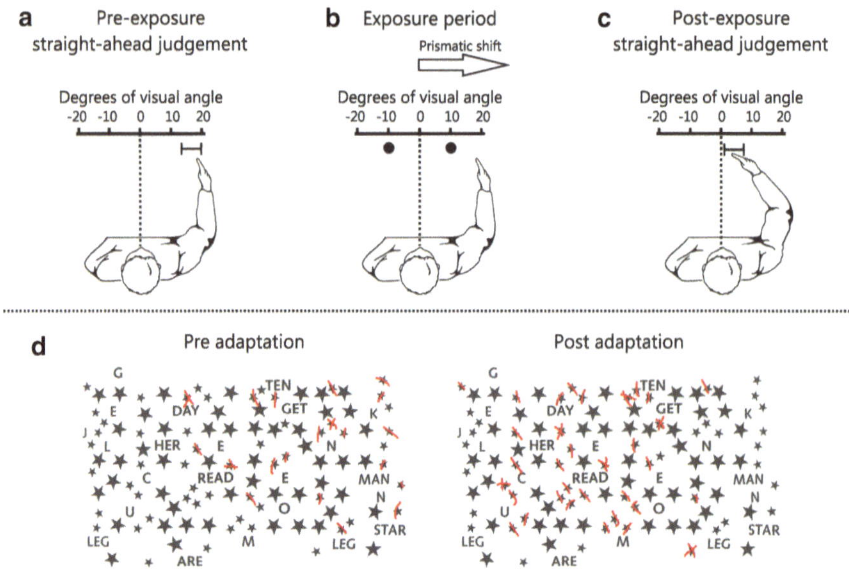

Fig. 5.7 Schematic representation of a typical prism adaptation procedure [119]. (**a**) Pre-exposure judgment of straight ahead while blindfolded. The typical patient will indicate a location to the right of true center. (**b**) Exposure period in which patients point to targets to the left and right of midline. Shown here is an initial pointing trial with a rightward error (i.e., in the direction of the prism shift). (**c**) Post-exposure judgments of straight ahead while blindfolded typically show an aftereffect in the direction opposite that of the prismatic shift, such that the patient now indicates a location closer to true center. (**d**) Pre- and postexposure cancellation performance. Note that while more targets are cancelled on the left after exposure to prisms, a concomitant number of right-sided targets are now omitted

We showed that prisms modified the way in which spatial attention was oriented [126, 128], such that after prisms, right brain-damaged patients with significant disengage deficits were faster to respond to left-sided targets after having first been cued to attend to right space (Fig. 5.8) [126, 129]. Critically, this work shows that one key aspect of neglect—deficient spatial reorienting—can be ameliorated by prisms. The same cannot be said for all aspects of the disorder [88]. Our initial demonstration of this made use of chimeric faces to explore the influence of prisms on perception versus action [124]. As discussed previously, neglect patients show a characteristic bias such that they claim chimeric faces shown to be smiling on the right appear happier (the opposite of the bias seen in controls [40, 41]). We explored this bias while monitoring fixations before and after prisms. Unsurprisingly, prior to prisms the patient fixated the right half of faces. After prisms, he now explored the left half of those same faces (Fig. 5.8). When asked which face was happier, even after prisms the patient chose the face shown smiling on the right despite the fact that prisms had allowed him to now explore previously neglected left-sided features of those faces (Fig. 5.8). This dissociation between altered attentional shifts and unchanged perceptual biases is not unique to faces [89, 130].

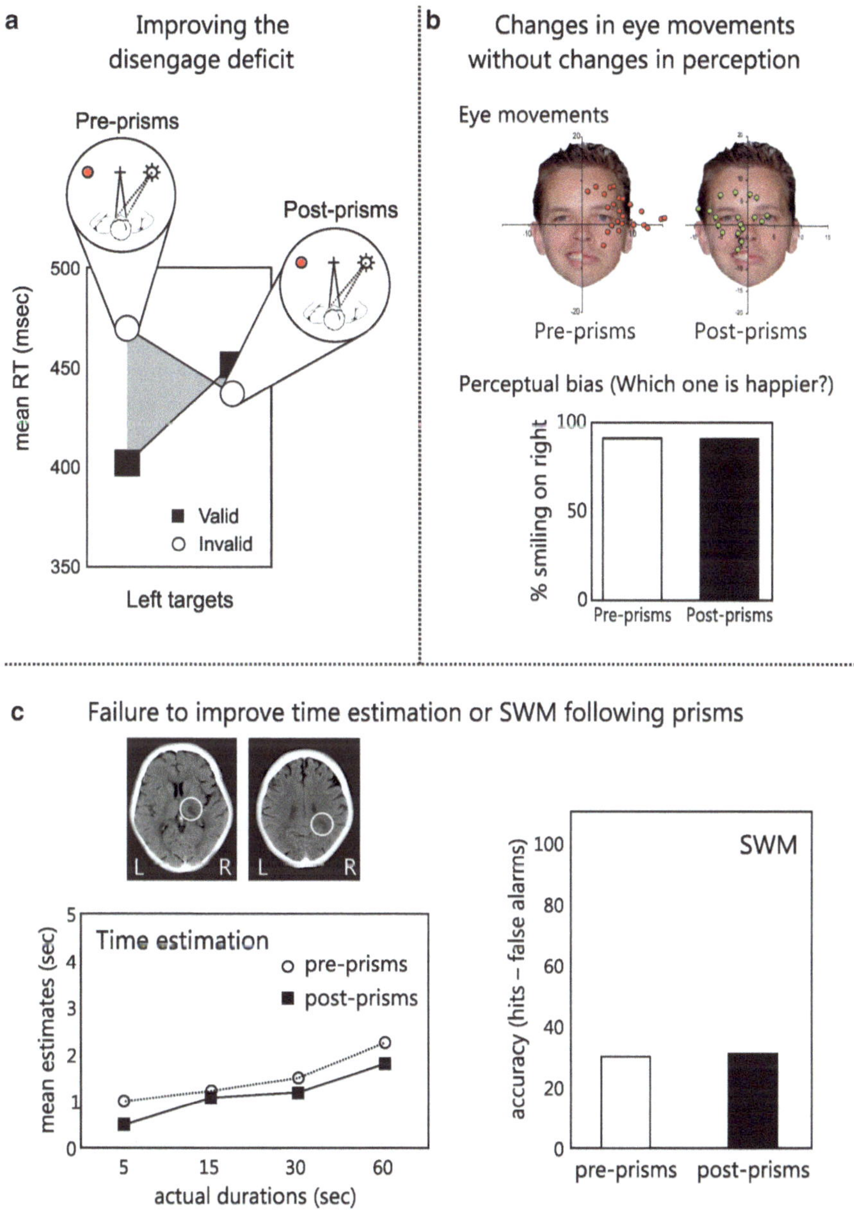

Fig. 5.8 (**a**) Improvement in the disengage deficit post-prisms in one patient with neglect. (**b**) Fixations made to chimeric faces prior to prisms (left face with *red circles*) and after prisms (right face with *green circles*). *Lower panel* shows the perceptual judgments of happiness (% of right smiling face chosen) before (*white bar*) and after (*black bar*) prisms. Note that while fixations shifted leftwards, there were no changes in perceptual bias. (**c**) *Left panel* shows time estimation performance in one neglect patient with a thalamic and parietal white matter lesion (*circles* on the CT scan show the lesions) before (*open circles*) and after prisms (*black squares*). The *right panel* shows SWM performance in the same patient before (*white bar*) and after (*black bar*) prisms. Note there is no change in either behavior

The failure of prisms to alter perceptual biases in neglect may reflect the fact that prisms primarily influence processes controlled by superior parietal cortices—prominent structures in the dorsal visual stream from V1 to posterior parietal cortex, known to be important for the control of visually guided actions and reorienting attention [35, 131]. The dorsal stream is typically undamaged in neglect and changes in behaviors normally controlled by this stream are unlikely to affect perceptual processing in the ventral pathway (V1 to inferotemporal cortex), especially given the fact that the two pathways are substantially disconnected from one another by IPL/STG lesions. Anecdotally, neglect patients continue to show many of the same problems in daily living (e.g., bumping into walls on their left) even after having shown improvements on clinical measures of the disorder following prisms. It may be the case that prisms alter exploratory motor behaviors and the ability to reorient attention, but do nothing to the impairments in generating and updating mental models (Fig. 5.5). In one patient we tested, we found no improvement in SWM and temporal estimation following prisms (Fig. 5.8). As previously suggested, the full neglect syndrome probably represents a constellation of symptoms including deficient spatial orienting of attention, impaired sustained attention, and a more generalized impairment in mental model creation and updating. Prisms may prove to be very effective in rehabilitating the former deficits, while failing to improve the latter.

Summary

Neglect is a common consequence of right-sided brain damage and presents in the acute phase after stroke with a heterogeneous range of symptoms. Classic models of neglect that focus on impaired spatial attention are inadequate to explain the full range of symptoms. Instead, neglect is likely a consequence of a constellation of symptoms including deficient spatial orienting, poor sustained attention, and impoverished use of mental models. Characterizing neglect deficits as impaired construction, use, or updating of mental models represents a fruitful way of explaining a broad swath of observed deficits. This does not suggest that deficits in mental model building or updating are sufficient to produce neglect. However, an inability to accurately create or update a representation of the environment and the rules that govern its operation would represent a severe impediment to fluid control of behavior. When coupled with a strong bias towards attending to right space, this deficit would explain many of the most prominent symptoms of neglect. It is likely that the degree to which updating deficits are present in any given neglect patient will depend not only on the site of cortical or subcortical damage, but also on the degree to which those regions have been disconnected from their target projection sites.

References

1. Bowen A, McKenna K, Tallis RC. Reasons for variability in the reported rate of occurrence of unilateral spatial neglect after stroke. Stroke. 1999;30:1196–202.
2. Ringman JM, Saver JL, Woolson RF, Clarke WR, Adams HP. Frequency, risk factors, anatomy, and course of unilateral neglect in an acute stroke cohort. Neurology. 2004;63:468–74.
3. Appelros P, Karlsson GM, Seiger A, Nydevik I. Neglect and anosognosia after first-ever stroke: incidence and relationship to disability. J Rehabil Med. 2002;34:215–20.
4. Cassidy TP, Lewis S, Gray CS. Recovery from visuospatial neglect in stroke patients. J Neurol Neurosurg Psychiatry. 1998;64:555–7.
5. Danckert J, Ferber S. Revisiting unilateral neglect. Neuropsychologia. 2006;44:987–1006.
6. Halligan PW, Fink GR, Marshall JC, Vallar G. Spatial cognition: evidence from visual neglect. Trends Cogn Sci. 2003;7:125–33.
7. Heilman KM, Watson RT, Valenstein E. Neglect and related disorders. In: Heilman KM, Valenstein E, editors. Clinical Neuropsychology (3rd edition). New York: Oxford University Press; 2003.
8. Mesulam M. A cortical network for directed attention and unilateral neglect. Ann Neurol. 1981;10:309–25.
9. Albert ML. A simple test of visual neglect. Neurology. 1973;23:658–64.
10. Ogden JA. Anterior-posterior interhemispheric differences in the loci of lesions producing visual hemineglect. Brain Cogn. 1985;4:59–75.
11. Wilson B, Cockburn J, Halligan P. Development of a behavioural test of visuospatial neglect. Arch Phys Med Rehabil. 1987;68:98–102.
12. Milner AD. Neglect, extinction, and the cortical streams of visual processing. In: Thier P, Karnath H-O, editors. Parietal lobe contributions to orientation in 3D space. Berlin: Springer; 1997. p. 3–22.
13. Rafal RD. Neglect. In: Parasuraman R, editor. The attentive brain. Cambridge, USA: MIT; 1996. p. 489–526.
14. Karnath HO, Himmelbach M, Küker W. The cortical substrate of visual extinction. Neuroreport. 2003;14:437–42.
15. Bisiach E, Luzzatti C. Unilateral neglect of representational space. Cortex. 1978;14:129–33.
16. Rode G, Rossetti Y, Boisson D. Prism adaptation improves representational neglect. Neuropsychologia. 2001;39:1250–4.
17. Rode G, Rossetti Y, Li L, Boisson D. Improvement of mental imagery after prism exposure in neglect: a case study. Behav Neurol. 1998;11:251–8.
18. Beschin N, Robertson IH. Personal versus extrapersonal neglect: a group study of their dissociation using a reliable clinical test. Cortex. 1997;33:379–84.
19. Bisiach E, Perani D, Vallar G, Berti A. Unilateral neglect: personal and extra-personal. Neuropsychologia. 1986;24:759–67.
20. Cowey A, Small M, Ellis S. Left visuo-spatial neglect can be worse in far than in near space. Neuropsychologia. 1994;32:1059–66.
21. Halligan PW, Marshall JC. Left neglect for near but not far space. Nature. 1991;350:498–500.
22. Berti A, Smania N, Allport A. Coding of far and near space in neglect patients. Neuroimage. 2001;14:98–102.
23. Pegna AJ, Petit L, Caldera-Schnetzer AS, Khateb A, Annoni JM, Sztajzel R, et al. So near yet so far: neglect in far or near space depends on tool use. Ann Neurol. 2001;50:820–2.
24. Driver J, Baylis GC, Goodrich SJ, Rafal RD. Axis-based neglect of visual shapes. Neuropsychologia. 1994;32:1353–65.
25. Tipper S, Behrmann M. Object-centered not scene-based visual neglect. J Exp Psychol Hum Percept Perform. 1996;22:1261–78.

26. Ladavas E, Shallice T, Zanella MT. Preserved semantic access in neglect dyslexia. Neuropsychologia. 1997;35:257–70.
27. Miceli G, Capasso R. Word-centred neglect dyslexia: evidence from a new case. Neurocase. 2001;7:221–37.
28. Vallar G, Guariglia C, Nico D, Tabossi P. Left neglect dyslexia and the processing of neglected information. J Clin Exp Neuropsychol. 1996;18:733–46.
29. Brain WR. Visual disorientation with special reference to lesions of the right cerebral hemisphere. Brain. 1941;64:224–72.
30. Paterson A, Zangwill OL. Disorders of visual space perception associated with lesions of the right cerebral hemisphere. Brain. 1944;67:331–58.
31. Mattingley JB. In: Code C, Wallesch CW, Joanette Y, et al., editors. Classic cases in neuropsychology. UK: Psychology Press; p. 173–188.
32. Adair JC, Barrett AM. Spatial neglect: clinical and neuroscience review: a wealth of information on the poverty of spatial attention. Ann N Y Acad Sci. 2008;1142:21–43.
33. Kerkhoff G. Spatial hemineglect in humans. Prog Neurobiol. 2001;63:1–27.
34. Corbetta M, Kincade JM, Shulman GL. Neural systems for visual orienting and their relationships to spatial working memory. J Cogn Neurosci. 2002;14:508–23.
35. Corbetta M, Shulman GL. Control of goal-directed and stimulus-driven attention in the brain. Nat Rev Neurosci. 2002;3:201–15.
36. Kinsbourne M. A model for the mechanism of unilateral neglect of space. Trans Am Neurol Assoc. 1970;95:143–6.
37. Kinsbourne M. Orienting bias model of unilateral neglect: Evidence from attentional gradients within hemispace. In: Robertson IH, Marshall JC, editors. Unilateral neglect: clinical and experimental studies. Hove: Lawrence Erlbaum; 1993. p. 63–86.
38. Posner MI, Walker JA, Friedrich FA, Rafal RD. Effects of parietal injury on covert orienting of attention. J Neurosci. 1984;4:1863–74.
39. Posner MI, Walker JA, Friedrich FA, Rafal RD. How do the parietal lobes direct covert attention? Neuropsychologia. 1987;25:135–45.
40. Levy J, Heller W, Banich MT, Burton LA. Asymmetry of perception in free viewing of chimaeric faces. Brain Cogn. 1983;2:404–19.
41. Heller W, Levy J. Perception and expression of emotion in right-handers and left-handers. Neuropsychologia. 1981;19:263–72.
42. Demaree HA, Everhart DE, Youngstrom EA, Harrison DW. Brain lateralization of emotional processing: historical roots and a future incorporating "dominance". Behav Cogn Neurosci Rev. 2005;4:3–20.
43. Mattingley JB, Bradshaw JL, Nettleton NC, Bradshaw JA. Can task specific perceptual bias be distinguished from unilateral neglect? Neuropsychologia. 1994;32:805–17.
44. Dehaene S. The neural basis of the Weber–Fechner law: a logarithmic mental number line. Trends Cogn Sci. 2003;7:145–7.
45. Umiltà C, Priftis K, Zorzi M. The spatial representation of numbers: evidence from neglect and pseudoneglect. Exp Brain Res. 2009;192:561–9.
46. Zorzi M, Priftis K, Meneghello F, Marenzi R, Umiltà C. The spatial representation of numerical and non-numerical sequences: evidence from neglect. Neuropsychologia. 2006;44:1061–7.
47. Zorzi M, Priftis K, Umiltà C. Brain damage: neglect disrupts the mental number line. Nature. 2002;417:138–9.
48. Behrmann M, Ghiselli-Crippa T, Sweeney JA, Di Matteo I, Kass R. Mechanisms underlying spatial representation revealed through studies of hemispatial neglect. J Cogn Neurosci. 2002;14:272–90.
49. De Renzi E, Gentilini M, Faglioni P, Barbieri C. Attentional shift towards the rightmost stimuli in patients with left visual neglect. Cortex. 1989;25:231–7.
50. Natale E, Marzi CA, Bricolo E, Johannsen L, Karnath HO. Abnormally speeded saccades to ipsilesional targets in patients with spatial neglect. Neuropsychologia. 2007;45:263–72.
51. Behrmann M, Watt S, Black SE, Barton JJ. Impaired visual search in patients with unilateral neglect: an oculographic analysis. Neuropsychologia. 1997;35:1445–58.

52. Bartolomeo P, Chokron S. Orienting of attention in left unilateral neglect. Neurosci Biobehav Rev. 2002;26:217–34.
53. Losier BJ, Klein RM. A review of the evidence for a disengage deficit following parietal lobe damage. Neurosci Biobehav Rev. 2001;25:1–13.
54. Robertson IH, Manly T, Beschin N, Daini R, Haeske-Dewick H, Hömberg V, et al. Auditory sustained attention is a marker of unilateral spatial neglect. Neuropsychologia. 1997;35:1527–32.
55. Robertson IH, Tegnér R, Tham K, Lo A, Nimmo-Smith I. Sustained attention training for unilateral neglect: theoretical and rehabilitation implications. J Clin Exp Neuropsychol. 1995;17:416–30.
56. Robertson IH, Ward A, Ridgeway V, Nimmo-Smith I. Test of everyday attention. Flempton: Thames Valley Test Company; 1994.
57. Robertson IH, Ward T, Ridgeway V, Nimmo-Smith I. The structure of normal human attention: the Test of Everyday Attention. J Int Neuropsychol Soc. 1996;2:525–34.
58. Husain M, Nachev P. Space and the parietal cortex. Trends Cogn Sci. 2007;11:30–6.
59. Husain M, Rorden C. Non spatially lateralised mechanisms in hemispatial neglect. Nat Rev Neurosci. 2003;4:26–36.
60. Robertson IH. Do we need the "lateral" in unilateral neglect? Spatially non-selective attention deficits in unilateral neglect and their implications for rehabilitation. Neuroimage. 2001;14:S85–90.
61. Robertson IH, Mattingley JB, Rorden C, Driver J. Phasic alerting of neglect patients overcomes their spatial deficit in visual awareness. Nature. 1998;395:169–72.
62. Husain M, Shapiro K, Martin J, Kennard C. Abnormal temporal dynamics of visual attention in spatial neglect patients. Nature. 1997;385:154–6.
63. Chun MM, Potter MC. A two-stage model for multiple target detection in rapid serial visual presentation. J Exp Psychol Hum Percept Perform. 1995;21:109–27.
64. Raymond JE, Shapiro KL, Arnell KM. Temporary suppression of visual processing in an RSVP task: an attentional blink? J Exp Psychol Hum Percept Perform. 1992;18:849–60.
65. Livesey EJ, Harris IM. Target sparing effects in the attentional blink depend on type of stimulus. Atten Percept Psychophys. 2011;73:2104–23.
66. Shapiro K, Hillstrom AP, Husain M. Control of visuotemporal attention by inferior parietal and superior temporal cortex. Curr Biol. 2002;12:1320–5.
67. Rizzo M, Akutsu H, Dawson J. Increased attentional blink after focal cerebral lesions. Neurology. 2001;57:795–800.
68. Schweizer TA, Alexander MP, Cusimano M, Stuss DT. Fast and efficient visuotemporal attention requires the cerebellum. Neuropsychologia. 2007;45:3068–74.
69. Ferber S, Danckert J. Lost in space—the fate of memory representations for non-neglected stimuli. Neuropsychologia. 2006;44:320–5.
70. Wojciulik E, Husain M, Clarke K, Driver J. Spatial working memory deficit in unilateral neglect. Neuropsychologia. 2001;39:390–6.
71. Wojciulik E, Rorden C, Clarke K, Husain M, Driver J. Group study of an "undercover" test for visuospatial neglect: invisible cancellation can reveal more neglect than standard cancellation. J Neurol Neurosurg Psychiatry. 2004;75:1356–8.
72. Parton A, Malhotra P, Nachev P, Ames D, Ball J, Chataway J, et al. Space re-exploration in hemispatial neglect. Neuroreport. 2006;17:833–6.
73. Malhotra P, Mannan S, Driver J, Husain M. Impaired spatial working memory: one component of the visual neglect syndrome? Cortex. 2004;40:667–76.
74. Basso G, Nichelli P, Frassinetti F, di Pellegrino G. Time perception in a neglected space. Neuroreport. 1996;7:2111–4.
75. Danckert J, Ferber S, Pun C, Broderick C, Striemer C, Rock S, et al. Neglected time: impaired temporal perception of multisecond intervals in unilateral neglect. J Cogn Neurosci. 2007;19:1706–20.
76. Merrifield C, Hurwitz M, Danckert J. Multimodal temporal perception deficits in a patient with left spatial neglect. Cogn Neurosci. 2010;4:244–53.

77. Becchio C, Bertone C. Time and neglect: abnormal temporal dynamics in unilateral spatial neglect. Neuropsychologia. 2006;44:2775–82.
78. Harrington DL, Haaland KY, Knight RT. Cortical networks underlying mechanisms of time perception. J Neurosci. 1998;18:1085–95.
79. Aboulafia-Brakha T, Christe B, Martory MD, Annoni JM. Theory of mind tasks and executive functions: a systematic review of group studies in neurology. J Neuropsychol. 2011;5:39–55.
80. Danckert J, Ferber S, Doherty T, Steinmetz H, Nicolle D, Goodale MA. Selective, non-lateralized impairment of motor imagery following right parietal damage. Neurocase. 2002;8:194–204.
81. Sirigu A, Duhamel J, Cohen L, Pillon B, Dubois B, Agid Y. The mental representation of hand movements after parietal cortex damage. Science. 1996;273(5281):1564–8.
82. Wolpert DM, Miall RC, Kawato M. Internal models in the cerebellum. Trends Cogn Sci. 1998;2:338–47.
83. Wolpert DM, Goodbody SJ, Husain M. Maintaining internal representations: the role of the human superior parietal lobe. Nat Neurosci. 1998;1:529–33.
84. Tenenbaum JB, Kemp C, Griffiths TL, Goodman ND. How to grow a mind: statistics, structure, and abstraction. Science. 2011;331:1279–85.
85. Pisella L, Mattingley JB. The contribution of spatial remapping to unilateral visual neglect. Neurosci Biobehav Rev. 2004;28:181–200.
86. Duhamel JR, Goldberg ME, Fitzgibbon EJ, Sirigu A, Grafman J. Saccadic dysmetria in a patient with a right fronto-parietal lesion: the importance of corollary discharge for accurate spatial behaviour. Brain. 1992;115:1387–402.
87. Heide W, Blankenburg M, Zimmermann E, Kömpf D. Cortical control of double-step saccades: implications for spatial orientation. Ann Neurol. 1995;38:739–48.
88. Striemer C, Danckert J. Through a prism darkly: re-evaluating prisms and neglect. Trends Cogn Sci. 2010;14:308–16.
89. Striemer C, Danckert J. Dissociating perceptual and motor effects of prism adaptation in spatial neglect. Neuroreport. 2010;21:436–41.
90. Druker M, Anderson B. Spatial probability AIDS visual stimulus discrimination. Front Hum Neurosci. 2010;4:63.
91. Craik K. The nature of explanation. New York: Cambridge University Press; 1943.
92. Johnson-Laird PN. Mental models and human reasoning. Proc Natl Acad Sci USA. 2010;107:18243–50.
93. Blakemore SJ, Goodbody SJ, Wolpert DM. Predicting the consequences of our own actions: the role of sensorimotor context estimation. J Neurosci. 1998;18:7511–8.
94. Desmurget M, Epstein CM, Turner RS, Prablanc C, Alexander GE, Grafton ST. Role of the posterior parietal cortex in updating reaching movements to a visual target. Nat Neurosci. 1999;2:563–7.
95. Frith CD, Blakemore SJ, Wolpert DM. Abnormalities in the awareness and control of action. Philos Trans R Soc Lond B Biol Sci. 2000;355:1771–88.
96. Danckert J, Stöttinger E, Quehl N, Anderson B. Right hemisphere brain damage impairs strategy updating. Cereb Cortex. 2012;22(12):2745–60. Epub 2011 Dec 15.
97. Vallar G, Perani D. The anatomy of unilateral neglect after right hemisphere stroke lesions. A clinical/CT-scan correlation study in man. Neuropsychologia. 1986;24:609–22.
98. Karnath HO, Ferber S, Himmelbach M. Spatial awareness is a function of the temporal not the posterior parietal lobe. Nature. 2001;411:950–3.
99. Mort DJ, Malhotra P, Mannan SK, Rorden C, Pambakian A, Kennard C, et al. The anatomy of visual neglect. Brain. 2003;126:1986–97.
100. Ferber S, Karnath HO. How to assess spatial neglect—line bisection or cancellation tasks? J Clin Exp Neuropsychol. 2001;23:599–607.
101. Buxbaum LJ, Ferraro MK, Veramonti T, Farne A, Whyte J, Ladavas E, et al. Hemispatial neglect: subtypes, neuroanatomy, and disability. Neurology. 2004;62:749–56.

102. Chechlacz M, Rotshtein P, Bickerton WL, Hansen PC, Deb S, Humphreys GW. Separating neural correlates of allocentric and egocentric neglect: distinct cortical sites and common white matter disconnections. Cogn Neuropsychol. 2010;27:277–303.
103. Rorden C, Fruhmann Berger M, Karnath HO. Disturbed line bisection is associated with posterior brain lesions. Brain Res. 2006;1080:17–25.
104. Verdon V, Schwartz S, Lovblad KO, Hauert CA, Vuilleumier P. Neuroanatomy of hemispatial neglect and its functional components: a study using voxel-based lesion-symptom mapping. Brain. 2010;133:880–94.
105. Bartolomeo P, Thiebaut de Schotten M, Doricchi F. Left unilateral neglect as a disconnection syndrome. Cereb Cortex. 2007;17:2479–90.
106. Doricchi F, Thiebaut de Schotten M, Tomaiuolo F, Bartolomeo P. White matter (dis)connections and gray matter (dys)functions in visual neglect: gaining insights into the brain networks of spatial awareness. Cortex. 2008;44:983–95.
107. Bird CM, Malhotra P, Parton A, Coulthard E, Rushworth MF, Husain M. Visual neglect following right posterior cerebral artery infarction. J Neurol Neurosurg Psychiatry. 2006;77:1008–12.
108. Doricchi F, Tomaiuolo F. The anatomy of neglect without hemianopia: a key role for parietal-frontal disconnection? Neuroreport. 2003;14:2239–43.
109. Thiebaut de Schotten M, Urbanski M, Duffau H, Volle E, Lévy R, Dubois B, et al. Direct evidence for a parietal frontal pathway subserving spatial awareness in humans. Science. 2005;309:2226–8.
110. Urbanski M, Thiebaut de Schotten M, Rodrigo S, Oppenheim C, Touzé E, Méder JF, et al. DTI-MR tractography of white matter damage in stroke patients with neglect. Exp Brain Res. 2011;208:491–505.
111. Urbanski M, Thiebaut de Schotten M, Rodrigo S, Catani M, Oppenheim C, Touzé E, et al. Brain networks of spatial awareness: evidence from diffusion tensor imaging tractography. J Neurol Neurosurg Psychiatry. 2008;79:598–601.
112. Farné A, Buxbaum LJ, Ferraro M, Frassinetti F, Whyte J, Veramonti T, et al. Patterns of spontaneous recovery of neglect and associated disorders in acute right brain-damaged patients. J Neurol Neurosurg Psychiatry. 2004;75:1401–10.
113. Adair JC, Na DL, Schwartz RL, Heilman KM. Caloric stimulation in neglect: evaluation of response as a function of neglect type. J Int Neuropsychol Soc. 2003;9:983–8.
114. Karnath HO, Christ K, Hartje W. Decrease of contralateral neglect by neck muscle vibration and spatial orientation of trunk midline. Brain. 1993;116:383–96.
115. Karnath HO, Fetter M, Dichgans J. Ocular exploration of space as a function of neck proprioceptive and vestibular input: observations in normal subjects and patients with spatial neglect after parietal lesions. Exp Brain Res. 1996;109:333–42.
116. Rubens AB. Caloric stimulation and unilateral visual neglect. Neurology. 1985;35:1019–24.
117. Eskes GA, Butler B, McDonald A, Harrison ER, Phillips SJ. Limb activation effects in hemispatial neglect. Arch Phys Med Rehabil. 2003;84:323–8.
118. Rossi PW, Kheyfets S, Reding MJ. Fresnel prisms improve visual perception in stroke patients with homonymous hemianopia or unilateral visual neglect. Neurology. 1990;40:1597–9.
119. Rossetti Y, Rode G, Pisella L, Farné A, Li L, Boisson D, Perenin MT. Prism adaptation to a rightward optical deviation rehabilitates left hemispatial neglect. Nature. 1998;395:166–9.
120. Nys GMS, de Haan EHF, Kunneman A, de Kort PL, Dijkerman HC. Acute neglect rehabilitation using repetitive prism adaptation. A randomized placebo-controlled trial. Restor Neurol Neurosci. 2008;26:1–12.
121. Turton AJ, O'Leary K, Gabb J, Woodward R, Gilchrist ID. A single blinded randomised controlled pilot trial of prism adaptation for improving self-care in stroke patients with neglect. Neuropsychol Rehabil. 2010;20:180–96.
122. Redding GM, Wallace B. Prism adaptation and unilateral neglect: review and analysis. Neuropsychologia. 2006;44:1–20.

123. Pisella L, Rode G, Farné A, Tilikete C, Rossetti Y. Prism adaptation in the rehabilitation of patients with visuo-spatial cognitive disorders. Curr Opin Neurol. 2006;19:534–42.
124. Ferber S, Danckert J, Joanisse M, Goltz HC, Goodale MA. Eye movements tell only half the story. Neurology. 2003;60:1826–9.
125. Tilikete C, Rode G, Rossetti Y, Pichon J, Li L, Boisson D. Prism adaptation to rightward optical deviation improves postural imbalance in left-hemiparetic patients. Curr Biol. 2001;11:524–8.
126. Striemer C, Danckert J. Prism adaptation reduces the disengage deficit in right brain damage patients. Neuroreport. 2007;18:99–103.
127. Frassinetti F, Angelini V, Meneghello F, Avanzi S, Làdavas E. Long-lasting amelioration of visuospatial neglect by prism adaptation. Brain. 2002;125:608–23.
128. Striemer C, Sablatnig J, Danckert J. Differential influences of prism adaptation on reflexive and voluntary covert attention. J Int Neuropsychol Soc. 2006;12:337–49.
129. Nijboer TC, McIntosh RD, Nys GM, Dijkerman HC, Milner AD. Prism adaptation improves voluntary but not automatic orienting in neglect. Neuroreport. 2008;19:293–8.
130. Dijkerman HC, McIntosh RD, Milner AD, Rossetti Y, Tilikete C, Roberts RC. Ocular scanning and perceptual size distortion in hemineglect: effects of prism adaptation and sequential stimulus presentation. Exp Brain Res. 2003;153:220–30.
131. Goodale MA, Milner AD. Separate visual pathways for perception and action. Trends Neurosci. 1992;15:20–5.
132. Karnath HO, Fruhmann-Berger M, Küker W, Rorden C. The anatomy of spatial neglect based on voxelwise statistical analysis: a study of 140 patients. Cereb Cortex. 2004;14:1164–72.
133. Karnath HO, Rorden C, Ticini LF. Damage to white matter fiber tracts in acute spatial neglect. Cereb Cortex. 2009;19:2331–7.
134. Karnath HO, Rennig J, Johannsen L, Rorden C. The anatomy underlying acute versus chronic spatial neglect: a longitudinal study. Brain. 2011;134:903–12.
135. Ptak R, Schnider A. The attention network of the human brain: relating structural damage associated with spatial neglect to functional imaging correlates of spatial attention. Neuropsychologia. 2011;49:3063–70.

Chapter 6
Post-stroke Aphasia

Marcelo L. Berthier, Guadalupe Dávila, Natalia García-Casares, and Ignacio Moreno-Torres

Introduction

Aphasia and Stroke

Aphasia is a generic term used to describe a range of impairments in language function following acquired brain damage typically involving the left hemisphere [1–3]. Aphasia may affect all modes of expressive and receptive communication including speaking, understanding, writing, reading, and gesturing [1–3]. This definition seems superficial because it merely describes the surface behavioral deficits of aphasia (e.g., word-finding difficulty and reduced fluency) and mentions nothing about the underlying source of functional impairments. Some scientists prefer to describe aphasia as a multimodal impairment of the integral constituents of language, such as phonology, syntax, morphology, and lexical semantics [4]. Aphasia should not be regarded as a domain-specific language disorder because other cognitive skills (e.g., attention, learning, and memory and executive function) essential for normal processing of language are usually impaired as well. A growing body of evidence indicates that the adequate functioning of these nonlanguage cognitive functions is crucial for the recovery of aphasia and communication deficits [5].

M.L. Berthier, M.D., Ph.D. (✉) • N. García-Casares, M.D., Ph.D.
Department of Medicine, Unit of Cognitive Neurology and Aphasia,
Centro de Investigaciones Médico-Sanitarias (CIMES), University of Malaga,
Marques de Beccaria 3, Malaga 29010, Spain
e-mail: mbt@uma.es

G. Dávila, Ph.D.
Department of Psychobiology, University of Malaga, Malaga, Spain

I. Moreno-Torres, Ph.D.
Department of Spanish Language, Philosophy, University of Malaga, Malaga, Spain

T.A. Schweizer and R.L. Macdonald (eds.), *The Behavioral Consequences of Stroke*,
DOI 10.1007/978-1-4614-7672-6_6, © Springer Science+Business Media New York 2014

This would imply that formal assessment of aphasia should not be restricted to the language domain; rather it needs to be expanded to include other cognitive and behavioral domains (see later). A new conceptualization of aphasia as a "discrete acquired disorder with a variety of aetiologies but specific characteristics" has been proposed as it provides a more informative viewpoint that may be useful to enhance public awareness of this still under-recognized condition [6]. In this chapter, we present an overview of acute and chronic post-stroke aphasia (PSA) covering epidemiology, pathophysiology, diagnosis, and treatment.

Aphasia is one of the most disabling cognitive deficits of stroke since it causes substantial functional disability and high psychological distress [7–11]. In acute rehabilitation settings, aphasic patients have longer length of stay and show poorer functional improvement than stroke patients without aphasia [12]. Moreover, individuals with chronic aphasia participate in fewer activities and have worse quality of life than stroke patients without aphasia [7]. Aphasia is a very frequent disorder. The National Aphasia Association (January 2012) reports that in the United States, aphasia affects one million inhabitants (i.e., 1 in 250 people) [13]. An annual incidence rate of 43–77.5 per 100,000 persons has been reported in cases of stroke [14–16] with an incidence of PSA following the first-ever episode ranging between 30 and 38 % in hospitalized patients [17–21]. These figures are even higher in the elderly population [14]. Despite its prevalence, PSA still remains an unknown condition for most people [2].

Aphasia is nearly always associated with left hemisphere lesions in right-handed individuals and in ~70 % of left-handers [1]. Right hemisphere damage can likewise provoke aphasia in left-handed subjects, but aphasia after right hemisphere involvement is exceptional in dextral subjects (crossed aphasia), except if they have left-handed relatives, personal history of developmental disorders (dyslexia, dysgraphia), or early acquired left hemisphere injury [22]. Acute aphasia due to ischemic stroke may be caused by atherothrombotic, hemodynamic, and cardioembolic mechanisms in vascular territories of the internal carotid artery and the middle cerebral artery perfusing the perisylvian language cortex [23, 24]. Less commonly, acute aphasias are the consequence of infarctions in borderzones between two major arteries (e.g., anterior cerebral and middle cerebral artery) [25, 26]. Infarction in these arterial territories is associated with different clinical types of aphasia [23, 24]. While ischemic infarctions are responsible for nearly 80 % of aphasic cases, hemorrhages are less frequent and their localization is not constrained by vascular territories [1]. Subcortical infarctions and hemorrhages involving the basal ganglia, internal capsule, or thalamus may also cause aphasia, yet the role of subcortical structures in language is poorly understood [23, 24, 27–30]. Various pathophysiological mechanisms have been suggested to account for subcortical aphasia including cortical diaschisis, deafferentiation of distant cortical areas connected with damaged white matter pathways or nuclei, or mass effect on intracerebral circulation [23, 24, 27–30]. For instance, aphasia after striato-capsular infarctions has been ascribed to sustained low perfusion in the overlying cortex and concomitant cortical infarction not visible in structural imaging studies [28].

Thalamic aphasia instead may result from direct damage to deep gray nuclei and white matter tracts, which interrupt fronto-thalamic neural networks implicated in attentional and lexical-semantic processing [23, 24, 27–30]. In light of the well-known role of subcortical structures in speech, language, and prosody, another likely mechanism underlying subcortical aphasia is direct damage to deep gray nuclei and pathways [30]. Accumulating evidence indicates that stroke lesions involving the right cerebellum may also induce speech disorders (e.g., mutism and foreign accent), aphasic signs (e.g., agrammatism), and genuine aphasic syndromes (e.g., anomic and transcortical motor aphasia) [31, 32]. Functional imaging studies in this condition revealed cerebellum-cerebral diaschisis, reflecting the functional inactivation of the language-dominant frontal lobe secondary to reduced input via cerebellum-cortical pathways [31].

Classification of Aphasic Syndromes

The current outline of aphasic syndromes derives from the pioneering studies of Broca, Wernicke, Lichtheim, and their contemporaries in the cerebral localization of vascular and traumatic lesions responsible for language dysfunction [23, 24, 27, 33–35]. Although these groundbreaking discoveries permitted the segregation of signs and symptoms into different syndromes (i.e., Broca's aphasia and Wernicke's aphasia) with reliable topographical correlates in acute cases (see next section) [33–35], they did not receive unanimous acceptance [36–39]. Hence, other classifications of aphasic disorders appeared and new syndromes were discovered (e.g., dynamic aphasia) [38], and new labels were also assigned to already well-defined aphasic syndromes. In this regard, it is remarkable that a lively debate on the classification and localization of aphasic syndromes remains 150 years after the publication of Broca's treatise on the brain substrates of language [33–43]. We admit that controversy many times sheds light on unsolved neuroscientific dilemmas, but learning the myriad of labels used for relatively analogous aphasic syndromes is a difficult endeavor for clinicians and may be one of the several reasons why aphasia does not receive as much attention as it may deserve. Therefore, in this chapter, a more practical approach to classifying aphasic syndromes and their neural correlates is adopted to help health professionals working in acute care settings to assist their patients.

Although some authors recommend against using taxonomic classifications of aphasic deficits [43, 44], this approach remains useful for clinical purposes because it allows lesion-symptom mapping and may inform prognosis [23, 24, 41]. For example, the presence of anterior perisylvian aphasias (e.g., Broca's aphasia) usually entails poorer prognosis for language recovery than anterior extrasylvian aphasias (e.g., transcortical motor aphasia) [22, 41]. Moreover, these distinct aphasic syndromes may have different pathophysiological mechanisms and complications. Classic Broca's aphasia is nearly always secondary to territorial infarctions in the middle cerebral artery branches, whereas transcortical motor aphasia frequently results from anterior borderzone infarctions, a mechanism that predispose to seizures more than territorial infarctions [26, 45].

Aphasic syndromes can be accurately established in more than two-thirds of cases with tasks tapping fluency in spontaneous speech, auditory comprehension, and repetition [23, 27, 41]. Naming deficits also need to be evaluated, but because word-finding difficulties and impaired naming are constant features in PSA, this domain lacks discriminative power to distinguish clinical profiles [27]. Traditional classifications of aphasia used dichotomous approaches (e.g., fluent versus nonfluent spontaneous speech and impaired versus spared auditory comprehension) [23, 24, 27, 40, 41]. A similar way to classify aphasias from a clinical perspective is evaluating the ability to repeat language, because performance on this variable is typically dichotomized (i.e., impaired versus spared repetition) in most patients with acute PSA and patently reflects the involvement or sparing of key cortical areas and white matter pathways (e.g., arcuate fasciculus) subserving auditory-verbal transcoding [23, 24, 27, 40, 41]. Indeed, some view the loss or preservation of word and sentence repetition in patients with acute PSA as a quite robust clinical sign, because it can provide valuable information about the involvement or sparing of the left perisylvian language area [46]. Acute aphasias caused by lesions involving the perisylvian core all share the impairment of repetition, whereas aphasias associated with acute lesions sparing the perisylvian region are chiefly characterized by the relative preservation of word and sentence repetition [24, 34]. Thus, the diagnosis of a specific acute aphasic syndrome may allow the prediction of the compromised vascular territory and is useful for clinical purposes [24, 47]. However, the clear-cut lesion-symptom mapping described in acute PSA is not applicable for patients with chronic aphasia, as the performance on language tasks in chronic PSA patients reflects the interplay between the conjoint activity of incompletely damaged neural tissue and dynamic compensation and reorganization processes underpinned by new neural connections [46–48]. Figure 6.1 shows an algorithm of aphasia classification based on repetition ability coupled with performance on fluency in spontaneous speech and auditory comprehension.

Six of the eight acute clinical aphasic syndromes can be broadly classified in two groups. Features common to one such group, usually described as "classic" aphasias (e.g., Broca's aphasia, Wernicke aphasia, and conduction aphasia), are *impaired* repetition and cortical involvement of the perisylvian language region irrigated by the middle cerebral artery [23, 24, 27, 40, 41]. Features characteristic of the other group (e.g., transcortical aphasias) are *preserved* repetition and cortical damage at or beyond the boundaries of the perisylvian language region [23, 24, 27, 34, 40, 41]. The cortical areas residing outside the perisylvian cortex are perfused by the anterior cerebral artery, the posterior cerebral artery, and the farthest end branches of two adjacent vascular territories [23, 24, 27, 34, 40, 41]. Of the remaining two aphasic syndromes, one (e.g., global aphasia) presents with *impaired* repetition, whereas repetition is *preserved* in the other (e.g., anomic aphasia). These two aphasias lack a consistent localizing value because they are associated with damage to different lesion sites [27].

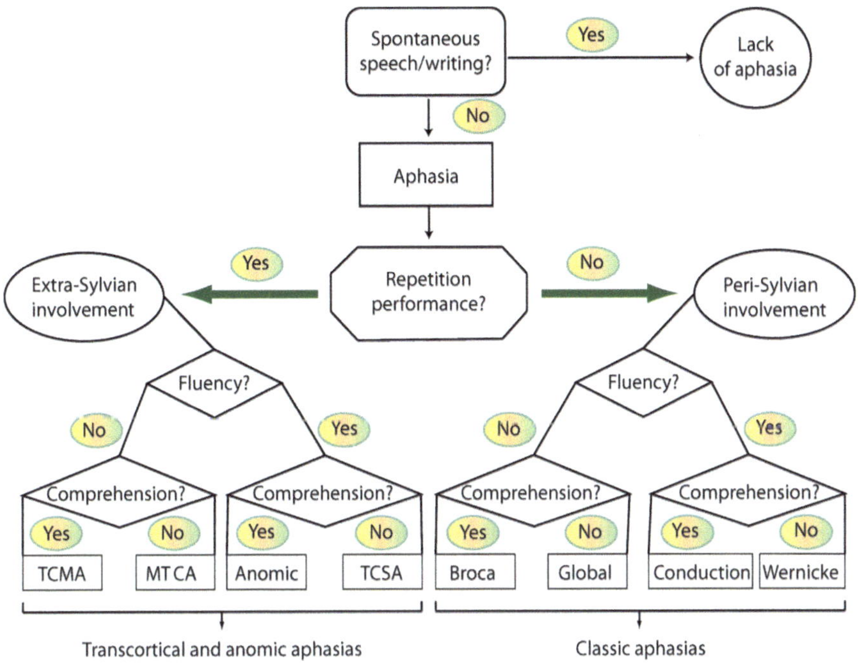

Fig. 6.1 Algorithm for different types of aphasia based on repetition ability coupled with performance on fluency in spontaneous speech and auditory comprehension. Adapted from [61]

Broca's Aphasia

Broca's aphasia is characterized by nonfluent spontaneous speech, naming, and repetition with relative sparing of auditory and reading comprehension except for syntactically complex sentences [23, 24, 27, 40, 41]. Nonfluent speech is characterized by reduced phrase length and word production (<50 words minute). Moreover, speech quality is also impaired by the presence of hesitant and effortful emissions, which contain abnormal rhythm, stress, and intonation. Grammar is impoverished (agrammatism) and many patients additionally display inconsistent articulation errors on repeated speech productions of the same utterance (apraxia of speech). Language comprehension is preserved for conversational, word-recognition tasks and simple commands, but understanding of syntactically complex sentences is abnormal. Repetition is impaired particularly for long words and short phrases, but sometimes is better than spontaneous speech. Naming is abnormal, but objects (nouns) are better named than actions (verbs). Oral reading and writing are grossly impaired. Patients with Broca's aphasia usually display depression and catastrophic reactions (e.g., outbursts of frustration and anger when confronted with a task) [49, 50].

Lesions responsible for Broca's aphasia are not restricted to the left frontal operculum; they usually exceed the boundaries of Broca's area extending deeply to lower motor cortex, temporal cortex, anterior insula, basal ganglia, and internal capsule and other white matter tracts [42].

Wernicke's Aphasia

Wernicke's aphasia is characterized by fluent speech with impaired comprehension and repetition [23, 24, 27, 40, 41]. Spontaneous speech is fluent, well-articulated, and melodic, but marred by the presence of neologisms, phonemic, and, less frequently, semantic paraphasias. Oral production may be abundant, copious, and sometimes adopt features of logorrhea (incoherent talkativeness) and pressure of speech, giving rise to an unintelligible jargon. Disinhibition of speech results, at least in part, from reduced self-monitoring of discourse, insight, and behavioral control (elation, irritability) [8, 49]. Repetition, naming, reading, and writing are grossly disrupted. Wernicke's aphasia is associated with large posterior cortical lesions involving the posterior temporal gyrus and neighboring parietal or occipital cortical areas [42].

Conduction Aphasia

Conduction aphasia is characterized by a disproportionate deficit in repetition in the context of fluent paraphasic verbal output and relative sparing of auditory comprehension [23, 24, 27, 40, 41]. Two main types of conduction aphasia have been described—*reproduction* and *repetition* [51]. The reproduction subtype is characterized by phonemic paraphasias in all verbal domains and recurrent production of sequential phonemic approximations to self-repair errors (conduite d'approche) [52, 53]. The reproduction conduction aphasia variety has been variously attributed to deficits in verbal praxis [54], disrupted speech programming [55], poor output phonological encoding [55, 56], or a combination of abnormal sensory-motor integration and reduced phonological short-term memory [57, 58], which may result from cortical damage without necessary compromise of the arcuate fasciculus. The repetition subtype shows virtually isolated repetition deficits that have been linked to a selective impairment in auditory-verbal short-term memory and cortical damage that extend deeply to affect the arcuate fasciculus [52, 53, 56].

Global Aphasia

Global aphasia is a severe condition that combines the deficits described in Broca's and Wernicke's aphasia [1, 23, 24, 27]. Thus, a grave reduction of verbal fluency (usually limited to word perseverations (e.g., "pa..pa..pa..pa")) coexists with severely impaired comprehension. Repetition, naming, reading, and writing are also profoundly impaired. As previously mentioned, acute global aphasia does not have a consistent localizing value because, although it usually results from involvement of both Broca's and Wernicke's area, large acute ischemic or hemorrhagic lesions in either cortical area can also produce the syndrome [27].

Transcortical Aphasia

Transcortical aphasia is the term used for syndromes in which the ability to repeat language is relatively preserved despite marked disturbances in other linguistic domains [34]. The transcortical syndromes are relatively similar to those describe previously, except for the preservation of repetition performance and a tendency to repeat words and utterances spoken by another person (echolalia) [24, 34]. Repetition is relatively preserved in the face of deficits in spontaneous speech (transcortical motor aphasia), auditory comprehension (transcortical sensory aphasia), or both (mixed transcortical aphasia) [34]. All clinical variants of transcortical aphasias have been attributed to lesions that spare the left perisylvian area and white matter tracts including the arcuate fasciculus. The most likely mechanism underpinning transcortical syndromes is the so-called isolation of speech area, which proposes that vascular lesions in watershed territories between major cerebral arteries or in the anterior cerebral artery and the posterior cerebral artery (extrasylvian cortical areas) disrupt the connectivity between frontal areas responsible for speech initiation or posterior conceptual-semantic representation areas and the intact perisylvian language core and its underlying white matter fiber pathways [23, 24, 34, 40]. Anterior lesions in watershed territories or in the anterior cerebral artery territory cause the motor variant, posterior lesions are responsible for the sensory type, and damage to both anterior and posterior watershed territories are associated with the mixed form.

Anomic Aphasia

Anomic aphasia is characterized by the inability to produce names in spontaneous speech and visual confrontation naming [23, 27]. Anomia is present in all types of aphasias, to the extent that the diagnosis of aphasia is uncertain in language-impaired patients showing preserved naming ability. Acute anomic aphasia is rare and may indicate damage to the left temporal cortex, whereas in chronic stroke, anomic aphasia is commonly the final classification for patients evolving from more severe fluent and nonfluent aphasias [58, 59].

Assessment: A Multidimensional Approach

A comprehensive evaluation of language deficits in PSA is essential for establishing neuroscientifically based recommendations for clinical practice of aphasia rehabilitation [1, 60, 61]. The evaluation of language deficits should be flexible and adjusted to the general, neurological, and functional status of the aphasic patient over the different phases of stroke recovery. For instance, in the early stages (i.e., hyperacute and acute periods) bedside language testing is preferable because most aphasic patients do not tolerate lengthy evaluations, whereas more in-depth evaluations are feasible in chronic patients. One simple manner to identify aphasia in the emergency room, particularly by clinicians unfamiliar with the assessment of language disorders, is testing spontaneous speech during an informal conversation or asking patients to write a sentence [61] (see Fig. 6.1). Aphasia is nearly always accompanied by deficits in spontaneous speech and writing, so that if the patient fails to do these simple tasks and commits a number of errors in both output modalities, the presence of aphasia is highly probable. Once the diagnosis of aphasia is established, it is necessary to assess the clinical profile and severity of language deficits for two important reasons. First, recent research has demonstrated that the initial aphasia severity is the best predictor of long-term outcome [62]. Second, data on aphasia type and severity are useful for managing recovery, devising tailor-made interventions, and informing the patient and relatives about prognosis as well as therapeutic possibilities [1].

An important aspect of the evaluation concerns the use of standardized aphasia tests. Figure 6.2 shows a diagram that can guide clinicians to determine the best assessment tools and timing for implementing different strategies of evaluation. In the acute PSA stage, testing of aphasic patients is often limited by short attention span, distractibility, fatigability, and lack of cooperation and other nonlanguage deficits affecting executive function and visual perception [59, 63–65]. Therefore, a bedside screening of language deficits using standardized scales (e.g., Mississippi Aphasia Screening Test, Frenchay Aphasia Screening Test) [66, 67] is recommended. These screening tests are not time-consuming and can be applied even in severely impaired patients. In less affected patients or in more chronic patients, a comprehensive evaluation is required. This can be done with standardized batteries (e.g., Western Aphasia Battery and Boston Diagnostic Aphasia Examination) [41, 68] that combine data from different language subtests (e.g., spontaneous, speech, auditory comprehension, repetition, and naming) to achieve a profile of aphasia (i.e., Broca's, Wernicke's, and others) and composite scores of aphasia severity (e.g., Aphasia Quotient of the Western Aphasia Battery) [41]. The individual test scores and composite scores of these batteries are useful to estimate longitudinal changes in the pattern and overall severity of aphasia occurring spontaneously or promoted by speech-language therapy (SLT) and other therapeutic interventions [64, 65]. Nevertheless, to plan rational and individualized SLT interventions, a more in-depth language evaluation in the frame of the cognitive neuropsychological approach is necessary. Tests like the Psycholinguistic Assessment of Language

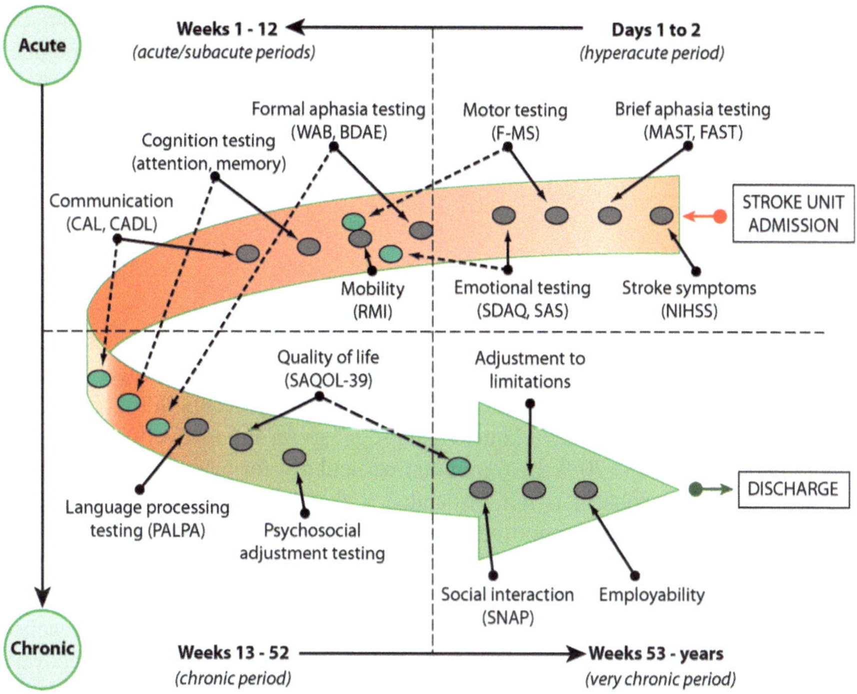

Fig. 6.2 The diagram shows the different post-stroke aphasia stages and testing tools, which can guide clinicians to select the appropriate tool for a specific function. *NIHSS* National Institutes of Health Stroke Scale [13], *F-MS* Fugl-Meyer Scale [111], *MAST* Mississippi Aphasia Screening Test, *FAST* Frenchay Aphasia Screening Test [66, 67], *SAQOL-39* Aphasia Quality of Life Scale-39 [112], *SDAQ* stroke aphasic depression questionnaire [113], *SAS* Starkstein's Apathy Scale [8, 9], *WAB* western aphasia battery [41], *BDAE* Boston Diagnostic Aphasia Examination [68], *PALPA* Psycholinguistic Assessment of Language Processing in Aphasia [69], *CAL* communicative activity log [98], *SNAP* social network with aphasia profile [114], *RMI* rivermead mobility index [115]. *Continuous lines* and *gray circles* indicate tasks that are preferentially administered in the period, whereas *dotted lines* and *turquoise circles* indicate tasks that may be administered in different aphasia periods

Processing in Aphasia (PALPA) [69] are particularly useful for identifying deficits in discrete cognitive systems as well as residual areas of strength that can be useful in the service of language recovery.

The aforementioned cognitive impairments in aphasic patients go beyond language deficits to affect other cognitive domains (e.g., attention, visuospatial working memory, processing speed, learning and memory, and executive function) essential for normal language functioning. Because the functioning of these nonlinguistic cognitive functions can predict recovery from language and communication deficits in PSA [5, 70, 71], formal assessment of aphasia should be expanded to other cognitive and noncognitive domains (e.g., mood and motivation). The rational manner to implement this methodology is by using a multidimensional approach [1].

In the past few years, the formal assessment of PSA has progressed from the evaluation of language deficits to a more comprehensive evaluation that includes testing of verbal and nonverbal communication, aphasia-related cognitive deficits (e.g., attention, executive function, and memory), mood and motivation, activities of daily living, social interaction, and the impact of aphasia on relatives and careers (see Fig. 6.2).

Prognosis and Mechanisms of Recovery

Recovery is an ongoing process and most patients continue to make progress several years after PSA onset. The greatest spontaneous recovery occurs in the first 2 or 3 months [1]. Some studies have claimed that prognosis for full recovery in the hyperacute period is good (>70 % by 6 months) particularly for patients suffering mild strokes and when aphasia is not associated with other neurologic deficits [72]. However, in some studies reporting good prognosis, language assessment carried out with testing tools (e.g., National Institutes of Health Stroke Scale) was unable to capture residual language deficits [13, 72]. In the remaining patients, the amount of improvement was less detectable in the following months, and most of them reached a plateau after 1 year [1]. Around 40 % of PSA patients showed complete recovery within 1 year. The type of aphasia always evolves to a less severe form during the first year and only certain aphasic patterns (i.e., anomic, Broca's, and conduction) prevail in chronic phases [19, 73, 74]. Figure 6.3 shows the possible patterns of recovery in PSA evolving from acute global aphasia to less severe aphasic syndromes. Similar patterns of recovery can also be observed in other acute aphasic syndromes (e.g., Wernicke's aphasia evolving to less severe conduction aphasia). By contrast, sometimes PSA worsens, indicating the occurrence of a new lesion, the aggravation of previous cognitive failure, or the development of post-stroke dementia in elderly patients [64, 75].

The mechanisms underlying recovery from PSA include restoration of cerebral blood flow in the area of the ischemic penumbra (a highly vulnerable peri-infarct tissue that may survive or die), release of remote cortical and subcortical areas from initial inhibition (diaschisis), dynamic network changes, unmasking of preexisting connections, activity-dependent synaptic changes, and restoration of neurotransmitter activity [76–78]. Restorative mechanisms may operate with different timetables, and their availability may depend upon the site and extent of the lesion as well as on the residual capacity of partially damaged and non-damaged neural tissue to ameliorate abnormal language functions [62, 79–82]. In this regard, the integrity of certain cortical areas (e.g., left hippocampus) [83] or white matter pathways (e.g., subcallosal fasciculus and arcuate fasciculus) [84, 85] seems to be crucial for recovering speech production in patients with large infarctions in the left middle cerebral artery (Fig. 6.4). On the other hand, PSA cases associated with infarcts in arterial border-zone areas between the left middle cerebral artery and either the anterior

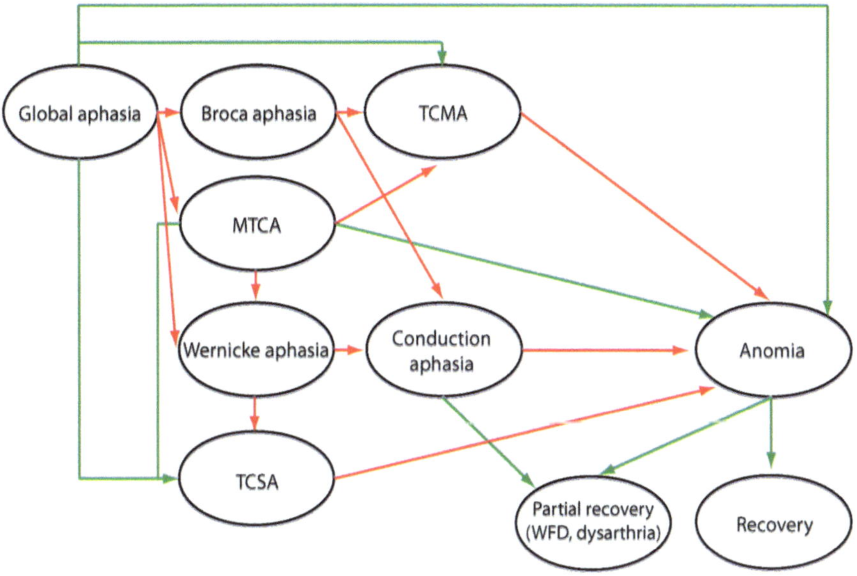

Fig. 6.3 Possible patterns of recovery in PSA evolving from acute global aphasia to less severe aphasic syndromes. *Red lines* show the usual transitions from one aphasic to others less severe following a logical sequence of resolving language deficits. For instance, improvement of auditory comprehension converts global aphasia in Broca's aphasia, and further recovery of repetition transforms Broca's aphasia in TCMA (transcortical motor aphasia). By contrast, *green lines* show possible, but less frequent, evolution of aphasic syndromes. Acute aphasic syndromes showing rapid resolution, as take place in hemorrhages, often circumvent the logical sequence of aphasia resolution, thus rapidly converting a severe aphasia to a very less severe syndrome or to partial/complete recovery. *MTCA* mixed transcortical aphasia, *TCSA* transcortical sensory aphasia, *WFD* word-finding difficulty

cerebral artery or the posterior cerebral artery territories sparing the perisylvian cortex show better long-term prognosis [86, 87]. Recovery also relies on the interplay between patient's characteristics (i.e., age, gender, education, handedness, premorbid representation of language, medical comorbidities, history of previous strokes, and regional atrophy in areas participating in recovery) and environmental contingencies (i.e., occupational and leisure activities and communication partners) [88]. Hence, it seems that individual differences on a number of factors accounts for the commonly observed heterogeneous patterns of spontaneous and therapy-induced recovery across aphasic individuals. Patients' attitudes toward their aphasic syndromes may influence the recovery process as well. For example, patients with Broca's aphasia are fully aware of their language deficits and, as a result, engage in more aphasia therapy than some Wernicke's aphasic patients who, due to lack of insight of their language problems, are unwilling to participate in language training [89]. Therefore, the search for new and better factors predictive of PSA recovery is growing very rapidly. During the past decades, most experts have agreed that variability

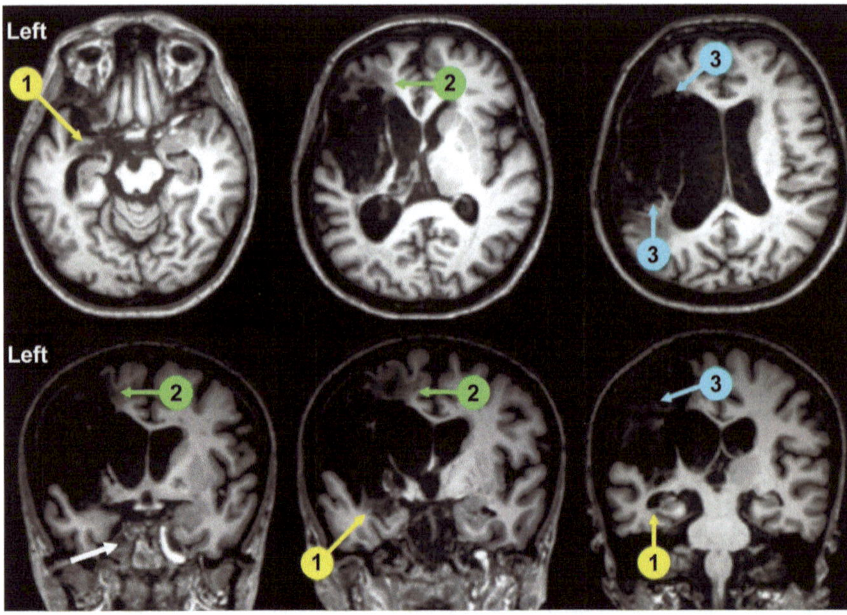

Fig. 6.4 Structural MRI of a patient with long-lasting (3 years) nonfluent aphasia (spontaneous speech limited to single words and perseverations) showing a large infarction in the left middle cerebral artery territory. Axial and coronal 3T MRI slices show the full extension of the lesion involving the fronto-temporoparietal cortex and extending deeply to insula, basal ganglia, and white matter tracts. Despite intensive aphasia therapy and a trial of memantine (20 mg/day during 3 months), no changes were noted in oral productive functions. Involvement of brain structures critical for recovery such as the hippocampus (*1*, *yellow circles* and *arrows*), subcallosum fasciculus (*2*, *green circles* and *arrows*), and arcuate fasciculus (*2*, *blue circles* and *arrows*) most likely accounted for the lack of recovery of verbal output

of language recovery after first-ever stroke is a frequent phenomenon and that the final outcome of aphasic deficits cannot be reliably predicted at 3 months post-onset by traditional demographic variables (e.g., age, sex, and handedness) and/or lesion characteristics (e.g., location and volume) [62, 90]. This is an important issue because a deeper understanding of these factors may improve the prediction of outcome in the acute stage and assist clinicians in the implementation of early therapeutic interventions [62, 90]. Some initiatives in this direction are emerging. Positron emission tomography and functional magnetic resonance imaging studies attempt to distinguish key regions of the brain that promote improvement of language deficits from others that may hinder full recovery through maladaptive reorganization [80, 82, 91, 92]. In complimentary terms, diffusion tensor imaging (i.e., white matter fiber tractography) reveals variability in the hemispheric lateralization of long white matter fiber pathways (e.g., arcuate fasciculus) [93] that may play a role in the clinical profile, prognosis, and recovery patterns of PSA [51]. Furthermore, tractography may also assist rehabilitation by identifying preserved structures participating in recovery as well as structural changes in white matter pathways triggered by intensive rehabilitation [94].

Treatment Interventions

Speech-Language Therapy

In light of what is known about spontaneous recovery from PSA [95], it can be expected that most aphasic patients will require therapeutic interventions to push recovery further. Aphasia rehabilitation begins during the acute hospitalization, as soon as patients regain ability to take part in language training. Although drugs can also be used in this phase (see later), speech-language therapy (SLT) remains the mainstay intervention for aphasia [1, 95]. Improvement of language deficits is highest in the initial 4 weeks after stroke [74], yet few studies of aphasia rehabilitation in the early post-stroke period have been performed. Several reasons can account for the paucity of rehabilitation trials in acute PSA. Recruitment of acute patients for early aphasia rehabilitation is not an easy enterprise, although a recent study demonstrated that randomized controlled intervention trials in acute PSA are feasible [96]. Reasons for exclusion or abandonment of SLT in acute PSA include medical complications (e.g., pneumonia and cardiac decompensation) or other factors inherent to the unstable neurological status including reduced alertness and attention span, aphasia severity, severe speech deficits (e.g., dysarthria, hypophonia, and apraxia of speech), lack of rehabilitation engagement due to fatigability, depression, apathy, aging (e.g., unwillingness to participate), or pre- and post-stroke dementia. Despite these disadvantages, recent studies in acute PSA have shown that 2 h/week of SLT is effective, but intensive interventions (i.e., 4 h/week) provide similar outcomes than less intensive regimes (i.e., 2 h/week), and most acute patients do not tolerate demanding regimes (i.e., 4 h/week) [89]. Another recent randomized controlled pilot study revealed that very early PSA therapy (range: 0–10 days after stroke onset) improved aphasia severity and communication deficits [97], but further studies are needed. The benefits provided by different aphasia therapies in patients with chronic PSA are undisputed, particularly when scientifically based interventions (e.g., constraint-induced aphasia therapy, errorless learning, and computer-based aphasia therapy) are implemented [59, 65, 98–100]. Most studies carried out in chronic PSA have demonstrated that aphasia therapy is efficacious when it is administered in an intensive and prolonged way. The fact that language deficits associated with PSA can be attenuated several years after onset demonstrates that the temporal window for aphasia recovery is much wider than previously estimated [59]. This new insight militates against the argument claiming that further benefits with aphasia rehabilitation cannot be expected

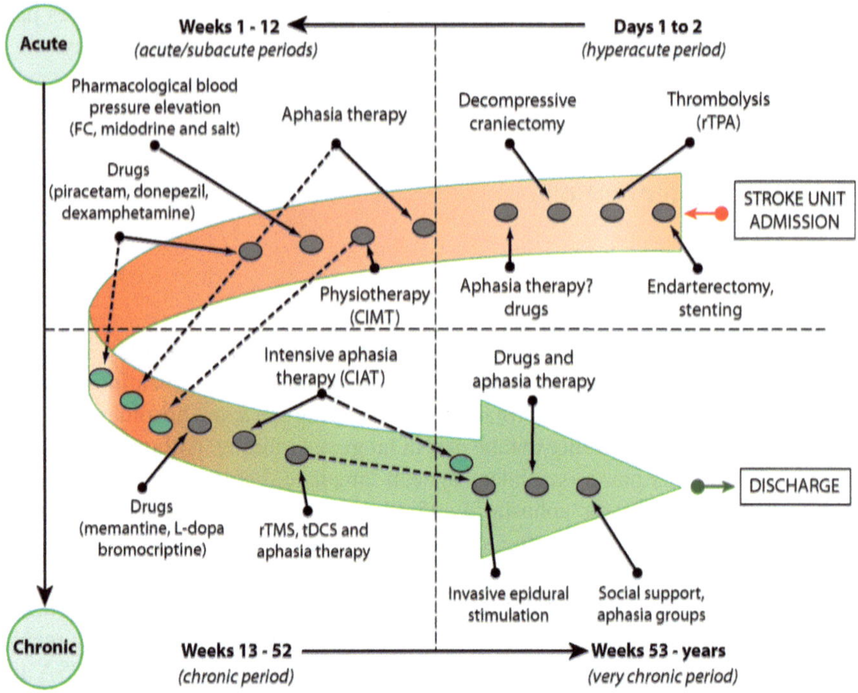

Fig. 6.5 The diagram shows various therapeutic strategies that can be implemented during the different evolution periods of post-stroke aphasia. *Continuous lines* and *gray circles* indicate interventions, which are preferentially used in the period, whereas *dotted lines* and *turquoise circles* indicate interventions that may be implemented in different aphasia periods

1 year after stroke onset. Figure 6.5 shows a diagram that will help clinicians and other health professionals (e.g., speech pathologists and neuropsychologists) to determine appropriate therapy options and timing for implementing different treatment strategies.

Pharmacotherapy and Other Biological Interventions

Although intensive SLT is effective, in the past few years efforts have been focused on developing complementary strategies to augment the gains provided by rehabilitation. Two main types of interventions have been used: *drugs* and *brain stimulation* [59, 64, 65, 94]. Drug therapy is emerging as a promising option to enhance cognitive function in healthy individuals and in brain-damaged patients [101], and a growing body of evidence also indicates that pharmacotherapy can significantly augment benefits provided by conventional and intensive SLT therapy in patients with acute and chronic PSA [59, 64, 65]. The motivation for using drugs to treat aphasia is to reestablish the activity of specific neurotransmitters in dysfunctional, but not irretrievably damaged brain areas and interconnecting networks [59, 64, 65]. Drugs acting on diverse neurotransmitters (i.e., glutamatergic, GABAergic, cholinergic, dopaminergic, serotoninergic, and norepinephrinergic systems) have been used with variable results (Table 6.1). Although more research is needed, well-designed and controlled studies showed that drugs commonly used to treat Alzheimer's disease and vascular dementia (e.g., donepezil and memantine) are safe and effective in patients with chronic PSA and that the obtained benefits are maintained at long-term follow-up [102, 103].

The advent of novel noninvasive methods of brain stimulation has increased the interest in applying these neurophysiological techniques as potential adjuvant treatments to language rehabilitation in PSA [94]. Repetitive transcranial magnetic stimulation (rTMS) and transcranial direct current stimulation (tDCS) approaches induce prolonged functional changes in the cerebral cortex [104–110]. Depending upon the type of stimulation, it is possible to suppress the maladaptive activity of cortical regions interfering with recovery or instead enhance cortical excitability promoting improvement of language function. Several studies have been performed in chronic aphasic patients with rTMS and tDCS [94, 105]. Positive effects were found on naming performance in patients with chronic, nonfluent [106], and fluent aphasias [105]. A pilot study used invasive epidural stimulation of the ipsilesional premotor cortex in patients with chronic nonfluent aphasia and extensive left hemisphere infarcts to improve language performance [109, 110]. A small sample of patients with severe and long-lasting, nonfluent PSA received a simultaneous treatment with intensive SLT and epidural stimulation during 6 weeks (3 h/day) [109, 110]. All treated patients showed significant benefits on aphasia testing. Although further studies are needed, epidural cortical stimulation seems to be safe and may be a viable option to enhance the limited benefits provided by SLT in patients with severe nonfluent aphasia [109, 110].

Summary

Aphasia after stroke is a frequent, but still neglected, condition. Like other neurologic disorders (e.g., Parkinson disease and multiple sclerosis), aphasia deserves the right to occupy a better position according to its high incidence, morbidity, and negative impact on patients and relatives. The diagnosis and classification of aphasic syndromes in acute stroke are easy to establish if three principal language domains (i.e., spontaneous speech, auditory comprehension and repetition) are examined. A correct initial diagnosis of aphasia type and severity is important to predict outcomes, implement rational interventions, and inform patients and relatives about evolution and therapeutic options. Comprehensive examinations of language deficits and related cognitive alterations are more appropriate in subacute and chronic phases. Formal evaluations are crucial to unravel the underlying functional

Table 6.1 Drug trials in acute and chronic post-stroke aphasia

Drug	Action mechanisms	Study design	Number of studies	Main outcomes: acute aphasia	Main outcomes: chronic aphasia
Bromocriptine	Dopamine agonist	Single cases Case series Open-label RCT	11	Not investigated	Positive effects in single cases, case series, and open-label trials mainly in transcortical motor aphasia, adynamic aphasia, and Broca's aphasia of mild to moderate severity. Improvement on production tasks in chronic patients. Variables outcomes in moderate Broca's aphasia and lack of positive effects in severe cases. Four of five RCTs negative. Few trials with concomitant SLT. Benefits in "crossed" nonfluent aphasia
Piracetam	Neuronal and vascular unspecific effects	RCT	8	Positive effects on overall language measures, spontaneous speech (communicative verbal behavior), object naming, comprehension, and written language. Language deficit improvement correlates with increases in blood flow in the left perisylvian cortex	Lack of long-term benefits in cases with large infarcts
Donepezil	IAChE	Single cases, case series, open-label RCT	7	Beneficial effects on acute aphasia on spontaneous speech, comprehension, repetition, and naming	Positive effects on aphasia severity and everyday functional communication. Significant benefits on spontaneous speech, comprehension, and naming in chronic aphasia. Efficacy maintained at long-term follow-up

Dexamphetamine	Catecholamine uptake inhibition	Open-label RCT, single-case RCT	4	Positive effects on overall performance in communication in subacute stroke	Positive effects on chronic stages. Variable outcomes on naming performance when paired with model-based naming therapy in chronic stages
Levodopa	Dopamine agonist	RCT	2	Improvement in acute/subacute anomic aphasia with levodopa and intensive computer-assisted therapy (CAT), but response similar to placebo plus intensive CAT	Positive effects in chronic patients when combined with SLT on naming and repetition especially in patients with frontal lobe damage
Amantadine	Dopamine agonist, uncompetitive NMDA receptor antagonist	Single case, open label	2	Useful in amotivational transcortical motor aphasia with improvement of verbal fluency when paired with SLT	Improvement of verbal fluency in a case of transcortical sensory aphasia secondary to diffuse brain hypoxia
Bifemelane	IAChE	RCT	2	Not investigated	Positive effects on naming, comprehension, and repetition, but not in fluency. Language improvements correlate with an increase in blood flow in the left perisylvian cortex
Physostigmine	IAChE	Open label	1	Not investigated	Positive effects in combination with lecithin on chronic anomia, but not on other cognitive and mood variables. Not used in clinical practice due to adverse event profile

(continued)

Table 6.1 (continued)

Drug	Action mechanisms	Study design	Number of studies	Main outcomes: acute aphasia	Main outcomes: chronic aphasia
Memantine	Moderate affinity, uncompetitive NMDA receptor antagonist with strong voltage-dependency and fast kinetics	RCT	1	Not investigated	Positive effects on aphasia severity and communication. Significant benefits on spontaneous speech, comprehension, and naming. Efficacy maintained at long-term follow-up. No adverse events
Fluvoxamine	SSRI	RCT	1	Not investigated	Positive effects on anomia
Moclobemide	IMAO-A	RCT	1	Not investigated	Negative effect
Zolpidem	GABA agonist interacting with Ω_1 receptor	Single case	1	Not investigated	Positive reversible effect on verbal fluency in Broca's aphasia
Propranolol	Central and peripheral beta-adrenergic antagonist	RCT	1	Not investigated	Positive effect on naming in Broca's aphasia
Vasopressin	Neurotrophic effect mediated by the V1 receptor	RCT	1	Not investigated	Positive effect on expressive and receptive language functions. Possible bi-hemispheric action with primary influence on the right hemisphere
Cerebrolysin	Mimics endogenous neurotrophic factors	RCT	1	Positive effect on spontaneous speech, repetition, and naming in acute Broca's aphasia at all study time points	Not investigated

Vasopressin is a neuropeptide and cerebrolysing a neurotrophic factor. See articles citation in ref. [1, 59, 65]
RCT randomized controlled trial

language and other cognitive deficits and devise individualized therapies. Speech-language therapy remains the most powerful therapeutic strategy for PSA in both acute and chronic phases. Gains obtained with aphasia therapy can be augmented with drugs and other new brain stimulation methods.

References

1. Berthier ML. Poststroke aphasia: epidemiology, pathophysiology and treatment. Drugs Aging. 2005;22:163–82.
2. Code C, Petheram B. Delivering for aphasia. Int J Speech Lang Pathol. 2011;13:3–10.
3. Varley R. Rethinking aphasia therapy: a neuroscience perspective. Int J Speech Lang Pathol. 2011;13:11–20.
4. McNeil MR, Pratt SR. Defining aphasia: some theoretical and clinical implications of operating from a formal definition. Aphasiology. 2001;15:901–11.
5. Lambon Ralph MA, Snell C, Fillingham JK, et al. Predicting the outcome of anomia therapy for people with aphasia post CVA: both language and cognitive status are key predictors. Neuropsychol Rehabil. 2010;20:289–305.
6. Sherratt S. Written media coverage of aphasia: a review. Aphasiology. 2011;25:1132–52.
7. Hilari K. The impact of stroke: are people with aphasia different to those without? Disabil Rehabil. 2011;33:211–8.
8. Starkstein SE, Robinson RG. Depression and aphasia. Aphaiology. 1988;2:1–20.
9. Starkstein SE, Fedoroff JP, Price TR, et al. Apathy following cerebrovascular lesions. Stroke. 1993;24:1625–30.
10. Jorge RE, Starkstein SE, Robinson RG. Apathy following stroke. Can J Psychiatry. 2010;55:350–4.
11. Mayo NE, Fellows LK, Scott SC, et al. A longitudinal view of apathy and its impact alter stroke. Stroke. 2009;40:3299–307.
12. Gialanella B, Prometti P. Rehabilitation length of stay in patients suffering from aphasia alter stroke. Top Stroke Rehabil. 2009;16:437–44.
13. National Institutes of Health Stroke Scale. National Institute of Neurological Disorders and Stroke. www.ninds.nih.gov.
14. Law J, Rush R, Pringle A-M, et al. The incidence of cases of aphasia following first stroke referred to speech and language therapy services in Scotland. Aphasiology. 2009;23:1266–75.
15. Engelter ST, Gostynski M, Papa S, et al. Epidemiology of aphasia attributable to first ischemic stroke: incidence, severity, fluency, etiology, and thrombolysis. Stroke. 2006; 37:1379–84.
16. Dickey L, Kagan A, Lindsay MP, et al. Incidence and profile of inpatient stroke-induced aphasia in Ontario,Canada. Arch Phys Med Rehabil. 2010;91:196–202.
17. Pedersen PM, Jørgensen HS, Nakayama H, et al. Aphasia in acute stroke: incidence, determinants, and recovery. Ann Neurol. 1995;38:659–66.
18. Laska AC, Hellblom A, Murray V, et al. Aphasia in acute stroke and relation to outcome. J Intern Med. 2001;249:413–22.
19. Pedersen PM, Vinter K, Olsen TS. Aphasia after stroke: type, severity and prognosis. The Copenhagen aphasia study. Cerebrovasc Dis. 2004;17:35–43.
20. Bersano A, Burgio F, Gattinoni M, et al. Aphasia burden to hospitalised acute stroke patients: need for an early rehabilitation program. Int J Stroke. 2009;4:443–7.
21. Tsouli S, Kyritsis AP, Tsagalis G, Virvidaki E, Vemmos KN. Significance of aphasia after first-ever acute stroke: impact on early and late outcomes. Neuroepidemiology. 2009; 33:96–102.

22. Berthier ML, Dávila G, García-Casares N, et al. Atypical conduction aphasia and the right hemisphere: cross-hemispheric plasticity of phonology in a developmentally dyslexic and dysgraphic patient with early left frontal damage. Neurocase. 2011;17:93–111.
23. Alexander MP. Aphasia: clinical and anatomical aspects. In: Feinberg TE, Farah MJ, editors. Behavioral neurology and neuropsychology. New York: McGraw-Hill; 1997. p. 133–49.
24. Hillis AE. Aphasia. Progress in the last quarter of century. Neurology. 2007;69:200–13.
25. Evrard S, Woimant F, Le Coz P, Polivka M, Cousin C, Haguenau M. Watershed cerebral infarcts: retrospective study of 24 cases. Neurol Res. 1992;14(2 Suppl):97–9.
26. Cauquil-Michon C, Flamand-Roze C, Denier C. Borderzone strokes and transcortical aphasia. Curr Neurol Neurosci Rep. 2011;11:570–7.
27. Albert ML, Goodglass H, Helm NA, et al. Clinical aspects of dysphasia. In: Arnold GE, Winckel F, Wyke BD, editors. Disorders of human communication 2. Wien New York: Springer; 1981.
28. Nadeau SE, Crosson B. Subcortical aphasia. Brain Lang. 1997;58:355–402; discussion 418–23.
29. Crosson B. Subcortical mechanisms in language: lexical-semantic mechanisms and the thalamus. Brain Cogn. 1999;40:414–38.
30. de Boissezon X, Démonet JF, Puel M, et al. Subcortical aphasia: a longitudinal PET study. Stroke. 2005;36:1467–73.
31. Marien P, Engelborghs S, Fabbro F, De Deyn PP. The lateralized linguistic cerebellum: a review and a new hypothesis. Brain Lang. 2001;79:580–600.
32. Ackermann H, Mathiak K, Riecker A. The contribution of the cerebellum to speech production and speech perception: clinical and functional imaging data. Cerebellum. 2007; 6:202–13.
33. Lazar RM, Mohr JP. Revisiting the contribution of Paul Broca to the study of aphasia. Neuropsychol Rev. 2011;21:236–9.
34. Berthier ML. Transcortical aphasias. In: Code C, Humphreys G, editors. Brain damage, behaviour and cognition series. Hove: Psychology Press; 1999.
35. Commentary KA. Ardila's attempt to alter aphasiology. Aphasiology. 2010;24:404–7.
36. Head H. Aphasia and kindred disorders of speech. London: Cambridge University Press; 1926.
37. Jakobson R. Toward a linguistic typology of aphasic impairments. In: DeReuck AVS, O'Connor M, editors. Disorders of language. Boston: Little Brown; 1964.
38. Luria A. Basic problems of neurolinguistics. New York: Mouton; 1976.
39. Ardila A. A proposed reinterpretation and reclassification of aphasic syndromes. Aphasiology. 2010;24:363–94.
40. Geschwind N. Disconnection syndromes in animals and man. Brain. 1965;88:237–94.
41. Kertesz A. Western aphasia battery. New York: Grune & Stratton; 1982.
42. Damasio H. Neuroanatomical correlates of the aphasias. In: Taylor Sarno M, editor. Acquired aphasia. New York: Academic; 1991. p. 45–71.
43. York III GK. Localization of language function in the twentieth century. J Hist Neurosci. 2009;18:283–90.
44. Caramazza A, McCloskey M. The case for single-patient studies. Cogn Neuropsychol. 1988;5:517–28.
45. Denier C, Masnou P, Mapoure Y, et al. Watershed infarctions are more prone than other cortical infarcts to cause early-onset seizures. Arch Neurol. 2010;67:1219–23.
46. Ross ED. Cerebral localization of functions and the neurology of language: facts versus fiction or is it something else? Neuroscientist. 2010;16:222–43.
47. Ochfeld E, Newhart M, Molitoris J, et al. Ischemia in broca area is associated with broca aphasia more reliably in acute than in chronic stroke. Stroke. 2010;41:325–30.
48. Lambon Ralph MA. Measuring language recovery in the underlying large-scale neural network: pulling together in the face of adversity. Ann Neurol. 2010;68:570–2.
49. Signer S, Cummings JL, Benson DF. Delusions and mood disorders in patients with chronic aphasia. J Neuropsychiatry Clin Neurosci. 1989;1:40–5.

50. Carota A, Rossetti AO, Karapanayiotides T, et al. Catastrophic reaction in acute stroke: a reflex behavior in aphasic patients. Neurology. 2001;57:1902–5.
51. Berthier ML, Lambon Ralph MA, Pujol J, et al. Arcuate fasciculus and repetition. The left sometimes can be right. Cortex. 2012;48:133–43.
52. Shallice T, Warrington EK. Auditory-verbal short-term memory and conduction aphasia. Brain Lang. 1977;4:79–491.
53. Nadeau SE. Phonology: a review and proposals from a connectionist perspective. Brain Lang. 2001;79:511–79.
54. Ardila A, Rosselli M. Conduction aphasia and verbal apraxia. J Neurolinguistics. 1990;5:1–14.
55. Bernal B, Ardila A. The role of the arcate fasciculus in conduction aphasia. Brain. 2009;132:2309–16.
56. Kohn SE. Conduction aphasia. Hove: Lawrence Erlbaum Associates; 1992.
57. Buchsbaum BR, Baldo J, Okada K, et al. Conduction aphasia, sensory-motor integration and phonological short-term memory: an aggregate analysis of lesion and fMRI data. Brain Lang. 2011;119:119–28.
58. Lazar M, Antoniello D. Variability in recovery from aphasia. Curr Neurol Neurosci Rep. 2008;8:497–502.
59. Berthier ML, Pulvermüller F, Dávila G, et al. Drug therapy of post-stroke aphasia: a review of current evidence. Neuropsychol Rev. 2011;21:302–17.
60. Cicerone KD, Langenbahn DM, Braden C, et al. Evidence-based cognitive rehabilitation: updated review of the literature from 2003 through 2008. Arch Phys Med Rehabil. 2011;92:519–30.
61. Othmer E, Othmer SC. The clinical interview using DSM-IV. Volume 1: fundamentals. Washington, DC: American Pyschiatric Association; 1994.
62. Lazar RM, Minzer B, Antoniello D, et al. Improvement in aphasia scores after stroke is well predicted by initial severity. Stroke. 2010;41:1485–8.
63. De Haan EH, Gudrun MN, Van Zandvoort MJV. Cognitive function following stroke and vascular cognitive impairment. Curr Opin Neurol. 2006;19:559–64.
64. Berthier ML, García-Casares N, Walsh SF, et al. Recovery from post-stroke aphasia: lessons from brain imaging and implications for rehabilitation and biological treatments. Discov Med. 2011;12:275–89.
65. Berthier ML, Pulvermüller F. Neuroscience insights improve neurorehabilitation of post-stroke aphasia. Nat Rev Neurol. 2011;7:86–97.
66. Nakase-Thompson R, Manning E, Sherer M, et al. Bedside screen of language disturbance among acute care admissions: initial psychometrics of the Mississippi aphasia screening test. Arch Clin Neuropsychol. 2002;17:848.
67. Enderby P, Crow E. Frenchay aphasia screening test: validity and comparability. Disabil Rehabil. 1996;18:238–40.
68. Goodglass H, Kaplan E. Assessment of aphasia and related disorders. Philadelphia: Lea & Febiger; 1976.
69. Kay J, Lesser R, Colheart M. Psycholinguistic assessment of language processing in aphasia (PALPA). Hove: Lawrence Erlbaum Associates; 1992.
70. Seniów J, Litwin M, Leśniak M. The relationship between non-linguistic cognitive deficits and language recovery in patients with aphasia. J Neurol Sci. 2009;15(283):91–4.
71. Yeung O, Law SP. Executive functions and aphasia treatment outcomes: data from an ortho-phonological cueing therapy for anomia in Chinese. Int J Speech Lang Pathol. 2010;12:529–44.
72. Maas MB, Lev MH, Ay H, et al. The prognosis for aphasia in stroke. J Stroke Cerebrovasc Dis. 2012;21(5):350–7.
73. Kertesz A. Recovery of aphasia. In: Feinberg TE, Farah MJ, editors. Behavioral neurology and neuropsychology. NewYork: McGraw-Hill; 1997. p. 167–82.
74. Bakheit AM, Shaw S, Carrington S, et al. The rate and extent of improvement with therapy from the different types of aphasia in the first year after stroke. Clin Rehabil. 2007;21:941–9.

75. Staekenborg SS, van der Flier WM, van Straaten EC, et al. Neurological signs in relation to type of cerebrovascular disease in vascular dementia. Stroke. 2008;39:317–22.
76. Johansson BB. Brain plasticity and stroke rehabilitation. The Willis lecture. Stroke. 2000;31:223–30.
77. Grafman J. Conceptualizing functional neuroplasticity. J Commun Disord. 2000;33:345–55.
78. May A. Experience-dependent structural plasticity in the adult human brain. Trends Cogn Sci. 2011;15:475–82.
79. Crosson B, Moore AB, Gopinath K, et al. Role of the right and left hemispheres in recovery of function during treatment of intention in aphasia. J Cogn Neurosci. 2005;17:392–406.
80. Heiss WD, Thiel A. A proposed regional hierarchy in recovery of post-stroke aphasia. Brain Lang. 2006;98:118–23.
81. Sharp DJ, Turkheimer FE, Bose SK, et al. Increased frontoparietal integration after stroke and cognitive recovery. Ann Neurol. 2010;68:753–6.
82. Turkeltaub PE, Messing S, Norise C, et al. Are networks for residual language function and recovery consistent across aphasic patients. Neurology. 2011;76:1726–34.
83. Meinzer M, Mohammadi S, Kugel H, et al. Integrity of the hippocampus and surrounding white matter is correlated with language training success in aphasia. Neuroimage. 2010;53:283–90.
84. Naeser MA, Palumbo CL, Helm-Estabrooks N, et al. Severe nonfluency in aphasia. Role of the medial subcallosal fasciculus and other white matter pathways in recovery of spontaneous speech. Brain. 1989;112(Pt 1):1–38.
85. Marchina S, Zhu LL, Norton A, et al. Impairment of speech production predicted by lesion load of the left arcuate fasciculus. Stroke. 2011;42:2251–6.
86. Berthier ML. Transcortical sensory aphasia: dissociation between naming and comprehension. Aphasiology. 1995;9:431–51.
87. Flamand-Roze C, Cauquil-Michon C, Roze E, et al. Aphasia in border-zone infarcts has a specific initial pattern and good long-term prognosis. Eur J Neurol. 2011;18:1397–401.
88. McClung JS, Rothi LJ, Nadeau SE. Ambient experience in restitutive treatment of aphasia. Front Hum Neurosci. 2010;4:183.
89. Bakheit AM, Shaw S, Barrett L, et al. A prospective, randomized, parallel group, controlled study of the effect of intensity of speech and language therapy on early recovery from post-stroke aphasia. Clin Rehabil. 2007;21:885–94.
90. Cloutman L, Newhart M, Davis C, et al. Acute recovery of oral word production following stroke: patterns of performance as predictors of recovery. Behav Neurol. 2009;21:145–53.
91. Crinion J, Price CJ. Right anterior superior temporal activation predicts auditory sentence comprehension following aphasic stroke. Brain. 2005;128(Pt 12):2858–71.
92. Crinion JT, Leff AP. Recovery and treatment of aphasia after stroke: functional imaging studies. Curr Opin Neurol. 2007;20:667–73.
93. Geva S, Correia M, Warburton EA. Diffusion tensor imaging in the study of language and aphasia. Aphasiology. 2011;25:543–58.
94. Schlaug G, Marchina S, Wan CY. The use of non-invasive brain stimulation techniques to facilitate recovery from post-stroke aphasia. Neuropsychol Rev. 2011;21:288–301.
95. Basso CS, Girelli L, et al. Treatment efficacy of language and calculation disorders and speech apraxia: a review of the literature. Eur J Phys Rehabil Med. 2011;47:101–21.
96. Laska AC, Kahan T, Hellblom A, et al. Design and methods of a randomized controlled trial on early speech and language therapy in patients with acute stroke and aphasia. Top Stroke Rehabil. 2008;15:256–61.
97. Godecke E, Hird K, Lalor EE, et al. Very early poststroke aphasia therapy: a pilot randomized controlled efficacy trial. Int J Stroke. 2012;7(8):635–44.
98. Pulvermüller F, Berthier ML. Aphasia therapy on a neuroscience basis. Aphasiology. 2008;22:563–99.
99. Meinzer M, Rodriguez AD, Gonzalez Rothi LJ. First decade of research on constrained-induced treatment approaches for aphasia rehabilitation. Arch Phys Med Rehabil. 2012;93 (1 Suppl):S35–45.

100. Basso A, Macis M. Therapy efficacy in chronic aphasia. Behav Neurol. 2011;24:317–25.
101. Husain M, Mehta MA. Cognitive enhancement by drugs in health and disease. Trends Cogn Sci. 2011;15:28–36.
102. Berthier ML, Green C, Higueras C, et al. A randomized, placebo-controlled study of donepezil in poststroke aphasia. Neurology. 2006;67:1687–9.
103. Berthier ML, Green C, Lara JP, et al. Memantine and constraint-induced aphasia therapy in chronic poststroke aphasia. Ann Neurol. 2009;65:577–85.
104. Holland R, Crinion J. Can tDCS enhance treatment of aphasia after stroke? Aphasiology. 2012;26(9):1169–91.
105. Naeser MA, Martin PI, Nicholas M, et al. Improved picture naming in chronic aphasia after TMS to part of right Broca's area: an open-protocol study. Brain Lang. 2005;93:95–105.
106. Naeser MA, Martin PI, Treglia E, et al. Research with rTMS in the treatment of aphasia. Restor Neurol Neurosci. 2010;28:511–29.
107. Barwood CH, Murdoch BE, Whelan BM, et al. The effects of low frequency repetitive transcranial magnetic stimulation (rTMS) and sham condition rTMS on behavioural language in chronic non-fluent aphasia: short term outcomes. NeuroRehabilitation. 2011;28:113–28.
108. Baker J, Rorden C, Fridriksson J. Using transcranial direct current stimulation (tDCS) to treat stroke patients with aphasia. Stroke. 2010;41:1229–36.
109. Cherney LR, Erickson RK, Small SL. Epidural cortical stimulation as adjunctive treatment for non-fluent aphasia: preliminary findings. J Neurol Neurosurg Psychiatry. 2010; 81:1014–21.
110. Cherney LR, Harvey RL, Babbitt EM, et al. Epidural cortical stimulation and aphasia therapy. Aphasiology. 2011;iFirst:1–26.
111. Fugl-Meyer AR, Jääskö L, Leyman I, et al. The post-stroke hemiplegic patient. 1. A method for evaluation of physical performance. Scand J Rehabil Med. 1975;7:13–31.
112. Hilari K, Byng S, Lamping DL, et al. Stroke and aphasia quality of life scale-39 (SAQOL-39): evaluation of acceptability, reliability, and validity. Stroke. 2003;34:1944–50.
113. Sutcliffe LM, Lincoln NB. The assessment of depression in aphasic stroke patients: the development of the stroke aphasic depression questionnaire. Clin Rehabil. 1998;12:506–13.
114. Code C. The quantity of life for people with chronic aphasia. Neuropsychol Rehabil. 2003;13:379–90.
115. Collen FM, Wade DT, Robb GF, et al. The rivermead mobility index: a further development of the rivermead motor assessment. Int Disabil Stud. 1991;13:50–4.

Chapter 7
Disorders of Emotional Communication After Stroke

Kenneth M. Heilman

Introduction

Emotions can be divided into at least three major domains: emotional experiences, emotional memories, and emotional behaviors. Stroke can induce changes in all three of these domains; however, in this chapter we will focus on emotional behaviors and primarily review the disorders of emotional communication that can be induced by stroke. There are several means by which emotions can be classified or categorized, including *valence* (positive, neutral, negative), *arousal* (high, moderate, low), and *activity* (approach, avoid, none) [1]. For example, fear would be an emotion characterized by negative valence, high arousal, and avoidance. Ross et al. [2] also divides emotions into primary (e.g., happy, sad, anger, fear) and social (e.g., embarrassed) categories. In this chapter, deficits of emotional communication will mainly address the primary emotions, which include happiness, sadness, anger, fear, disgust, surprise, and neutrality.

Since emotions can be communicated by propositional speech and written language, many patients with aphasia, alexia, and agraphia will have some problems with emotional communication. Aphasic disorders will not be fully discussed in this chapter as it is covered in Chap. 6 of this book. There are many other means by which emotions can be expressed including facial movements and postures as well as nonverbal speech such as prosody. These means of emotional expression are often the primary means by which we express emotions and these forms of communication are often impaired in patients who have had a stroke. This review will be limited to these modalities of communication. For each of these forms of communication, there are both receptive and expressive elements and each will be discussed independently.

K.M. Heilman, M.D. (✉)
Department of Neurology, University of Florida College of Medicine,
Box 100236, Gainesville, FL 32610, USA
e-mail: heilman@neurology.ufl.edu

T.A. Schweizer and R.L. Macdonald (eds.), *The Behavioral Consequences of Stroke*,
DOI 10.1007/978-1-4614-7672-6_7, © Springer Science+Business Media New York 2014

Disorders of Emotional Speech

Comprehension of Emotional Speech Prosody

Almost all people who have been in a close relationship have heard their significant other state, “It was not what you said, but how you said it.” When people speak, and when people hear other people speak to them, there are at least two sounds that people may hear. One is a sequence of speech sounds called *phonemes*, which form words, and the other is *prosody*, which includes the pitch of the voice, the amplitude or loudness, the speed and rhythm or tempo. When hearing someone speak, these elements of prosody may change.

When people speak, they also convey at least two types of messages: propositional messages and emotional messages. Propositions are messages that can either be correct or incorrect. Propositional messages are usually conveyed by a series of words, the comprehension of which requires hearing the phonemes that comprises the words, recognizing the sequence of phonemes that comprises a specific previously heard word (lexical processing), understanding the meaning of each of the words (semantic processing), and understanding how the sequence and relationships between these words influence meaning (syntax) (e.g., “John hit Mary” versus “Mary hit John”).

Although prosody in English can be used to express a question, command, or statement, prosody is often used along with propositional speech to express emotions [3]. About 40 years ago I saw a woman in the clinic who had a right hemispheric temporal parietal lesion. Her husband was considering leaving her, because according to him, they no longer had a meaningful relationship. When I asked him to explain, he told me that his business was not doing well and that he was very depressed, but she did not appear to understand his mood. I tested her ability to understand emotions expressed by speech prosody and found that she was terribly impaired. Hence, we subsequently studied a group of patients that had either left or right-hemisphere strokes [4, 5]. These patients were asked to identify the emotional tone of the speaker based not on what was said, but rather how it was said. We presented recorded sentences that had a neutral propositional content, but when these recorded sentences were played, each had one of four different emotional prosodies (happy, sad, angry, and neutral-indifferent). Although the patients with the left hemispheric strokes were aphasic, all had sufficient comprehension to understand the task instructions. We found that the non-aphasic patients with right-hemisphere strokes performed worse on this emotional prosody task than those with left-hemisphere lesions, suggesting that the right hemisphere is dominant in comprehending emotionally intonated speech [6]. Similar findings were reported by Ross [7].

As we mentioned, in English, speech prosody can also be used to label sentences as questions, commands, or statements. Weintraub et al. [8] reported that, relative to normal control participants, patients with right-hemisphere strokes were impaired in determining whether intoned sentences were statements, commands, or questions.

Based on these results, they suggested that the prosodic deficit induced by right-hemisphere damage (RHD) was not specific to emotions. Since a left-hemisphere damage (LHD) group was not tested by Weinstraub et al. [8], they could not determine if this deficit was specific to RHD. Therefore, Heilman et al. [9] compared RHD and LHD patients for comprehension of emotional (happy, sad, angry, neutral) and nonemotional prosody (questions, commands, statements). We found that both the RHD and LHD subjects were equally impaired on the nonemotional (syntactic-propositional) prosody comprehension task when compared to healthy participants. In contrast, in regard to comprehension of emotional prosody, the patients with RHD performed significantly worse than the LHD and the normal control participants. This hemispheric dissociation between deficits in propositional and emotional prosody suggests that the right hemisphere does indeed play a dominant role in the comprehension of emotional prosody, although both hemispheres may be important in comprehending syntactic-propositional prosody.

Since these early studies, a large body of clinical research has shown that impairments in the comprehension of emotional prosody is primarily associated with damage to the right cerebral hemisphere [10–13]. Although these studies do reveal that patients with a right hemispheric stroke may have impairments in the comprehension of spoken emotional prosody, the neuropsychological mechanism or mechanisms that may be responsible for this deficit is not entirely known. Geschwind [14] noted that strokes of the right hemisphere can not only destroy portions of the cortex that store information and process stimuli, but these lesions can also involve the pathways that connect the right hemisphere's cortex to the left-verbal cortex. Thus, patients with right hemispheric strokes may not be able to correctly name emotional intonations in addition to being unable to understand them, because this information cannot access the verbal left hemisphere.

An alternate hypothesis is that these patients' right-hemisphere injuries induced a disorder that, in some respects, parallels Wernicke's aphasia, which occurs due to left-hemisphere lesions, such that the patients cannot properly discriminate affective intonations in speech. To learn if patients with right-hemisphere lesions and a failure to comprehend emotional prosody could discriminate between affective intonations of speech without having to name or verbally denote these intonations, we had patients listen to recorded pairs of sentences [5]. Sometimes these two sentences were spoken with the same emotional prosody and sometimes with different emotional prosodies. In this testing paradigm all the patients had to do was to indicate to the examiner if the prosody in these two sentences were the same or different. We found that the patients with right-hemisphere strokes performed more poorly on this task than did patients with left-hemisphere strokes. Based on these findings, we posited that the posterior temporal-inferior parietal region in the right hemisphere contains representations of affective prosodic expressions and destruction of these representations or an inability to access them impairs both comprehension and discrimination of emotional prosody.

There is, however, another possible explanation for the emotional prosodic comprehension disturbance associated with right-hemisphere temporoparietal infarctions. One of the reasons why the left hemisphere processes propositional speech and the right-hemisphere emotional prosody is that having these two hemispheres analyze different forms of speech allows for parallel processing, so that we can hear what is said and how it is said at the same time. It is possible that the intact left hemisphere can comprehend emotional prosody, but like all cognitive systems it has a limited capacity, cannot simultaneous process incoming words together with emotional prosody, and has a preference for attending to words rather than emotional prosody. To test this hypothesis, we tested RHD and LHD subjects with an emotional prosody task that varied in the degree of "conflict" between the emotional prosodic message and the message that was conveyed by the propositional content of the speech [15]. If this "left hemisphere distraction" hypothesis has merit, RHD stroke patients should be easily "distracted" by conflicting emotional propositional and prosodic content. Thus, their comprehension of emotional prosody should be worse when the propositional content and prosody messages are strongly conflicting (i.e., "All the puppies are dead" said in a happy tone of voice) than when the there is less conflict (e.g., "The boy walked to the store" said with a happy tone of voice). Analysis of our results revealed that the RHD group was more disrupted when the propositional and prosodic emotional messages were highly conflicting than when they were less conflicting. In contrast, LHD stroke patients were unaffected by increasing the discrepancy between the verbal and prosodic messages. The results of this study indicate that, to some degree, the deficit RHD patients experience in comprehending emotional prosody is related to verbal "distraction." This verbal distraction hypothesis cannot entirely explain why right-hemisphere injury patients are impaired in comprehending emotional prosody, since even when speech is altered by a filtering technique such that words become incomprehensible the patients with RHD are still impaired in comprehending emotional prosody [15].

Comprehension of Verbal Stimuli

Patients with strokes of the left hemisphere who have *pure word deafness*, Wernicke's aphasia, *transcortical sensory aphasia*, *mixed transcortical aphasia*, or *global aphasia* are impaired in comprehending spoken verbal stimuli, and patients with *alexia* are impaired in understanding written verbal stimuli. Although these LHD aphasic or word-deaf patients may have difficulty comprehending propositional speech, several studies have revealed that some of these patients can still comprehend emotional intonations, and their comprehension of propositional speech may even be aided by these intonations [16, 17]. For example, I was examining a young woman with severe global aphasia from a left carotid artery dissection and during this examination her husband and children came to visit her, but were in the hallway where she could not see them. One of the children's voice was very sad when she told her father that she was worried about crying when she saw her mother. The patient could hear her child's voice and looked to the door and after hearing how sad the child sounded this young globally aphasic woman started to cry.

There is also evidence that aphasic patients with comprehension disorders can understand emotional words relatively better than words that do not express any emotions [18–20]. In addition, Landis et al. [19] reported that LHD stroke patients were able to read emotional words better than nonemotional words. A study of normal subjects supports the postulate that emotional words are more likely to be processed by the right hemisphere and that this ability of the right hemisphere to comprehend emotional words may account for the preserved comprehension of words in patients with aphasia from left-hemisphere infarctions [21].

If the right hemisphere were important for comprehending emotional words, would patients with strokes of the right hemisphere have impaired comprehension of these words? To test this hypothesis, Borod et al. [22] assessed participants with strokes in either their right or left hemisphere by presenting sentences and words with emotional and nonemotional content. They found that subjects with RHD were more impaired than those with LHD when attempting to understand these emotional sentences and words [22]. Other investigators such as Morris et al. [23] found no differences between RHD and LHD patients' ability to comprehend the denotative or connotative meaning of emotional versus nonemotional words. In addition, Blonder et al. [24] presented sentences describing a situation that would induce an emotion in the person described in the sentence and asked patients with RHD what they thought was the emotion this person would experience. They found that patients with RHD were not impaired in their emotional comprehension. Although several investigators have found that some patients with RHD may be impaired in understanding the affective-emotional content of stories [25, 26], these problems do not appear to be specific to "emotion" per se. Instead, they are related to more general difficulties that RHD patients have in drawing inferences, understanding figures of speech, metaphor, and logical reasoning [26–29].

Expression of Emotional Prosody

After learning that patients with right hemispheric strokes were impaired in comprehending emotional prosody, we wanted to learn if patients with right-hemisphere strokes were also impaired in expressing emotional speech prosody. In our first study, patients with right hemispheric strokes were asked to produce semantically neutral sentences such as, "The boy went to the store," intoned with different forms of emotional prosody such as sadness, happiness, anger, and neutrality [5]. The participants with right hemispheric strokes were not impaired when producing the words in these sentences, but were impaired when attempting to produce emotional prosody such that, when assessed by judges, almost all the sentences produced sounded neutral. Subsequently, Borod replicated our study [30], but in addition Ross and Mesulam [31] described two patients with right anterior hemispheric lesions who could also not express emotionally intoned speech prosody; however, these two patients were able to normally comprehend the emotion expressed by speech prosody. In addition, Ross [7] described patients who could not comprehend the emotionally intoned prosody, but could nevertheless normally repeat emotionally intoned speech. Based on these observations, he postulated that right-hemisphere lesions might disrupt the comprehension, repetition, or production of emotional prosody in a manner that is parallel to those patients with the aphasic syndromes (propositional speech disorders) associated with left hemispheric lesions. For example, just as there are patients who have non-fluent verbal propositional speech, with impaired speech repetition (Broca's aphasia) the patients reported by Ross and Mesulam had similar problems with emotional prosody. There are also patients with Wernicke's aphasia who have a verbal comprehension deficit with impaired word–sentence repetition and these are similar to the patients reported by Tucker et al. [5]. There are aphasic patients who cannot comprehend propositional speech, but who can repeat it normally. This disorder is called *transcortical sensory aphasia*. Ross described a patient with a right-hemisphere stroke who was not aphasic, but who could not comprehend emotional prosody. This patient, however, could normally repeat the words and emotional prosody that they heard. Patients with left-hemisphere medial frontal lesions often have an impairment in producing spontaneous propositional speech but have intact repetition of propositional phrases and comprehension of propositional speech, a disorder called *transcortical motor aphasia*. We have even described a patient with a right medial frontal lesion who was not aphasic and who could not express emotional prosody spontaneously, but could repeat this prosody and comprehend emotional prosody [32]. This patient appeared to have a prosodic defect that is parallel to transcortical motor aphasia, thereby providing support for the "parallel propositional-prosody speech hypothesis" posited by Ross. Acoustic analyses of the speech produced by patients with right hemispheric strokes have also demonstrated prosodic abnormalities such as decreased variation in fundamental frequency [13, 33–36].

Normally when people speak they combine their propositional message with their emotional prosodic message. Since propositional speech is primarily mediated by the left hemisphere and emotional prosody by the right hemisphere, when normal people are speaking the two hemispheres would have to be interactive. Speedie et al. [37] studied stroke patients with LHD who were able to normally repeat propositional sentences. These patients' imaging data demonstrated that their lesions interrupted the callosal pathways that interconnect the left-hemisphere speech areas with the right hemisphere. When attempting to repeat emotionally intoned sentences, these patients correctly repeated the propositional verbal message but without emotional intonations or prosody, thereby providing evidence that this propositional and affective prosodic mixing occurs inter-hemispherically.

Verbal Expression of Emotions

Since Paul Broca's paper where he described a series of right-handed individuals who developed aphasia from left-hemisphere injury, it has repeatedly been demonstrated that the left hemisphere is dominant for processing propositional speech; however, in the last several decades investigators have provided some evidence that patients with RHD have alterations when expressing emotions even when using propositional speech. For example, Cimino et al. [38] had RHD patients as well as matched controls produce emotional autobiographical memories. The judges that assessed these patients' production rated the memories described by the RHD participants as less emotional than those of normal volunteers. Bloom et al. [39] reported that RHD patients used fewer words when denoting emotions in their spontaneous speech. Another study done by Borod et al. [40] had raters judge the emotionality of monologues produced by RHD, LHD, as well as normal volunteers and found that the monologues of the RHD patients were rated as significantly less emotional than the monologues of normal controls. There was even a trend for RHD patients' monologues to be rated as less emotional than LHD patients. In contrast, when interviewing patients Blonder et al. [41] found that RHD patients who had emotional aprosodic speech produced a greater percentage of emotion words relative to total words than did the aphasic LHD patients. The reason for these conflicting results remains to be determined, but perhaps these RHD participants were aware of their prosodic disability and attempted to verbally compensate.

Some patients with LHD and aphasia may have difficulty expressing emotions when these emotions are expressed using a spoken or written propositional message; however, some aphasics, and especially those who are non-fluent, may become very fluent when using expletives. In addition, Landis et al. [19] and Roeltgen et al. [42] demonstrated that aphasic patients with agraphia were able to write emotional words better than nonemotional words. This preserved ability may also be related to right-hemisphere function, but needs further investigation.

Disorders of Facial Emotional Communication

Comprehension of Facial Emotional Expressions

One of the most important means of communicating an emotion or a mood is by the display of facial expressions. Thus, in order to understand other people's emotional responses and moods a person has to be able to perceive and comprehend other people's emotional facial expressions.

More than 3 decades ago we wanted to learn if strokes could impair the ability of patients to understand facial emotional expressions and the location of strokes that could produce this deficit. In our first study, we assessed participants who had left- or right hemispheric strokes, as well as control subjects without hemispheric disorders, for their ability to perceive, comprehend, and name emotional facial expressions [43]. In this study, the participants were shown pictures of people displaying emotional faces and asked to name the emotions being expressed, or asked to point to faces expressing the specific emotions named by the examiner. These participants were also presented with a series of paired pictures—in which an actor's face either expressed the same emotion twice or two different emotions—and were asked if the emotions shown were the same or different. We found that the patients with right-hemisphere infarctions were impaired in naming emotional facial expressions, pointing to faces expressing an emotion named by the examiner, as well as determining whether two faces were expressing the same emotion or different emotions. This study revealed that right hemispheric strokes cause a loss of the ability for patients to perceive and understand emotional faces.

This study did not reveal if the deficit was specific to emotions or if it was simply a disorder of facial recognition. Benton and Van Allen had demonstrated that patients with right-hemisphere strokes often have impaired visuospatial functions including facial recognition [44]. For example, when shown a series of pictures containing two faces—at times belonging to the same person, and at other times, belonging to different people—the RHD patients were impaired in determining whether the two pictures were of the same person or of different people. Thus, it is possible that the reason our patients with right-hemisphere strokes could not normally process emotional faces (nor name, point and match) was because their right-hemisphere strokes caused a visuospatial facial recognition deficit.

When we covaried for neutral facial discrimination (a visuoperceptual, but non-emotional facial recognition task), we no longer found that our RHD stroke subjects were more impaired in recognizing (naming and pointing) than the other subject groups [43]. However, when the RHD patients were compared to the other groups, they remained impaired in determining same and different emotional faces [43]. These results suggest that one of the primary disabilities of these RHD patients is a visuospatial facial perceptual disorder and it is this disability that might account for their inability to recognize and name emotional facial expressions. However, when we reviewed the performance of the individual participants, we found that about one-third of the patients with RHD performed poorly on both the neutral facial matching (same or different people) and all emotional faces tasks. Another one-third performed well on both. The remaining patients with RHD, however, performed relatively well on the neutral facial discrimination task but poorly on the emotional faces tasks. This observation suggests that a visuospatial facial perceptual deficit cannot account for the impairment in emotional facial recognition in all RHD patients and that some of these patients do have an impairment of processing emotional facial expressions.

After we completed the emotional facial recognition study, we became aware that there was a weakness in the design of the study performed by DeKosky et al. [43], such that the subjects could have accurately made same–different determinations based on physiognomic/structural configurations, without any regard to emotional expressions displayed on these faces. We performed a second study that corrected for these methodological problems and again assessed stroke patients with hemispheric lesions and control subjects using a series of "perceptual" and "associative" facial recognition and facial emotion tasks [45]. This revised technique allowed us to measure the participants' visuospatial abilities and statistically equate for their visuoperceptual ability when analyzing their ability to recognize emotional faces. In this study we found that the RHD stroke group was impaired in their ability to name, select (point), and discriminate (same/different) facial emotional expressions even when accounting for their nonemotional facial recognition abilities. These results provided strong evidence that the RHD patients do in fact have impaired recognition of facial emotional expressions and that this recognition deficit cannot be solely attributed to a general deficit in visuoperceptual facial processing. In addition, these results also suggest that the right hemisphere's dominance for recognizing–comprehending facial emotional expressions is at least in part independent from its dominance in processing faces. Studies from other laboratories have also reported that RHD stroke patients are more impaired than their LHD counterparts in recognizing or categorizing facial emotions [10–12, 24, 46–53]. In a meta-analysis, Borod et al. [50] reported that 20 out of 23 published studies found that deficits in facial affect recognition was associated with focal lesions in the right hemisphere.

Studies using different types of participants have provided further evidence for a right-hemisphere dominance in recognizing emotional facial expressions. For example, studies of patients undergoing Wada testing, studies of split brain patients, and even visual half field studies in normal subjects provide evidence that the right hemisphere is dominant for the recognition of emotional facial expressions [54–59].

Blonder et al. [24] presented RHD and LHD stroke patients with verbal descriptions of nonverbal emotional gestures and found that the patients with right-hemisphere strokes were even impaired at recognizing the verbal descriptions of emotional gestures. In this study the poor performance of the RHD group could not be attributed to a visual–perceptual disturbance. These same participants were also given another task in which they had to make inferences about emotions that are linked to various situational contexts (e.g., "The children track dirt over your white carpet"). In this task the RHD stroke patients were not impaired in recognizing the emotions that these situations would induce. Thus, these results suggest that the impairment of these RHD subjects in recognizing emotional gestures is not conceptual. Based on these and other studies, we posited that the poor performance of the RHD patients in assigning an emotion to verbally described nonverbal signals is induced by a facial emotional gestural representational deficit [24]. But how are these facial gestures represented?

We posited that the right hemisphere contains a store of facial emotional iconic representations [60]. Support for this iconic hypothesis, that the right hemisphere contains representations of species-specific typical emotional expressions, comes from a study we performed in which RHD and LHD patients were evaluated on two tasks where they were required to use visual imagery [61]. In one of these tasks the subjects were asked to imagine a target emotional face (i.e., a face expressing anger). The examiner then asked a series questions regarding the structural characteristics of the face (i.e., "Are the outer lips pulled down?") and the participant had to supply yes or no answers. In the other task—the object imagery task—the subjects were asked to imagine common objects and were then asked questions about the visual characteristics of these objects and also had to provide yes or no answers. We found that the RHD group was more impaired in the emotional facial imagery task than in the object imagery task and the LHD was more impaired in object imagery than in facial emotional imagery. These RHD patients were also more impaired at recognizing visible facial emotions than were the LHD group. These results suggest that the right hemisphere normally contains a store of typical facial emotional representations and that right-hemisphere strokes can damage these iconic representations. Bowers et al. [61] also described a patient with a ventral temporal-occipital lesion of the right hemisphere who was not impaired in recognizing emotional facial expressions, but was impaired in imagining facial emotional expressions. This dissociation suggests that the emotional facial imagery generation is at least in part mediated by the posterior ventral area. However, the facial emotional representations may be stored in an anatomically distinct area. But where are these representations stored?

Harciarek and Heilman [52] compared facial affect recognition of patients with right-hemisphere strokes who had anterior versus posterior lesions and found that while both groups were impaired relative to normal controls, the patients with anterior lesions performed significantly worse than those with posterior lesions. Based on these observations, it is possible that mirroring the observed emotional facial expression with activation of the somatosensory cortex is crucial for the recognition of facial affect. Harciarek and Heilman's report appears to support the findings of Adolphs et al. [48] for the role of the somatosensory cortex in the recognition of facial expressions. This mirror-somatosensory hypothesis could also help explain the failure of RHD patients to correctly imagine facial emotional expressions [61], and to recognize the verbal description of emotional expressions [24].

Generating Facial Emotion

Expressions

One of the first studies examining stroke patients' ability to express facial emotions was performed by Buck and Duffy who had RHD and LHD stroke patients view slides of people with whom they were familiar and unpleasant scenes [62]. They reported that the RHD patients were less facially expressive than LHD patients when viewing these slides. These impairments of facial emotional expressions induced by right-hemisphere strokes have been replicated by other investigators [30, 35, 49, 63, 64].

A reduction of facial emotional expressions has also been observed in more "naturalistic" settings. Blonder et al. videotaped interviews they had with stroke patients and their spouses in their homes [41, 65]. These videotapes were then viewed by judges who were unaware of the patients' clinical status and it was found that the judges rated patients with RHD as less facially expressive than LHD patients and normal controls.

Using the facial action scoring system (FACS) of Ekman and Friesen [66], investigators have reported finding no differences in facial emotional expressiveness between RHD and LHD stroke patients [67, 68]. Other investigators have found that intra-hemispheric lesion location rather than hemispheric laterality is critical for inducing this emotional facial expressive deficit. According to these investigators, strokes that extend into the frontal lobes, regardless of laterality, induce a reduction of facial expression [69, 70].

Whereas the basis for these discrepant findings in studies of laterality remains unclear [71], in a meta-analysis of emotional processing deficits in patients with unilateral brain damage, Borod and colleagues reported that 9 out of 13 studies of spontaneous facial expressions found patients with RHD to be more impaired than those with LHD [50]. In addition, several studies of normal subjects indicated that they express emotions more intensely on the left side of the face than on the right, providing evidence for right hemispheric dominance [58, 59, 72, 73]. However, Triggs and colleagues demonstrated that there is an asymmetry of corticobulbar projections to muscles of facial expression in the lower face, with the left having stronger connectivity than the right [74].

Summary

Humans have two hemispheres affording us the ability to perform parallel processing. When we are communicating, it is important for us to comprehend and express not only propositional messages but also emotional messages. For more than 150 years it has been known that in most people, it is the left hemisphere that is dominant for mediating propositional communication. In the last several decades we have learned that it is the right hemisphere that appears to mediate many forms of emotional communication, including expressing and comprehending emotional facial expressions as well as expressing and understanding emotional speech prosody. Thus, damage to the right hemisphere by stroke may cause impairments in emotional communication. Studies of aphasic patients have revealed that although left anterior hemispheric strokes are more likely to cause expressive deficits, posterior lesions are more likely to cause comprehension deficits and this rule also appears to hold true for expressions of emotions such as emotional prosody. Whereas disorders of propositional communication are more often apparent to clinicians, disorders of emotional communication can cause profound disabilities more noticeable to the general public. Patients should therefore be tested for these disorders and treated when able [75–77].

References

1. Heilman KM, Blonder LX, Bowers D, Valenstein E. Emotional disorders associated with neurological diseases. In: Heilman KM, Valenstein E, editors. Clinical neuropsychology. 5th ed. New York: Oxford University Press; 2012. p. 466–503.
2. Ross ED, Homan RW, Buck RW. Differential hemispheric lateralization of primary and social emotions. Neuropsychiatry Neuropsychol Behav Neurol. 1994;7:1–19.

3. Monrad-Krohn G. The prosodic quality of speech and its disorders. Acta Psychol Scand. 1947;22:225–65.
4. Heilman KM, Scholes R, Watson RT. Auditory affective agnosia: disturbed comprehension of affective speech. J Neurol Neurosurg Psychiatry. 1975;38:69–72.
5. Tucker DM, Watson RT, Heilman KM. Affective discrimination and evocation in patients with right parietal disease. Neurology. 1977;17:947–50.
6. Broca P. Anatomie comparee des enconvolutions cerebrales: le grand lobe limbique et al scissure limbique dans la seire des mammiferes. Rev Antrop. 1878;1:385–498.
7. Ross ED. The aprosodias: functional-anatomic organization of the affective components of language in the right hemisphere. Ann Neurol. 1981;38:561–89.
8. Weintraub S, Mesulam MM, Kramer L. Disturbances in prosody. Arch Neurol. 1981;38:742–4.
9. Heilman KM, Bowers D, Speedie L, Coslett HB. Comprehension of affective and nonaffective speech. Neurology. 1984;34:917–21.
10. Charbonneau S, Scherzer BP, Aspirot D, Cohen H. Perception and production of facial and prosodic emotions by chronic CVA patients. Neuropsychologia. 2003;41:605–13.
11. Harciarek M, Heilman KM, Jodzio K. Defective comprehension of emotional faces and prosody as a result of right hemisphere stroke: modality versus emotion-type specificity. J Int Neuropsychol Soc. 2006;12(6):774–81.
12. Kucharska-Pietura K, Phillips ML, Gernand W, David AS. Perception of emotions from faces and voices following unilateral brain damage. Neuropsychologia. 2003;41:1082–90.
13. Pell MD. Fundamental frequency encoding of linguistic and emotional prosody by right hemisphere-damaged speakers. Brain Lang. 1999;69(2):161–92.
14. Geschwind N. Disconnexion syndromes in animals and man. Brain. 1965;88(237–94):585–644.
15. Bowers D, Coslett HB, Bauer RM, Speedie LJ, Heilman KM. Comprehension of emotional prosody following unilateral hemispheric lesions: processing defect vs. distraction defect. Neuropsychologia. 1987;25:317–28.
16. Heilman KM, Gold MS, Tucker DM. Improvement of aphasics' comprehension by use of novel stimuli. Trans Am Neurol Assoc. 1975;100:201–2.
17. Coslett HB, Brasher HR, Heilman KM. Pure word deafness after bilateral primary auditory cortex infarcts. Neurology. 1984;34:347–52.
18. Boller F, Cole M, Vtunski P, Patterson M, Kim Y. Paralinguistic aspects of auditory comprehension in aphasia. Brain Lang. 1979;7:164–74.
19. Landis T, Graves R, Goodglass H. Aphasic reading and writing: possible evidence for right hemisphere participation. Cortex. 1982;18:105–22.
20. Reuterskiold C. The effects of emotionality on auditory comprehension in aphasia. Cortex. 1991;27:595–604.
21. Abbassi E, Kahlaoui K, Wilson MA, Joanette Y. Processing the emotions in words: the complementary contributions of the left and right hemispheres. Cogn Affect Behav Neurosci. 2011;11(3):372–85.
22. Borod JC, Andelman F, Obler LK, Tweedy JR, Welkowitz J. Right hemisphere specialization for the identification of emotional words and sentences: evidence from stroke patients. Neuropsychologia. 1992;30(9):827–44.
23. Morris MK, Bowers D, Verfaellie M, Blonder L, Cimino C, Bauer R, et al. Lexical denotation and connotation in right and left hemisphere damaged patients. J Clin Exp Neuropsychol. 1992;14:105.
24. Blonder LX, Bowers D, Heilman KM. The role of the right hemisphere on emotional communication. Brain. 1991;114:1115–27.
25. Gardner H, Ling PK, Flam I, Silverman J. Comprehension and appreciation of humorous material following brain damage. Brain. 1975;98:399–412.
26. Brownell H, Michel D, Powelson J, Gardner H. Surprise but not coherence: sensitivity to verbal humor in right hemisphere patients. Brain Lang. 1983;18:20–7.

27. Brownell H, Potter H, Birhle A. Inferences deficits in right brain damaged patients. Brain Lang. 1986;27:310–21.
28. McDonald S, Wales R. An investigation of the ability to process inferences in language following right hemisphere brain damage. Brain Lang. 1986;29:68.
29. Cheang HS, Pell MD. A study of humour and communicative intention following right hemisphere stroke. Clin Linguist Phon. 2006;20(6):447–62.
30. Borod JC, Koff E, Lorch MP, Nicholas M. Channels of emotional communication in patients with unilateral brain damage. Arch Neurol. 1985;42:345–8.
31. Ross ED, Mesulam MM. Dominant language functions of the right hemisphere? Prosody and emotional gesturing. Arch Neurol. 1979;36:144–8.
32. Heilman KM, Leon SA, Rosenbek JC. Affective aprosodia from a medial frontal stroke. Brain Lang. 2004;89(3):411–6.
33. Behrens SJ. Characterizing sentence intonation in a right hemisphere-damaged population. Brain Lang. 1989;36(2):181–200.
34. Blonder LX, Pickering JE, Heath RL, Smith CD, Butler SM. Prosodic characteristics of speech pre- and post-right hemisphere stroke. Brain Lang. 1995;51(2):318–35.
35. Kent RD, Rosenbek JC. Prosodic disturbance and neurologic lesion. Brain Lang. 1982;15(2):259–91.
36. Ross ED, Edmondson JA, Seibert GB, Homan RW. Acoustic analysis of affective prosody during right-sided Wada Test: a within-subjects verification of the right hemisphere's role in language. Brain Lang. 1988;33(1):128–45.
37. Speedie LJ, Coslett HB, Heilman KM. Repetition of affective prosody in mixed transcortical aphasia. Arch Neurol. 1984;41:268–70.
38. Cimino CR, Verfaellie M, Bowers D, Heilman KM. Autobiographical memory with influence of right hemisphere damage on emotionality and specificity. Brain Cogn. 1991;15:106–18.
39. Bloom R, Borod JC, Obler L, Gerstman LJ. Impact of emotional content on discourse production in patients with unilateral brain damage. Brain Lang. 1992;42:153–64.
40. Borod JC, Rorie KD, Haywood CS, Andelman F, Obler LK, Welkowitz J, et al. Hemispheric specialization for discourse reports of emotional experiences: relationships to demographic, neurological, and perceptual variables. Neuropsychologia. 1996;34(5):351–9.
41. Blonder LX, Heilman KM, Ketterson T, Rosenbek J, Raymer A, Crosson B, et al. Affective facial and lexical expression in aprosodic versus aphasic stroke patients. J Int Neuropsychol Soc. 2005;11(6):677–85.
42. Roeltgen DP, Sevush S, Heilman KM. Ponological agraphia: writing by the lexical semantic route. Neurology. 1983;33:755–65.
43. DeKosky ST, Heilman KM, Bowers D, Valenstein E. Recognition and discrimination of emotional faces and pictures. Brain Lang. 1980;9:206–14.
44. Benton AL, Van Allen MW. Impairment in facial recognition in patients with cerebral disease. Cortex. 1968;4:344–58.
45. Bowers D, Bauer RM, Coslett HB, Heilman KM. Processing of faces by patients with unilateral hemispheric lesions. I. Dissociation between judgments of facial affect and facial identity. Brain Cogn. 1985;4:258–72.
46. Cicone M, Wapner W, Gardner H. Sensitivity to emotional expressions and situations in organic patients. Cortex. 1980;16:145–58.
47. Adolphs R, Damasio H, Tranel D, Damasio AR. Cortical systems for the recognition of emotion in facial expressions. J Neurosci. 1996;16:7678–87.
48. Adolphs R, Damasio H, Tranel D, Cooper G, Damasio AR. A role for somatosensory cortices in the visual recognition of emotion as revealed by three-dimensional lesion mapping. J Neurosci. 2000;20:2683–90.
49. Borod JC, Koff E, Perlman Lorch M, Nicholas M. The expression and perception of facial emotions in brain damaged patients. Neuropsychologia. 1986;24:169–80.
50. Borod JC, Bloom RL, Brickman AM, Nakhutina L, Curko EA. Emotional processing deficits in individuals with unilateral brain damage. Appl Neuropsychol. 2002;9:23–36.

51. Etcoff N. Perceptual and conceptual organization of facial emotions. Brain Cogn. 1984;3:385–412.
52. Harciarek M, Heilman KM. The contribution of anterior and posterior regions of the right hemisphere to the recognition of emotional faces. J Clin Exp Neuropsychol. 2008;9:1–9.
53. Mandal MK, Borod JC, Asthana HS, Mohanty A, Mohanty S, Koff E. Effects of lesion variables and emotion type on the perception of facial emotion. J Nerv Ment Dis. 1999;187:603–9.
54. Benowitz LI, Bear DM, Mesulam MM, Rosenthal R, Zaidal EE, Sperry RW. Nonverbal sensitivity following lateralized cerebral injury. Cortex. 1983;19:5–12.
55. Ahern G, Schumer D, Kleefield J, Blume H, Cosgrove GR, Weintraub S, et al. Right hemisphere advantage in evaluating emotional facial expressions. Cortex. 1991;27:193–202.
56. Suberi M, McKeever W. Differential right hemisphere memory storage of emotional and nonemotional faces. Neuropsychologia. 1977;15:757–68.
57. Ley R, Bryden M. Hemispheric differences in recognizing faces and emotions. Brain Lang. 1979;1:127–38.
58. Campbell R. Asymmetries in interpreting and expressing a posed facial expression. Cortex. 1978;14:327–42.
59. Heller W, Levy J. Perception and expression of emotion in right handers and left handers. Neuropsychologia. 1981;19:263–72.
60. Bowers D, Heilman KM. Dissociation of affective and nonaffective faces: a case study. J Clin Neuropsychol. 1984;6:367–79.
61. Bowers D, Blonder LX, Feinberg T, Heilman KM. Differential impact of right and left hemisphere lesions on facial emotion and object imagery. Brain. 1991;114:2593–609.
62. Buck R, Duffy RJ. Nonverbal communication of affect in brain damaged patients. Cortex. 1980;16:351–62.
63. Borod J, Koff E, Perlman-Lorch M, Nicholas M, Welkowitz J. Emotional and nonemotional facial behavior in patients with unilateral brain damage. J Neurol Neurosurg Psychiatry. 1988;51:826–32.
64. Richardson C, Bowers D, Eyeler L, Heilman KM. Asymmetrical control of facial emotional expression depends on the means of elicitation. Presented at the meeting of the International Neuropsychology Society, San Diego. 1992.
65. Blonder LX, Burns A, Bowers D, Moore RW, Heilman KM. Right hemisphere facial expressivity during natural conversation. Brain Cogn. 1993;21(1):44–56.
66. Ekman P, Friesen WV. Facial action coding system. Palo Alto, CA: Consulting Psychologists Press; 1978.
67. Mammucari A, Caltagirone C, Ekman P, Friesen W, Gainotti G, Pizzamiglio L, et al. Spontaneous facial expression of emotions in brain damaged patients. Cortex. 1988;24:521–33.
68. Caltagirone C, Ekman P, Friesen W, Gainotti G, Mammucari A, Pizzamiglio L, et al. Posed emotional facial expressions in brain damaged patients. Cortex. 1989;25:653–63.
69. Kolb B, Milner B. Observations on spontaneous facial expression after focal cerebral excisions and after intracarotid injection of sodium amytal. Neuropsychologia. 1981;19:505–14.
70. Weddell R, Miller R, Trevarthen C. Voluntary emotional facial expressions in patients with focal cerebral lesions. Neuropsychologia. 1990;28:49–60.
71. Buck R. Using FACS versus communication scores to measure spontaneous facial expression of emotion in brain damaged patients. Cortex. 1990;26:275–80.
72. Sackeim H, Gur R, Saucy M. Emotions are expressed more intensely on the left side of the face. Science. 1978;202:434–6.
73. Moreno CR, Borod J, Welkowitz J, Alpert M. Lateralization for the expression and perception of facial emotion as a function of age. Neuropsychologia. 1990;28:119–209.
74. Triggs WJ, Ghacibeh G, Springer U, Bowers D. Lateralized asymmetry of facial motor evoked potentials. Neurology. 2005;65(4):541–4.

75. Rosenbek JC, Crucian GP, Leon SA, Hieber B, Rodriguez AD, Holiway B, et al. Novel treatments for expressive aprosodia: a phase I investigation of cognitive linguistic and imitative interventions. J Int Neuropsychol Soc. 2004;10(5):786–93.
76. Rosenbek JC, Rodriguez AD, Hieber B, Leon SA, Crucian GP, Ketterson TU, et al. Effects of two treatments for aprosodia secondary to acquired brain injury. J Rehabil Res Dev. 2006;43(3):379–90.
77. Leon SA, Rosenbek JC, Crucian GP, Hieber B, Holiway B, Rodriguez AD, et al. Active treatments for aprosodia secondary to right hemisphere stroke. J Rehabil Res Dev. 2005;42(1):93–101.

Chapter 8
Dysexecutive Syndrome After Stroke

Olivier Godefroy, Pierre Yves Garcia, Jean Marc Bugnicourt, Claire Leclercq, and Martine Roussel

Introduction

Disorders of executive functions represent a leading form of post-stroke cognitive impairment, regardless of stroke mechanism, and are important contributors to post-stroke disability. This chapter reviews executive functions and impairment of these functions and then discusses the main disorders observed in stroke and their assessment.

Executive Functions and Disorders

Based on Luria's approach [1], the term "executive function" was coined by Lezak [2] and was initially defined as brain function related to goal setting, action initiation, planning, shifting, and verification. Baddeley et al. [3] coined the term "dysexecutive syndrome" to describe the disorders of executive function, especially marked in the behavioral domain, observed in a patient with bilateral frontal damage. Recent aspects of social and emotional processes, theory of mind, attentional processes, strategic processes of episodic memory, and metacognition have been variably incorporated in the domain of executive functions. These somewhat

O. Godefroy, M.D., Ph.D. (✉) • M. Roussel, Ph.D.
Department of Neurology, University Hospital, Amiens 80054, France
e-mail: Godefroy.olivier@chu-amiens.fr

P.Y. Garcia, Ph.D.
Department of Neurology, Centre Hospitalier Compiegne, Compiegne, France

J.M. Bugnicourt, M.D.
Department of Neurology, Amiens University Hospital, Amiens, France

C. Leclercq, M.D.
Department of Neurology, CHU Amiens—Hôpital Nord, Amiens, France

T.A. Schweizer and R.L. Macdonald (eds.), *The Behavioral Consequences of Stroke*,
DOI 10.1007/978-1-4614-7672-6_8, © Springer Science+Business Media New York 2014

Table 8.1 Main diagnostic criteria of behavioral and cognitive dysexecutive disorders

Behavioral disorders[a]	Cognitive disorders[a]
Indicated by reliable informant using validated inventory	Shown by validated performance index on normalized tests
Significant changes compared to premorbid behavior	Not attributable to developmental disorder
Significant changes in activities of daily living, social life, or work	
Highly suggestive	
Global hypoactivity with apathy and/or abulia	Response inhibition (e.g., Stroop test)
Global hyperactivity with distractibility and/or psychomotor instability and/or disinhibition	Rule deduction and generation (e.g., Card sorting test)
	Shifting of sets (e.g., perseveration)
Stereotyped and perseverative behavior	Information generation and strategic search (e.g., fluency tasks)
Supportive deficits and developing areas	
Disorders of emotional control (euphoria, emotional lability)	Planning (e.g., Tower of London test)
Disorders of social behavior including loss of empathy	Coordination of dual tasks (e.g., dual task)
Disorders of sexual, eating, and urinary behavior	Action initiation and sustained alertness (e.g., Simple Reaction Time test)
Spontaneous confabulations, reduplicative paramnesia	Episodic memory strategic processes (retrieval and memory selection)
Anosognosia, anosodiaphoria	Theory of mind and metacognitive processes

[a]To be considered as dysexecutive, the disorder should not be more readily explained by perceptuomotor, psychiatric (depression, manic state, or obsessive-compulsive disorder), sleep (hypersomnia) or other cognitive (language, memory, visuospatial) disturbances

disparate definitions of executive disorders complicate clinical assessment and experimental studies, emphasizing the need for a consensus on diagnostic criteria. The Groupe de Réflexion pour l'Evaluation des Fonctions EXécutives (GREFEX) study group has proposed [4] and validated criteria for dysexecutive syndrome [5]. Separate criteria for behavioral and cognitive dysexecutive syndromes have been proposed, as behavioral disorders (i.e., changes observed clinically or on a behavior inventory) may be observed in patients without cognitive deficit (i.e., deficits observed on tests) and vice versa. The need for separate criteria was supported by results of the GREFEX study showing that dissociated disorders (i.e., behavioral disorders without test impairment and test deficit without behavioral disorder) were more frequent than combined disorders affecting both behavioral and cognitive domains [5]. Dysexecutive disorders (Table 8.1) include eight behavioral changes and nine cognitive disorders. The criteria were strictly defined and operationalized using well-defined performance indices. Recent studies have refined the criteria for impairment of strategic processes underlying episodic memory and response initiation (for review, see [6]). Frontal lobe damage has been shown to impair strategic processes involved in retrieval as shown by cueing benefits (i.e., impaired free recall and normal recall following cueing) [7] and memory selection, as shown by false recognitions [8]. This profile of episodic memory impairment was found to be specific to a frontal location of the stroke [9, 10] and was also found to improve the prediction of frontal stroke compared to the use of conventional executive tests only [10].

Initiation has long been regarded as the rapidity to initiate action and is frequently confounded with action speed. Impairment of action speed is a leading deficit in stroke patients [11–13] and may be observed in the long term [14, 15]. It is frequently labeled "psychomotor slowing." However, the term "action slowing" appears to be more appropriate, as it has been shown to involve, to varying degrees, perceptual, motor, and cognitive aspects of action speed [16–18]. In addition, these results indicate that action slowing cannot be attributed solely to an initiation deficit, as it may also be due to perceptual or motor impairment. Initiation of action is especially involved in situations requiring repeated activation of a sequence of simple actions. This function is typically assessed by Reaction Time tests. Following experimental validation, Reaction Time [19] and benefits of a warning cue [20, 21] are now used as an index of initiation processes. This method shows that initiation depends on the activation of a thalamo-parieto-prefronto-cingulate network predominantly in the right hemisphere. Lesion studies indicate that the mediofrontal region is especially critical for initiation [17, 20]. These converging results indicate that initiation can be differentiated from action speed and other executive processes [20, 22]. In clinical practice, action speed is usually assessed by the simple condition of chronometric tests (such as part A of the Trail Making Test and naming and reading Stroop subtests) although this does not provide a pure index of initiation (Table 8.1).

The use of a comprehensive cognitive battery and behavioral inventory shows that impairment (usually mild) may be observed in one test or one behavioral domain in normal subjects. For this reason, it is essential to adjust criteria of impairment according to the number of performance indices. In the GREFEX study, this adjustment was performed by differentiating dysexecutive disorder (i.e., impairment in one test or one behavioral domain) from dysexecutive syndrome, which requires the association of several impairments on the behavioral inventory (behavioral dysexecutive syndrome) or tests (cognitive dysexecutive syndrome). The usefulness of this approach is supported by the demonstration of a relationship between dysexecutive impairment and loss of autonomy in dysexecutive syndrome [5].

Disorders of Executive Functions in Stroke

The present review will focus on studies of consecutive patients (hospital- or population-based studies) using a validated battery of tests assessing executive functions and reporting data with sufficient details to determine the profile of cognitive impairment. Published studies provide a consistent picture of the high frequency of executive deficit across various types of stroke with about 35 % of patients suffering from action slowing and about 20 % of patients suffering from a deficit of rule deduction, abstract reasoning, or set shifting. Action slowing is assessed by various tests, such as the Digit-Symbol substitution subtest, Trail Making Test, naming and reading Stroop subtests, and more rarely, Reaction Time tests. Executive deficits are reported with frequencies ranging from approximately 10 to 50 % [11, 23–30]. On tests assessing abstract reasoning, such as the similarities subtest of the revised Wechsler Adult Intelligence Scale or Modified Card Sorting Test, a deficit was observed in 15–30 % of patients [11, 23, 27]. Verbal fluency tests are reported to be impaired in 15–40 % of patients [11, 24, 25]. A study that used a global score of executive functions summarizing performance on verbal fluency, Wisconsin Card Sorting, Stroop interference tests, and Part B of the Trail Making Test reported a deficit in 41 % of patients [31]. These high prevalences of dysfunction tend to decrease on retesting performed 6 months to 4 years post-stroke, suggesting that the deficits improve in about 25–30 % of patients [25, 27]. However, a deficit in executive functions, motor speed, and episodic memory is still frequently observed in studies with long-term stroke survivors [14, 15], indicating that a large percentage of patients still suffer from attentional and executive disorders in the long term. The presence of dysexecutive disorders was associated, in several studies, with a poor functional outcome [13, 15, 27, 31, 32]. This has been shown for both behavioral [33] and cognitive disorders [5] (Table 8.2).

Profile of Executive Deficits

Studies including consecutive patients usually have different test batteries and usually do not assess behavioral changes, and thus preclude a precise description for the profile of dysexecutive disorders. In addition, most studies report crude test performance results that fail to distinguish between deficits due to true executive disorders or to associated cognitive impairment (e.g., fluency test impairment may be due to language or executive disorders). The GREFEX study assessed 152 stroke patients (infarct: $n=47$; complicated ruptured intracranial aneurysm: $n=76$; cerebral venous thrombosis: $n=11$) referred for cognitive assessment. This study design is not appropriate to determine the prevalence of a deficit, but is useful to assess the relative frequencies of certain disorders. Behavioral and cognitive dysexecutive syndromes were observed in 25 % and 29 % of patients, respectively [5], indicating that the failure to systematically assess behavioral changes results in an underestimation of dysexecutive disorders. The prominent behavioral changes were hypoactivity with apathy (observed in 20 % of patients) and anosognosia. Apathy without depression has been observed with a frequency ranging between 20 and 30 % in stroke patients [33, 34]. The prominent cognitive disorders involved information generation (i.e., fluency tests), rule deduction (i.e., number of categories achieved on the Modified Card Sorting Test and correct responses on the Brixton Test), and shifting (i.e., perseveration on Trail Making Test B and the Modified Card Sorting Test) [5]. The specific dysexecutive impairment profile has not yet been defined according to stroke type.

Table 8.2 Summary of findings concerning cognitive impairment after stroke

References	#Pts	Type of stroke	Time of assessment post-admission	Impairment in	% Impaired
[11]	227	Ischemic	3 Months	Abstract reasoning	16–20
				Attention	20–39
				Language	13–33
				Memory	10–25
				Visuospatial	17–25
[15]	307	Ischemic and hemorrhagic	5 Years	Executive function	33
				Information processing	33
				Language	24
				Verbal memory	8
				Visual memory	9
				Visuospatial	3
[23]	37	Hemorrhagic	6 Months	Executive function	19
				Memory	14
[24]	150	Unspecified	>3 Months	Memory	26
				Numerical working memory	43–47
				Spatial working memory	14–21
[25]	196	Ischemic and hemorrhagic	1 Month	Attention	32
				Executive function	32
				Information processing	60
				Language	35
				Memory	24
				Orientation	10
				Visuospatial	37
[26]	47	Unspecified	<10 Days	Executive function	61
[27]	111	Ischemic and hemorrhagic	<3 Weeks	Abstract reasoning	24
				Executive function	30
				Language	21
				Verbal memory	21
				Visual memory	16
				Visuospatial	31
[28]	74	Hemorrhagic	20 Months	Episodic memory	42
				Executive function	14
				Language	5
				Verbal memory	36
				Visual memory	19
				Visuospatial	5
				Working memory	5
[29]	55	Ischemic and hemorrhagic	<3 Weeks	Executive function	46
				Language	18
				Long-term memory	26
				Visuospatial	27
[30]	48	Hemorrhagic	40 Months	Episodic memory	52
				Executive function	37
				Language	35
				Visuospatial	19
[31]	256	Ischemic	3 Months	Executive function	41

Determinants of Dysexecutive Disorders

Stroke location is usually considered to be the main determinant of the type of cognitive disorder. Executive functions are mainly supported by the frontal lobes, basal ganglia (especially the caudate nucleus, pallidum, anterior and median thalamus), and their connections (especially the anterior centrum semiovale, anterior limb, and genu of the internal capsule). Executive disorders are therefore essentially expected in strokes affecting these structures. This classical view has been challenged by many studies. Firstly, several studies have shown that anterior *and* posterior strokes result in impaired performance on tests of executive functions [12, 35]. This probably accounts for the contrast between the high prevalence of dysexecutive impairment and the lower frequency of anterior stroke. Secondly, stroke location is not the sole determinant of dysexecutive disorders; white matter abnormalities [36–38], microbleeds [39], medial temporal atrophy [40], and pre-stroke cognitive disorders [41] also contribute to post-stroke executive impairments. White matter abnormalities are frequently correlated with executive test performance, especially on tests with time constraint and chronometric performance [42]. Accordingly, it has been demonstrated that the volume of white matter abnormalities is related to action slowing [43]. Cerebral microbleeds have been associated with deficits on executive tests and action speed [39, 44, 45] independent of white matter abnormalities [39, 44]. Additionally, patients with impaired executive tests had a greater number of microbleeds in frontal and basal ganglia regions [39, 44]. Pre-stroke cognitive deficits are more frequent in patients after the age of 75 [41]. They may be caused by previous strokes or by nonvascular lesions, especially Alzheimer's disease [41]. Preexisting dementia has been observed in patients with spontaneous cerebral hemorrhage with a prevalence similar to that reported in ischemic stroke patients [46]. It may be caused by previous vascular lesions (especially in subcortical hemorrhages) or by Alzheimer's disease (especially in lobar hemorrhages) [46].

Dysexecutive Disorders According to Stroke Type and Location

The high frequency of executive disorders relative to other cognitive deficits seems to concern all types of stroke, as it has been reported in ischemic [11, 13, 24, 27, 31], hemorrhagic [30], and mixed ischemic-hemorrhagic strokes [15, 25, 29]. The majority of studies determining the prevalence of dysexecutive disorders were performed by focusing on the most frequent stroke type (cerebral infarction). They showed that action slowing is observed in about one-third of patients, and a cognitive dysexecutive deficit is observed in about one-fifth of patients. The profile of dysexecutive disorders has been assessed in patients with various infarct types and locations.

Infarcts involving the frontal lobes, basal ganglia, and their connections are mainly due to occlusions of the anterior/superior branch of the middle cerebral artery and/or the deep perforating branches (lateral lenticulostriate arteries). Infarcts in the territory of the anterior cerebral artery are rare (accounting for 2–5 % of infarcts), but they can cause frontal (medial part) and deep (ventral part of caudate nucleus) cerebral infarctions. Finally, infarcts of the anterior part of the thalamus are mainly located in the territory of the tuberothalamic artery, a branch of the posterior communicating artery, and the anterior choroidal artery. Infarcts of the middle part of the thalamus are in the territory of the paramedian thalamic-subthalamic arteries, which originate from the first segment of the posterior cerebral artery.

Infarcts in the territory of the anterior/superior branch of the middle cerebral artery usually induce cognitive dysexecutive disorders that are reported to be especially marked on rule deduction, shifting of rules, information generation, and maintenance of information in working memory. Conversely, behavioral dysexecutive changes are usually reported to be moderate and can even be absent. Global hypoactivity with abulia-apathy is the most frequent disorder. Dysexecutive disorders are usually associated with other cognitive deficits (aphasia in left-sided infarct and visuo-constructional disorders in right-sided infarct—see Chaps. 5 and 6 for more information). Pure unilateral lenticulostriate infarcts may induce dysexecutive disorders, although the deficits are usually mild. In the behavioral domain, global hypoactivity with abulia-apathy or hyperactivity with distractibility-impulsivity may be observed [47–49]. Cognitive executive deficits concern tests that assess response activation and suppression, information generation, and rule deduction [49, 50]. They are usually associated with memory disturbances [10, 51, 52] and, in left-sided infarcts, language disorders [10, 50, 53]. Bilateral striatal lesions may be observed in anoxia, which usually also causes lesions outside the striatal region. Such a condition may induce "loss of psychic autoactivation" [54] characterized by severe global hypoactivity with abulia-apathy (usually associated with "empty mind"), which is reversed by heteroactivation (usually following instructions from the examiner or the caregiver). This disorder has also been reported after bilateral thalamic infarction [55].

In anterior cerebral artery infarcts, the profile of dysexecutive disorders has been only partly documented. Akinetic mutism (which is an extreme form of global hypoactivity with apathy) may be as a result of bilateral infarcts followed by severe dysexecutive disorders persisting after the acute phase [56, 57]. Transient akinetic mutism and acute confusional states have been reported in patients following left-sided infarcts [56–58]. Acute action slowing has also been reported in patients after right-sided infarcts [56–58].

Although *thalamic infarcts* in the territory of the tuberothalamic and paramedian arteries are frequently reported to cause executive deficits, few studies have assessed patients using validated test batteries. Behavioral disorders are a classical component of the clinical picture during the acute phase of stroke. In unilateral lesions, they are usually mild and mainly consist of global hypoactivity with abulia-apathy [59–61]. A bilateral paramedian infarct is associated with more severe clinical features with impaired consciousness at the acute phase and presence of akinetic mutism. After the acute stage, mild behavioral disorders may persist with more severe disorders in the case of bilateral lesions, which may induce "loss of psychic auto-activation" [55]. In patients with global hypoactivity, the possible role of hypersomnia must be assessed, especially in bilateral paramedian infarcts. Cognitive dysexecutive disorders have been observed with a frequency ranging from one-half to one-third of patients according to a review by Van der Werf et al. [62]. Impairment of executive functions and response speed were found to depend on several structures including the mediodorsal nucleus and internal medullary lamina [60]. Executive disorders are usually associated with episodic memory deficits and aphasia or hemineglect, all of which must be taken into account when interpreting test results. Follow-up examinations have shown that spectacular improvement is possible within a few months, even in patients with bilateral infarcts. However, bilateral infarcts causing persistent behavioral and cognitive disorders—thus satisfying the criteria of the so-called thalamic dementia—may be quite severe.

Patients with *lacunar infarcts* are frequently reported to have dysexecutive disorders. This impairment has been supported by recent studies comparing subcortical ischemic vascular disease with either normal controls [63] or other stroke types [64]. The presence of cognitive dysexecutive deficit in patients with lacunes has initially been attributed to their prominence in the basal ganglia and thalamus [40, 65]. However, small vessel disease, the usual cause of lacunes, is also associated with other cerebral abnormalities, such as white matter signal hyperintensities, microbleeds, reduced volume of cortical gray matter, and medial temporal atrophy [66]. In subcortical ischemic vascular disease, white matter lesions and cerebral atrophy were found to be independent predictors of executive deficits [37]. These results indicate that lacunes are a marker of a disease process responsible for various lesions that contribute to executive impairments.

In *subarachnoid hemorrhage* (see Chap. 10), the frequency of executive deficits varies considerably across studies. A review of 61 published studies assessing cognitive outcome reported a frequency of executive deficits ranging from 7 to 76 % [67]. This wide frequency range is likely due to the heterogeneity of assessment and patient selection. Studies assessing executive function in large samples of consecutive patients have reported dysexecutive deficits in 18–51 % of patients [23, 68–73]. The profile of cognitive dysfunction is usually only minimally influenced by aneurysm location [23, 69] and seems more related to diffuse cerebral damage and cerebral edema [70, 71]. However, cognitive deficits have been found to be more frequent in patients with left-sided lesions [71]. In addition, a few studies using a validated method of clinical-anatomical correlation analysis reported that lesion location influenced the profile of cognitive deficit [28, 74]. Assessment of executive disorders is especially important after rupture of anterior communicating artery aneurysms. The usual clinical presentation is comprised of prominent behavioral changes with mild cognitive dysexecutive impairment and memory deficit [75], while perceptuomotor abilities and instrumental cognitive functions (e.g., language, visuospatial, and constructive abilities) are typically spared. Behavioral executive changes include hypoactivity with abulia-apathy, hyperactivity with distractibility-disinhibition, and, in more severe cases, stereotypy, anosognosia, disturbances of social behavior, confabulations, and reduplicative paramnesia. These disorders may be observed in the same patient. A common situation is the observation of hypoactivity (and akinetic mutism in very severe cases) in the acute phase, followed by the progressive development of hyperactivity. Cognitive dysexecutive syndrome is usually mild (and may be absent) and mainly concerns action initiation and response suppression. A deficit of episodic memory is usually observed, but may present with various patterns. Apart from the typical impairment of strategic processes (see above), a mediotemporal profile (characterized by impairment of free and cued recalls and recognition) and, very rarely, selective impairment of recognition have been reported [7, 9, 10, 76].

Cognitive outcome after *intracerebral hemorrhage* (see Chap. 11) is the subject of several major ongoing studies. A cross-sectional study of intracerebral hemorrhage, with comprehensive neuropsychological assessment, revealed a picture similar to that of infarct, with impaired episodic memory in 52 % of patients, psychomotor speed in 44 % of patients, and executive functions in 37 % of patients [30]. A study in patients undergoing surgery for arteriovenous malformations reported the presence of attentional and executive deficits during the first year post-surgery [77].

The cognitive outcome of cerebral venous thrombosis remains poorly examined [78–80]. A cross-sectional study of cerebral venous thrombosis, with comprehensive neuropsychological assessment, showed an overall lower frequency of cognitive impairment as compared to arterial stroke [81], in close keeping with the relatively good prognosis of this condition [82]. Cognitive impairment was prominent in the domain of episodic memory, executive function, and action speed, a similar pattern to that observed in arterial stroke [81].

Clinical Implications

Assessment of Executive Disorders

Clinical examination is not very sensitive to dysexecutive disorders but is essential in analyzing associated deficits in sensorimotor, cognitive (especially transcortical motor aphasia and hemineglect), and psychiatric (especially depression) domains, as they influence prognosis and the interpretation of neuropsychological assessment. In the acute stage, clinical examination of stroke patients is usually based on the National Institutes of Health (NIH) stroke scale score, which includes a short assessment of orientation, language, and hemineglect. However, this scale misses mild deficits and does not assess action speed, memory, and executive function. Clinical examination and interview of the patient and an informant should be conducted to assess significant changes in personality and behavior, as this information is very important for diagnosis. In patients without behavioral changes, dysexecutive disorders should be suspected when an apparently good clinical outcome contrasts with persisting difficulties in occupational activities or instrumental activities of daily living. These difficulties are more frequently detected when a questionnaire is used to score the Rankin scale [83]. A careful search for depression as a symptom is important considering its frequency in stroke patients [84] (see Chap. 12). Clinical examination looks for the presence of grasp and sucking reflexes, as well as abnormal imitation and utilization of objects. Gaze "grasping" (i.e., the inability to avoid orienting one's gaze toward salient stimuli) is the only gaze abnormality observed on routine clinical examination and is attributed to bilateral lesions of the frontal eye fields. Oppositional (increased tone in response to passive movement) and facilitatory (when the patient acts in the same direction as the passive movement) paratonia are due to frontal lobe damage and are especially observed in the context of diffuse pathology [85]. The examination of motor sequences (fist-palm-edge), crossed tapping ("tap once in response to a double tone and tap twice in response to a single tone") or Go/No-go ("tap once in response to a double tone and do not respond to a single tone"), and verbal fluency may be useful for diagnosis of executive disorders. Among the various screening tests, the Mini-Mental State Examination is not very sensitive to executive deficit. The Frontal Assessment Battery [86] and Montréal Cognitive Assessment [87] include more subtests assessing executive functions, although their superiority to detect cognitive impairment in stroke patients remains questionable [29].

In most cases, neuropsychological examination is necessary to diagnose a dysexecutive syndrome. It should include a validated inventory of dysexecutive behavioral changes and, when possible, a test battery. Several questionnaires examine behavioral changes in various brain-damaged populations and more or less specifically assess dysexecutive disorders. Personality changes can be assessed by the Neuropsychiatric Inventory (NPI) [88] and Frontal Lobe Personality Scale [89]. The Behavioral Assessment of the Dysexecutive Syndrome battery includes a questionnaire [90] and has been used in mixed pathologies [91]. The Frontal Behavioral Inventory [92] and Frontotemporal Behavioral Scale [93] have been especially used in dementia. The Inventory of Behavioral Dysexecutive Syndrome has been normalized [94] and has been used in stroke patients [5].

Cognitive assessment of executive disorders is based on a battery of several tests. Verbal fluency, Trail Making, Stroop, and the Wisconsin Card Sorting tests are the most commonly used. However, this classic battery has shown certain limitations as compared with a more complete battery including the Six Elements test, Brixton test, and a dual-task test [94]. A verbal episodic memory test should be used to examine strategic encoding and retrieval processes [9, 10]. Action speed can be assessed with various tests such as the Digit-Symbol substitution subtest, Trail Making Test Part A, naming and reading Stroop subtests, and, more rarely, Reaction Time tests. Test interpretation should take into account sensorimotor deficit and deficit of instrumental functions and memory. However, these tests may fail to detect a selective or more subtle impairment. Additional tests may be used in these patients, especially to assess attention, emotional, and social behavior, as well as metacognitive processes. Deficits of nonspatial attention observed in this context mainly involve problems in sustained alertness and monitoring of information. Sustained alertness, defined as the ability to maintain a high level of preparedness during a task [95], is close to our concept of initiation (see above) and can be assessed with Reaction Time tests using individual distributions of the results [19] and examining the benefit of warning cues [20, 21]. Sustained alertness is impaired in superomedial frontal damage [17, 20]. Monitoring of information is assessed by the reaction time advantage conferred by long intertrial intervals and has been found to be impaired following right lateral frontal damage [20]. Regulation of emotional and social behavior has been extensively investigated. It can be assessed by testing the identification of facial emotion, using social questionnaires (e.g., Faux pas questionnaire [96]) and theory of mind tests (e.g., [97, 98]), all of which can be impaired after frontal damage. The ability to use emotional information to guide behavior can be assessed by a gambling task [99]. Assessment of metacognitive processes includes testing reactions to verbal and cartoon humor [100, 101] and comparing performance on the remember-know memory task [102].

Summary

Dysexecutive disorders are frequent in stroke and they contribute to disability. Although classically attributed to stroke involving the frontal lobe and basal ganglia, they may be observed in strokes affecting the cerebral hemispheres. Dysexecutive disorders are under-recognized due to the moderate sensitivity of usual screening tests for these disorders. Diagnosis may be improved by the systematic assessment of dysexecutive disorders examining both behavioral (using a validated questionnaire) and cognitive (using a validated battery of tests) components. Interpretation of performance on both questionnaire and cognitive tests needs to take into account elementary disorders, especially sensorimotor, depression, and cognitive disorders (especially transcortical motor aphasia and hemineglect). Test batteries should also assess action speed and episodic memory and provide validated indices to disentangle dysexecutive/strategic impairment from other deficits. Selected patients with behavioral or cognitive dysexecutive disorders may be referred to rehabilitation programs; however, sophisticated randomized controlled trials assessing the efficacy of pharmacological and rehabilitation-based treatments remain limited (see Chap. 16).

References

1. Luria AR. Higher cortical functions in man. New York: Basic Books; 1966.
2. Lezak MD. The problem of assessing executive functions. Int J Psychol. 1982;17:281–97.
3. Baddeley A, Wilson B. Frontal amnesia and the dysexecutive syndrome. Brain Cogn. 1988;7:212–30.
4. Godefroy O. Frontal syndromes and disorders of executive functions. J Neurol. 2003;250:1–6.
5. Godefroy O, Azouvi P, Robert P, Roussel M, LeGall D, Meulemans T, Groupe de Réflexion sur l'Evaluation des Fonctions Exécutives Study Group. Dysexecutive syndrome: diagnostic criteria and validation study. Ann Neurol. 2010;68:855–64.

6. Godefroy O, Stuss DT. Dysexecutive syndromes. In: Godefroy O, editor. Vascular cognitive impairment: the cognitive and behavioural neurology of stroke. Cambridge: Cambridge University Press; 2013. p. 255–76.
7. Jetter W, Poser U, Freeman Jr RB, Markowitsch HJ. A verbal long term memory deficit in frontal lobe damaged patients. Cortex. 1986;22:229–42.
8. Incisa Della Rocchetta A, Milner B. Strategic search and retrieval inhibition: the role of frontal lobes. Neuropsychologia. 1993;31:503–24.
9. Stuss DT, Alexander MP, Palumbo CL, Buckle L, Sayer L, Pogue J. Organizational strategies of patients with unilateral or bilateral frontal lobe injury in word list learning tasks. Neuropsychology. 1994;8:355–73.
10. Godefroy O, Roussel M, Leclerc X, Leys D. Deficit of episodic memory: anatomy and related patterns in stroke patients. Eur Neurol. 2009;61:223–9.
11. Tatemichi TK, Desmond DW, Stern Y, Paik M, Sano M, Bagiella E. Cognitive impairment after stroke: frequency, patterns, and relationship to functional abilities. J Neurol Neurosurg Psychiatr. 1994;57:202–7.
12. Leskelä M, Hietanen M, Kalska H, Ylikoski R, Pohjasvaara T, Mäntylä R, et al. Executive functions and speed of mental processing in elderly patients with frontal or nonfrontal ischemic stroke. Eur J Neurol. 1999;6:653–61.
13. Sachdev PS, Brodaty H, Valenzuela MJ, Lorentz L, Looi JC, Wen W, et al. The neuropsychological profile of vascular cognitive impairment in stroke and TIA patients. Neurology. 2004;62:912–9.
14. Engstad T, Almkvist O, Viitanen M, Arnesen E. Impaired motor speed, visuospatial episodic memory and verbal fluency characterize cognition in long-term stroke survivors: the Tromso Study. Neuroepidemiology. 2003;22:326–31.
15. Barker-Collo S, Feigin VL, Parag V, Lawes CM, Senior H. Auckland Stroke Outcomes Study. Part 2: cognition and functional outcomes 5 years poststroke. Neurology. 2010;75:1608–16.
16. Godefroy O, Rousseaux M. Binary choice in patients with prefrontal or posterior brain damage. A relative judgement theory analysis. Neuropsychologia. 1996;34:1029–38.
17. Godefroy O, Lhullier-Lamy C, Rousseaux M. SRT lengthening: role of an alertness deficit in frontal damaged patients. Neuropsychologia. 2002;40:2234–41.
18. Godefroy O, Spagnolo S, Roussel M, Boucart M. Stroke and action slowing: mechanisms, determinants and prognostic value. Cerebrovasc Dis. 2010;29:508–14.
19. Godefroy O, Roussel M, Despretz P, Quaglino V, Boucart M. Age-related slowing: perceptuomotor, decision or attention decline? Exp Aging Res. 2010;36:169–89.
20. Stuss DT, Alexander MP, Shallice T, Picton TW, Binns MA, Macdonald R, et al. Multiple frontal systems controlling response speed. Neuropsychologia. 2005;43:396–417.
21. Périn B, Godefroy O, Fall S, de Marco G. Exploration of an attentional network in the right hemisphere: an fMRI study of brain interactivity enhanced during phasic alertness. Brain Cogn. 2010;72:271–81.
22. Godefroy O, Cabaret M, Petit-Chenal V, Pruvo JP, Rousseaux M. Control functions of the frontal lobes. Modularity of the central-supervisory system? Cortex. 1999;35:1–20.
23. Tidswell P, Dias PS, Sagar HJ, Mayes AR, Battersby RD. Cognitive outcome after aneurysm rupture: relationship to aneurysm site and perioperative complications. Neurology. 1995;45:875–82.
24. Ballard C, Stephens S, Kenny R, Kalaria R, Tovee M, O'Brien J. Profile of neuropsychological deficits in older stroke survivors without dementia. Dement Geriatr Cogn Disord. 2003;16:52–6.
25. Rasquin SM, Lodder J, Ponds RW, Winkens I, Jolles J, Verhey FR. Cognitive functioning after stroke: a one-year follow-up study. Dement Geriatr Cogn Disord. 2004;18:138–44.
26. Zinn S, Bosworth HB, Hoenig H, Swartzwelder HS. Executive function deficits in acute stroke. Arch Phys Med Rehabil. 2007;88:173–80.

27. Nys GM, van Zandvoort MJE, de Kort PML, van der Worp HB, Jansen BP, Algra A, et al. The prognostic value of domain-specific cognitive abilities in acute first-ever stroke. Neurology. 2005;64:821–7.
28. Martinaud O, Perin B, Gérardin E, Proust F, Bioux S, Gars DL, et al. Anatomy of executive deficit following ruptured anterior communicating artery aneurysm. Eur J Neurol. 2009;16:595–601.
29. Godefroy O, Fickl A, Roussel M, Auribault C, Bugnicourt JM, Lamy C, et al. Is the Montreal Cognitive Assessment superior to the Mini-Mental State Examination to detect post-stroke cognitive impairment? A study with neuropsychological evaluation. Stroke. 2011;42:1712–6.
30. Garcia PY, Roussel M, Lamy C, Lamy C, Canaple S, Peltier J, et al. Cognitive impairment and dementia following intracerebral hemorrhage: a cross-sectional study of a university hospital-based series. J Stroke Cerebrovasc Dis. 2013;22:80–6.
31. Pohjasvaara T, Leskelä M, Vataja R, Kalska H, Ylikoski R, Hietanen M, et al. Post-stroke depression, executive dysfunction and functional outcome. Eur J Neurol. 2002;9:269–75.
32. Hommel M, Miguel ST, Naegele B, Gonnet N, Jaillard A. Cognitive determinants of social functioning after a first ever mild to moderate stroke at vocational age. J Neurol Neurosurg Psychiatr. 2009;80:876–80.
33. Mayo NE, Fellows LK, Scott SC, Cameron J, Wood-Dauphinee S. A longitudinal view of apathy and its impact after stroke. Stroke. 2009;40:3299–307.
34. Starkstein SE, Fedoroff JP, Price TR, Leiguarda R, Robinson RG. Apathy following cerebrovascular lesions. Stroke. 1993;24:1625–30.
35. Anderson SW, Damasio H, Jones RD, Tranel D. Wisconsin Card Sorting Test performance as a measure of frontal lobe damage. J Clin Exp Neuropsychol. 1991;13:909–22.
36. Pohjasvaara T, Mäntylä R, Salonen O, Aronen HJ, Ylikoski R, Hietanen M, et al. How complex interactions of ischemic brain infarcts, white matter lesions, and atrophy relate to post-stroke dementia. Arch Neurol. 2000;57:1295–300.
37. Mungas D, Jagust WJ, Reed BR, Kramer JH, Weiner MW, Schuff N, et al. MRI predictors of cognition in subcortical ischemic vascular disease and Alzheimer's disease. Neurology. 2001;57:2229–35.
38. O'Brien JT, Wiseman R, Burton EJ, Barber B, Wesnes K, Saxby B, et al. Cognitive associations of subcortical white matter lesions in older people. Ann N Y Acad Sci. 2002;977:436–44.
39. Werring DJ, Frazer DW, Coward LJ, Losseff NA, Watt H, Cipolotti L, et al. Cognitive dysfunction in patients with cerebral microbleeds on T2*-weighted gradient-echo MRI. Brain. 2004;127:2265–75.
40. Vataja R, Pohjasvaara T, Mäntylä R, Ylikoski R, Leppävuori A, Leskelä M, et al. MRI correlates of executive dysfunction in patients with ischaemic stroke. Eur J Neurol. 2003;10:625–31.
41. Hénon H, Pasquier F, Durieu I, Godefroy O, Lucas C, Lebert F, et al. Preexisting dementia in stroke patients. Baseline frequency, associated factors, and outcome. Stroke. 1997;28:2429–36.
42. Longstreth Jr WT, Manolio TA, Arnold A, Burke GL, Bryan N, Jungreis CA, et al. Clinical correlates of white matter findings on cranial magnetic resonance imaging of 3301 elderly people. The Cardiovascular Health Study. Stroke. 1996;27:1274–82.
43. Wright CB, Festa JR, Paik MC, Schmiedigen A, Brown TR, Yoshita M, et al. White matter hyperintensities and subclinical infarction: associations with psychomotor speed and cognitive flexibility. Stroke. 2008;39:800–5.
44. van Norden AG, van den Berg HA, de Laat KF, Gons RA, van Dijk EJ, de Leeuw FE. Frontal and temporal microbleeds are related to cognitive function: the Radboud University Nijmegen Diffusion Tensor and Magnetic Resonance Cohort (RUN DMC) Study. Stroke. 2011;42:3382–6.

45. Poels MM, Ikram MA, van der Lugt A, Hofman A, Niessen WJ, Krestin GP, et al. Cerebral microbleeds are associated with worse cognitive function: the Rotterdam Scan Study. Neurology. 2012;78:326–33.
46. Cordonnier C, Leys D, Dumont F, Deramecourt V, Bordet R, Pasquier F, et al. What are the causes of pre-existing dementia in patients with intracerebral hemorrhages? Brain. 2010;133:3281–9.
47. Mendez MF, Adams NL, Lewandowski KS. Neurobehavioral changes associated with caudate lesions. Neurology. 1989;39:349–54.
48. Caplan LR, Schmahmann JD, Kase CS, Feldmann E, Baquis G, Greenberg JP, et al. Caudate infarcts. Arch Neurol. 1990;47:133–43.
49. Godefroy O, Rousseaux M, Leys D, Destée A, Scheltens P, Pruvo JP. Frontal lobe dysfunction n patients with unilateral lenticulo-striate infarcts. Prominent role of cortical lesion. Arch Neurol. 1992;49:1285–9.
50. Weiller C, Ringelstein EB, Reiche W, Thron A, Buell U. The large striatocapsular infarct. A clinical and pathophysiological entity. Arch Neurol. 1990;47:1085–91.
51. Godefroy O, Rousseaux M, Pruvo JP, Cabaret M, Leys D. Neuropsychological changes related to unilateral lenticulostriate infarcts. J Neurol Neurosurg Psychiatr. 1994;57:480–5.
52. Exner C, Weniger G, Irle E. Implicit and explicit memory after focal thalamic lesions. Neurology. 2001;57:2054–63.
53. Damasio AR, Damasio H, Rizzo M, Varney N, Gersh F. Aphasia with non hemorrhagic lesions in the basal ganglia and internal capsule. Arch Neurol. 1982;39(1):15–24.
54. Laplane D, Widlocher D, Pillon B, Baulac M, Binoux F. Comportement compulsif d'allure obsessionnelle par nécrose circonscrite bilatérale pallido-striatale. Rev Neurol. 1981;137(4):269–76.
55. Bogousslavsky J, Regli F, Delaloye B, Delaloye-Bischof A, Assal G, Uske A. Loss of psychic self-activation with bithalamic infarction: neurobehavioural, CT, MRI and SPECT correlates. Acta Neurol Scand. 1991;83:309–16.
56. Bogousslavsky J. Frontal stroke syndromes. Eur Neurol. 1994;34:306–15.
57. Kumral E, Bayulkem G, Evyapan D, Yunten N. Spectrum of anterior cerebral artery territory infarction: clinical and MRI findings. Eur J Neurol. 2002;9:615–24.
58. Alonso A, Gass A, Rossmanith C, Kern R, Griebe M, Binder J, et al. Clinical and MRI patterns of pericallosal artery infarctions: the significance of supplementary motor area lesions. J Neurol. 2011;259:944–51.
59. Bogousslavsky J, Regli F, Uske A. Thalamic infarcts: clinical syndromes, etiology, and prognosis. Neurology. 1988;38:837–48.
60. Van der Werf YD, Scheltens P, Lindeboom J, Witter MP, Uylings HB, Jolles J. Deficits of memory, executive functioning and attention following infarction in the thalamus; a study of 22 cases with localised lesions. Neuropsychologia. 2003;41:1330–44.
61. Nishio Y, Hashimoto M, Ishii K, Mori E. Neuroanatomy of a neurobehavioral disturbance in the left anterior thalamic infarction. J Neurol Neurosurg Psychiatr. 2011;82:1195–200.
62. Van der Werf YD, Witter MP, Uylings HB, Jolles J. Neuropsychology of infarctions in the thalamus: a review. Neuropsychologia. 2000;38:613–27.
63. Kramer JH, Reed BR, Mungas D, Weiner MW, Chui HC. Executive dysfunction in subcortical ischaemic vascular disease. J Neurol Neurosurg Psychiatr. 2002;72:217–20.
64. Pohjasvaara T, Mäntylä R, Ylikoski R, Kaste M, Erkinjuntti T. Clinical features of MRI-defined subcortical vascular disease. Alzheimer Dis Assoc Disord. 2003;17:236–42.
65. Ishii N, Nishihara Y, Imamura T. Why do frontal lobe symptoms predominate in vascular dementia with lacunes? Neurology. 1986;36:340–5.
66. Du AT, Schuff N, Laakso MP, Zhu XP, Jagust WJ, Yaffe K, et al. Effects of subcortical ischemic vascular dementia and AD on entorhinal cortex and hippocampus. Neurology. 2002;58:1635–41.
67. Al-Khindi T, Macdonald RL, Schweizer TA. Cognitive and functional outcome after aneurysmal subarachnoid hemorrhage. Stroke. 2010;41:e519–36.

68. Ljunggren B, Sonesson B, Säveland H, Brandt L. Cognitive impairment and adjustment in patients without neurological deficits after aneurysmal SAH and early operation. J Neurosurg. 1985;62:673–9.
69. Ogden JA, Mee EW, Henning M. A prospective study of impairment of cognition and memory and recovery after subarachnoid hemorrhage. Neurosurgery. 1993;33:572–86.
70. Hütter BO, Kreitschmann-Andermahr I, Mayfrank L, Rohde V, Spetzger U, Gilsbach JM. Functional outcome after aneurysmal subarachnoid hemorrhage. Acta Neurochir Suppl. 1999;72:157–74.
71. Kreiter KT, Copeland D, Bernardini GL, Bates JE, Peery S, Claassen J, et al. Predictors of cognitive dysfunction after subarachnoid hemorrhage. Stroke. 2002;33(1):200–8.
72. Mayer SA, Kreiter KT, Copeland D, Bernardini GL, Bates JE, Peery S, et al. Global and domain-specific cognitive impairment and outcome after subarachnoid hemorrhage. Neurology. 2002;59:1750–8.
73. Hütter BO. Neuropsychological sequelae of subarachnoid hemorrhage and its treatment. Wien: Springer; 2000.
74. Rousseaux M, Godefroy O, Cabaret M, Bernati T. [Dysexecutive syndrome and disorders of motor control in prefrontal mediobasal and cingulate lesions]. Rev Neurol. 1996;152:517–27.
75. Godefroy O, Lhullier C, Rousseaux M. Non-spatial attention disorders in patients with prefrontal or posterior brain damage. Brain. 1996;119:191–202.
76. Delbecq-Derouesné J, Beauvois MF, Shallice T. Preserved recall versus impaired recognition: a case study. Brain. 1990;113:1045–74.
77. Mahalick DM, Ruff RM, Heary RF, U HS. Preoperative versus postoperative neuropsychological sequelae of arteriovenous malformations. Neurosurgery. 1993;33:563–70.
78. de Bruijn SF, Budde M, Teunisse S, de Haan RJ, Stam J. Long-term outcome of cognition and functional health after cerebral venous sinus thrombosis. Neurology. 2000;54:1687–9.
79. Buccino G, Scoditti U, Patteri I, Bertolino C, Mancia D. Neurological and cognitive long-term outcome in patients with cerebral venous sinus thrombosis. Acta Neurol Scand. 2003;107:330–5.
80. Koopman K, Uyttenboogaart M, Vroomen PC, van der Meer J, De Keyser J, Luijckx GJ. Long-term sequelae after cerebral venous thrombosis in functionally independent patients. J Stroke Cerebrovasc Dis. 2009;18:198–202.
81. Bugnicourt JM, Roussel-Pieronne M, Dupuy-Sonntag D, Canaple S, Godefroy O. [Cerebral venous thrombosis: long-term functional and cognitive outcome in 16 patients]. Rev Neurol. 2008;164(2):131–7. Epub 2008 Feb 20.
82. Ferro JM, Canhão P, Stam J, Bousser MG, Barinagarrementeria F, ISCVT Investigators. Prognosis of cerebral vein and dural sinus thrombosis: results of the International Study on Cerebral Vein and Dural Sinus Thrombosis (ISCVT). Stroke. 2004;35:664–70.
83. Godefroy O, Just A, Ghitu A, Leclercq C, Garcia PY, Lamy C, et al. Rankin scale with revised structured interview: effect on reliability, grading of disability and detection of dementia. Int J Stroke. 2012;7(2):183.
84. Hackett ML. Depression after stroke and cerebrovascular disease. In: Godefroy O, editor. Vascular cognitive impairment: the cognitive and behavioural neurology of stroke. Cambridge: Cambridge University Press; 2013. p. 363–74.
85. Beversdorf DQ, Heilman KM. Facilitory paratonia and frontal lobe functioning. Neurology. 1998;51:968–71.
86. Dubois B, Slachevsky A, Litvan I, Pillon B. The FAB: a frontal assessment battery at bedside. Neurology. 2000;55:1621–6.
87. Nasreddine ZS, Phillips NA, Bédirian V, Charbonneau S, Whitehead V, Collin I, et al. The Montreal Cognitive Assessment, MoCA: a brief screening tool for mild cognitive impairment. J Am Geriatr Soc. 2005;53:695–9.

88. Cummings JL, Mega M, Gray K, Rosenberg-Thompson S, Carusi DA, Gornbein J. The Neuropsychiatric Inventory: comprehensive assessment of psychopathology in dementia. Neurology. 1994;44:2308–14.
89. Grace J, Stout JC, Malloy PF. Assessing frontal lobe behavioral syndromes with the frontal lobe personality scale. Assessment. 1999;6:269–84.
90. Wilson BA, Evans JJ, Emslie H, Alderman N, Burgess P. The development of an ecologically valid test for assessing patients with a dysexecutive syndrome. Neuropsychol Rehabil. 1998;8:213–28.
91. Burgess PW, Alderman N, Evans J, Emslie H, Wilson BA. The ecological validity of tests of executive function. J Int Neuropsychol Soc. 1998;4:547–58.
92. Kertesz A, Davidson W, Fox H. Frontal behavioral inventory: diagnostic criteria for frontal lobe dementia. Can J Neurol Sci. 1997;24:29–36.
93. Lebert F, Pasquier F, Souliez L, Petit H. Frontotemporal behavioral scale. Alzheimer Dis Assoc Disord. 1998;12:335–9.
94. Godefroy O; and the GREFEX investigators. Fonctions executives et pathologies neurologiques et psychiatriques. Evaluation en pratiqueclinique. Marseille: Solal; 2008.
95. Posner MI, Petersen SE. The attention system of the human brain. Annu Rev Neurosci. 1990;13:25–42.
96. Stone VE, Baron-Cohen S, Knight RT. Frontal lobe contributions to theory of mind. J Cogn Neurosci. 1998;10:640–56.
97. Stuss DT, Gallup GG, Alexander MP. The frontal lobes are necessary for 'theory of mind'. Brain. 2001;124:279–86.
98. Shamay-Tsoory SG, Aharon-Peretz J. Dissociable prefrontal networks for cognitive and affective theory of mind: a lesion study. Neuropsychologia. 2007;45:3054–67.
99. Bechara A, Damasio H, Tranel D, Anderson SW. Dissociation of working memory from decision making within the human prefrontal cortex. J Neurosci. 1998;18:428–37.
100. Goel V, Dolan RJ. The functional anatomy of humor: segregating cognitive and affective components. Nat Neurosci. 2001;4:237–8.
101. Shammi P, Stuss DT. Humour appreciation: a role of the right frontal lobe. Brain. 1999;122:657–66.
102. Wheeler MA, Stuss DT. Remembering and knowing in patients with frontal lobe injuries. Cortex. 2003;39:827–46.

Chapter 9
Memory Loss After Stroke

Signy Sheldon and Gordon Winocur

Introduction

Memory loss is one of the most common types of cognitive impairment affecting patients recovering from stroke. Memory and related deficits are most severe in the first few weeks following stroke, but can persist for years after the episode [1–3]. The result is a reduction in quality of life, including decreased abilities in performing daily activities and in returning to the workplace [4, 5]. In one quarter of stroke survivors, impairments worsen and develop into dementia, a neurodegenerative condition marked by a range of cognitive deficits [6, 7]. Recent data show that the risk of developing dementia after ischemic stroke is higher than age- and education-matched healthy controls, suggesting common risk factors or some underlying common pathophysiologic mechanisms.

Memory is not a unitary construct. There are many different types of memory, each serving a different purpose and each mediated by different cognitive processes and underlying brain systems. This means that post-stroke memory problems will vary with the nature of the stroke. Understanding the diversity of memory impairments following stroke has important implications for the assessment of overall function as well as for developing effective rehabilitation for stroke survivors.

S. Sheldon, Ph.D. (✉)
Rotman Research Institute, Baycrest Centre for Geriatric Care, Toronto, ON, Canada
e-mail: signy.sheldon@utoronto.ca

G. Winocur, Ph.D.
Rotman Research Institute, Baycrest Centre for Geriatric Care, Toronto, ON, Canada

Department of Psychiatry, University of Toronto, Toronto, ON, Canada

Department Psychology, University of Toronto, Toronto, ON, Canada

T.A. Schweizer and R.L. Macdonald (eds.), *The Behavioral Consequences of Stroke*,
DOI 10.1007/978-1-4614-7672-6_9, © Springer Science+Business Media New York 2014

Definitions of Memory Types

1. *Explicit Memory*: the conscious recall of knowledge and experiences.
   (a) *Episodic*: recalling events or experiences that occurred in a single time and place.
      (i) *Recollection*: Specific recall of a past event.
      Example: "I remember meeting Sarah at Victoria's party on Friday night. We talked about politics while eating cheese fondue."
      (ii) *Familiarity*: Knowledge that something is old.
      Example: "I know I have met Sarah somewhere before. I have a feeling it was at a party."
   (b) *Semantic*: Knowledge of general facts about the world or self.
   Example: "Cheese fondue is a Swiss delicacy."
2. *Implicit Memory*: Remembering previously experienced information without conscious awareness.
   (a) *Priming*: bias of task performance by a previous experience without conscious awareness (not discussed in this chapter).
   (b) *Procedural*: remembering or learning procedures without conscious awareness.
   Example: Knowing how to prepare cheese fondue without following a recipe or instructions.

This chapter reviews different subtypes of memory, but focuses specifically on episodic memory, the conscious recall and recognition of information. Processes underlying episodic memory and the neural regions that support them are particularly vulnerable to stroke. In considering the relationship between stroke and episodic memory, we will review the impact of infarct location on the expression of impairment. Given the propensity of strokes to be associated with later development of dementia, we will appraise the relationship between stroke and dementia to illustrate vascular conditions and factors that promote the conversion of memory impairment to dementia. We begin with a review of the different types of memory and the common methods for memory assessment.

Multiple Forms of Memory

A basic distinction in memory is between short-term and long-term memory. Short-term memory involves holding information in consciousness for a limited time, in the range of seconds to minutes; the permanent storage of this information is

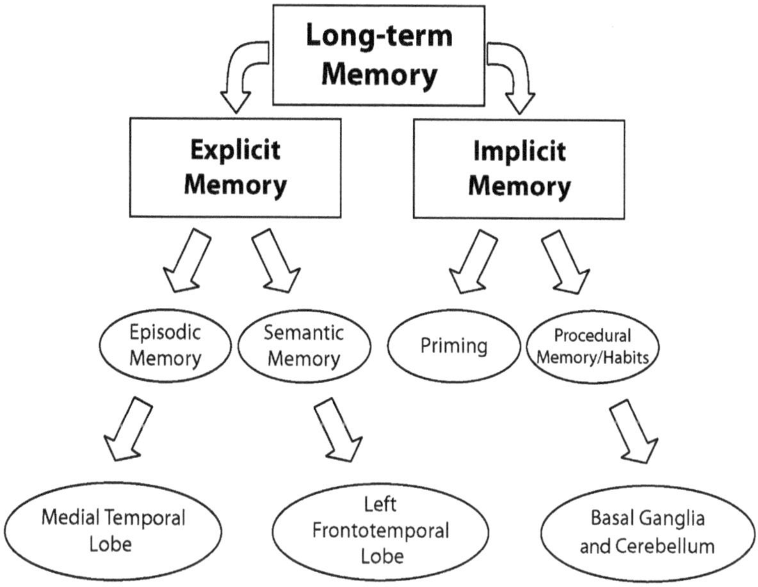

Fig. 9.1 The distinctions in long-term memory that are relevant to stroke and the primary supporting neural structures

supported by long-term memory. The transition of information from short-term to long-term memory begins with successful encoding. After a memory is encoded it undergoes a process of consolidation during which it is represented in the brain as a stable and durable long-term memory. Later retrieval of a memory depends on how well it was encoded and consolidated and on the availability of suitable cues at retrieval. Without delving into the complexity of these processes, it is important to appreciate that impairment of any combination of encoding, consolidation, or retrieval of information can impair memory functioning.

Within long-term memory, there are further divisions (see previous Definitions of Memory Types and Fig. 9.1 for how the distinctions in memory are made). Memories that can be consciously recalled, such as memories for specific events ("I remember my 18th birthday party") and facts ("The CN tower is located in Toronto") are called *declarative* or *explicit* memories. Within explicit memory, the distinction is made between *episodic* and *semantic* memory. Episodic memory refers to the conscious recollection of an experienced event and its associated contextual details, like the time and place that the event occurred. The retrieval of an episodic memory often entails a feeling of going back and reexperiencing the event that is being recalled.

In the clinic, episodic memory is measured with tasks that require one to learn a list of stimuli (e.g., words, objects) or more complex materials (e.g., short stories or figures) and, after a delay, recall or recognize as much information as possible (see Table 9.1 for examples of common neuropsychological tests used to examine episodic memory and other types of memory). These methods are useful for studying

Table 9.1 Some common examples of neuropsychological measures used to assess memory function

Test	Age cut-off	Administration time (min)	Test description	Processes measured
Implicit memory measures				
Mirror-tracing task	–	–	Participant is asked to trace a pattern through the reflection in a mirror. Examine practice effects	Procedural memory via the improvement of mirror-tracing over time
Serial reaction task	–	–	A dot is shown on a computer screen and the participant indicates the position. Sometimes it is random, sometimes it is predictable	Implicit learning via the time advantage between the blocks that are predictable and those that are random
Semantic memory measures				
Boston naming test (BNT)	89	5	60 line drawings of objects and the participant is asked to name each drawing. If unable to do so, semantic and phonemic cues are provided	Integrity of the semantic network; vocabulary
Semantic fluency tasks	95	5	Participant is asked to generate words other than proper names that belong to categories such as animals	Integrity of the semantic network; vocabulary
Peabody picture vocabulary test	90	15	Out of four pictures, participant points to one that is described by a given word	Vocabulary
Episodic memory measures				
California Verbal Learning Tests (CVLT)	89	20	Learning, recall and recognition of a word list of related items over multiple trials	Verbal learning and retrieval strategy; acquisition rate; immediate and delay recall; interference

Hopkins Verbal Learning Test (HVLT)	80	15	Learning, recall and recognition of a word list of related items over multiple trials	Verbal learning and retrieval strategy; acquisition rate; immediate and delay recall; interference
Brief visual memory test (BVMT)	70	15	Recall of six designs and locations over multiple trials, recognition trial included at delay	Visual-spatial learning, recall and recognition
Rey-Osterrieth complex figure	93	15	Construction, recall and recognition of a complex figure	Visual-spatial construction skills, learning, recall and recognition
Doors and people test	80	40	Recall and recognition of verbal and visual information and associations	Visual and verbal memory and associative memory processes
Autobiographical memory interview	80	30	Recall of remote personal and factual events	Episodic and semantic components of autobiographical memory
Weschler memory scale (WMS)-III: Logical memory subtest	89	15	Immediate and Delayed recall of two short passages and delayed recognition of story details	Verbatim and gist recall and recognition of verbal narratives
Weschler memory scale (WMS)-III: Paired associates subtest	89	20	Learning and delayed recall and recognition of a list of unrelated word pairs	Associative verbal learning, recall and recognition
Weschler memory scale (WMS)-III: Design memory subtest	89	15	Recall item information and location of eight designs on a grid for immediate and delayed periods	Visual-spatial immediate and delayed recall of item and location information

anterograde memory, the ability to form new episodic memories. *Retrograde memory*, the ability to remember previous events, can be reliably assessed with tests like the Autobiographical Memory Interview [8], which evaluate the number and type of details in older episodic memories.

Whereas episodic memory entails the conscious recall of contextual events, semantic memory refers to general, context-independent knowledge. This includes memory for facts about the world (e.g., Paris is the capital of France) as well as general knowledge about ourselves (personal preference for spaghetti with tomato sauce versus alfredo sauce). Neuropsychological measures of semantic memory assess the ability to recover these factual details in formal tests (Table 9.1). For example, the category fluency task requires the generation of many items that belong to specific categories (e.g., animals) in a limited time period. The Boston Naming Test requires one to name a series of pictured objects, beginning with common objects (e.g., canoe) and ending with less common objects (e.g., abacus). In general, semantic impairment following a stroke will present as deficits in comprehension or retrieval of factual information, such as the name of an object, the president of the United States, etc. Such impairment is typically seen following damage to the left prefrontal cortex or the left posterior temporal lobe and inferior parietal cortex [9–11] (Table 9.1).

Implicit Memory

Implicit memory is a form of memory that involves the expression of knowledge without conscious awareness of how that knowledge was acquired. Remembering acquired skills, habits, and rules, such as cooking techniques, riding a bicycle, or playing chess are examples of implicit or procedural memories. Strokes in regions such as the basal ganglia and the cerebellum can result in deficits in remembering established procedural memories and in forming new ones [12]. For example, in the commonly used mirror-tracing task, an individual is asked to trace shapes using only visual feedback from a mirror. This task may be initially hard, but it becomes easier with practice. Stroke survivors with deficits in implicit memory do not benefit from practice when mirror tracing [13–15] (see Table 9.1 for other examples of measures of implicit memory). While all forms of memory are important, episodic memory processes and structures are the most likely to be affected by stroke [16]. Accordingly, the remainder of this chapter will focus on the effect of stroke on episodic memory.

Episodic Memory

Strokes causing damage to the neural network that supports episodic memory can impair the learning, storage, and retrieval of episodic memories. The nature of this impairment is dependent on where in the network damage occurs (Table 9.2).

Using a number of different methodologies, neuropsychological investigations have determined that the core of the episodic memory neural network is the hippo-

Table 9.2 A brief summary of predicted memory impairments resulting from neuronal damage to regions of the episodic memory network

Neural damage	Vascular supply	Side	Process affected
Medial temporal: Hippocampus	Posterior communicating artery; anterior choroidal artery	Left	Verbal memory; recall impaired more than recognition
		Right	Nonverbal memory; recall impaired more than recognition
Medial temporal: Hippocampus and Parahippocampal regions	Posterior communicating artery; anterior choroidal artery	Left	Verbal memory; recall impairment = recognition impairment
		Right	Nonverbal memory; recall impairment = recognition impairment
Thalamus: Extended hippocampal system	Polar (tuberothalamic/premammillary) artery; anterior choroidal artery	Left	Verbal memory; recall impaired more than recognition
		Right	Nonverbal memory; recall impaired more than recognition
Thalamus: Mediodorsal-perirhinal circuit	Paramedian (thalamo-perforating) artery	Left	Recall impairment = recognition impairment; executive deficits
		Right	Recall impairment = recognition impairment; executive deficits
Frontal: Oribtofrontal cortex/basal forebrain	Perforating branches from anterior communicating artery, orbitofrontal artery branches of anterior (medial) and middle (lateral) cerebral arteries	Bilateral	Impaired source recall; context confusion; confabulation
Frontal: Prefrontal cortex	Anterior cerebral artery	Left	Memory encoding and general impairment in strategic and organizational (executive) processing
		Right	Memory retrieval and general impairment in strategic and organizational (executive) processing
Posterior association areas	Posterior cerebral artery, middle cerebral artery	Bilateral	Impairment in encoding and retrieving particular sensory/perceptual details

campus [12, 17, 18]. Subcortical and cortical structures, including the thalamus, the prefrontal cortex and posterior cortical regions, maintain dense connections with the hippocampus and other medial temporal lobe (MTL) structures. These regions mediate specific processes, which, if affected by vascular damage, bias the presentation of memory impairment in specific ways.

The Hippocampus and Episodic Memory

Appreciation of the central importance of the hippocampus (hippocampus proper, dentate gyrus, and subiculum) for episodic memory dates back to the pioneering work of Milner and her colleagues with the classic patient HM. HM suffered from intractable epilepsy and underwent bilateral MTL surgery to remove the hippocampus, parahippocampus, and related regions. The surgery relieved his epilepsy but left him with profound anterograde and retrograde amnesia for episodic events. Significantly, HM's memory for semantic information and his implicit memory were intact [19, 20]. Since HM, considerable work has confirmed the importance of the hippocampus in episodic memory [21–25].

From these behavioral findings, theoretical frameworks, such as the Multiple Trace Theory, have implicated the hippocampus in both the encoding and retrieval stages of episodic memory [17, 26]. At encoding, the hippocampus serves the critical role of binding all the relevant memory-related details into a coherent structure across multiple brain regions. Over time, episodic memories become represented in a distributed network of subcortical and cortical brain regions. During retrieval, hippocampal neurons act as a pointer to these storage regions, initiating the process of bringing the memory into awareness.

The Hippocampus: Recollection Versus Familiarity

As indicated previously, the hippocampus is believed to be crucial for *conscious recollection*, a process that entails the re-experiencing and recall of an event and its associated details. Clinically, recollection is crucial for tests such as free recall and those that require the retrieval of associative information, such as where and when an event was experienced. The hippocampus is not required for memory that is based solely on familiarity. In contrast to recollection, *familiarity-based memory* is less detailed and usually involves remembering that an object or item has been encountered before, but without necessarily being able to recall anything else about it. Recognition tasks that require decisions based on matching memory to present information can be successfully completed with familiarity-based assessments.

Recollection impairments are present after stroke in the MTL due to, for example, hemorrhage or occlusion of the vascular supply to the hippocampus (involving

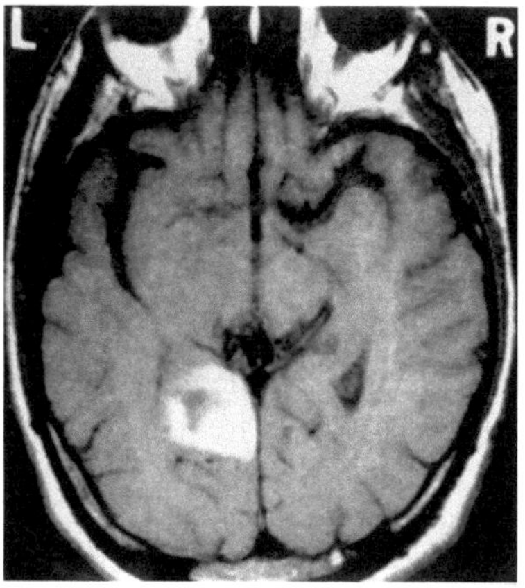

Fig. 9.2 A representative scan of one of the patients that presented with left hippocampal and parahippocampal gyrus damage. Reprinted with permission from Ott BR, Saver JL. Unilateral amnesic stroke. Six new cases and a review of the literature. Stroke. 1993 Jul;24(7):1033–42

the anterior and temporal artery, and to a lesser degree the anterior choroidal artery). Most strokes involving the hippocampus and MTL are ischemic; hemorrhages isolated to the hippocampus are uncommon and those in the temporal lobe are usually more lateral. Typical causes would be amyloid angiopathy or rupture of a middle cerebral artery aneurysm into the adjacent temporal lobe. Ott and Saver [27] reported cases of hippocampal lesions from strokes that were sufficient to induce amnesia (see Fig. 9.2 for a representative computed tomography [CT] scan of such a case).

Other investigations [24] have examined memory performance of patients who suffered from mild hypoxic ischemic encephalopathy, which is a form of transient global brain injury. Although diffuse brain regions can be affected, *hypoxic ischemic encephalopathy* can also selectively damage the hippocampus. Compared to healthy control participants, these patients showed memory impairment on recall and recognition memory tests; however, their recall deficit was significantly greater than their recognition deficit. Experimental measures designed to dissociate recollection from familiarity identified a selective recollection deficit in these patients [28, 29].

In 2004, Mayes and colleagues [30] described patient YR, who suffered an ischemic stroke that selectively affected both hippocampi. Consistent with the previously mentioned findings [28, 29], YR's primary deficit reflected a loss of hippocampus-dependent memory. Her short-term, implicit and semantic memories, as well as familiarity-based recognition memory, were all spared. Her performance on tests that required recollection, such as free recall and paired-associate learning, however, was impaired.

Many of the studies that have reported anterograde amnesia after hippocampal damage also demonstrated deficits in the patients' ability to recall details from past personal experiences (e.g., describe the first day of a job) [31]. These findings suggest that memory loss following hippocampal strokes extend to events of the distant past (retrograde amnesia). Further evidence of retrograde amnesia comes from a study that tested 20 patients with unilateral strokes, including patients who sustained damage to the hippocampus [32]. The patients were given tests of autobiographical memory in which they were asked to describe personal events from three remote time periods. They were also given tests of public events in which they answered questions about well-known events from different decades (e.g., "Who was the first person to set foot on the moon?"). The patients with hippocampal damage generated fewer details and had poorer recall of past personal events, particularly those from remote times. They also answered fewer correct responses to questions about well-known public events compared to their matched counterparts.

While recollection-specific deficits are a cardinal feature of hippocampal damage due to stroke, for example, familiarity-based memory impairment can supplement these deficits if the lesions are accompanied by damage to other MTL structures, namely the entorhinal, perirhinal, and the parahippocampal cortices. In their review, Aggleton and Shaw [33] reported, that, while focal hippocampal lesions selectively disrupt free recall (recollection), damage that extends beyond the hippocampus into other MTL regions results in recall *and* recognition deficits (also see [18]; but see [34] for a different view of recollection and familiarity).

The Hippocampus: Left Versus Right

Conventional laterality-specific functions are represented in the hippocampus. The left hippocampus is involved predominately in verbal episodic memory tasks whereas the right hippocampus is involved more in nonverbal episodic memory tasks, including the encoding and retrieval of spatial relations and location information. For example, poor verbal memory, as measured by story recall and tests that assess memory for a previously learned list of unrelated words, is associated with left hemisphere strokes [35]. Patients with ischemic right posterior hippocampal lesions present with relatively intact verbal memory, but have a specific impairment in nonverbal and spatial memory [36]. Szabo and colleagues [29] studied more than 50 patients using a neuropsychological battery that included a list-learning task, an auditory learning task, and tests of nonverbal memory, such as the Rey-Osterreith complex figure test. Strokes affecting the left hippocampus resulted in worse performance on the verbal memory tests (e.g., list-learning test) when tested at both short and long delays, and better performance on nonverbal memory tests (e.g., the Rey-Osterreith complex figure test). In contrast, strokes affecting right hippocampal regions resulted in the opposite pattern: poorer nonverbal than verbal memory.

The Hippocampus: Summary

The hippocampus is integral for the formation and recall of episodic memories, irrespective of when they are acquired. When the hippocampus is damaged as part of stroke-induced pathology, both anterograde and retrograde amnesia are experienced, as they are in other disease processes. If hippocampal lesions from stroke are accompanied by damage to additional regions of the MTL, familiarity-based recognition memory is also likely to be affected. Finally, the pattern of impairment from hippocampal strokes is dependent on the side of damage. Left-sided damage results in verbal memory impairments and right-sided damage results in nonverbal impairments.

The Thalamus and Episodic Memory

The role of the thalamus in episodic memory is related to its multiple connections with the MTL (Fig. 9.3). Two main circuits connect the anterior and medial dorsal thalamic nuclei to the hippocampus and parahippocampus, respectively. The anterior (and latero-dorsal) thalamic nuclei are densely connected to the hippocampus via the mammothalamic tract and the mammillary bodies forming the *extended hippocampal system* [33, 37]. These structures are supplied by the tuberothalamic artery, a branch of the posterior communicating artery (which also supplies anterior regions of the thalamus, including the mammillary bodies and part of the mammillothalamic tract) and the anterior choroidal arteries (the anterior nuclei complex). The medial dorsal thalamic nucleus is linked to the amygdala and the perirhinal cortex in the parahippocampal gyrus via connections that pass through the internal medullary lamina [38, 39] and also maintains reciprocal connections with the prefrontal cortices.

The thalamus is supplied by perforating arteries arising from the posterior communicating artery (polar, tuberothalamic, anterior internal optic, or premamillary artery), paramedian thalamic-subthalamic arteries from first segment of the posterior cerebral artery, inferolateral (thalamogeniculate) arteries from the second segment of the posterior cerebral artery and posterior (medial and lateral) choroidal arteries usually also from the second segment of the posterior cerebral artery. These are all derived from the vertebrobasilar arterial system, with the exception of the polar artery. Vascular supply to the medial dorsal thalamic nucleus also comes from the polar and paramedian arteries [40]. The thalamus is frequently damaged by stroke, including ischemic stroke with occlusion of the perforating arteries supplying it due to hypertension. Intracerebral hemorrhage secondary to hypertension also is common in the thalamus (Fig. 9.3).

A pattern of memory loss that is reminiscent of hippocampal amnesia occurs following damage to the extended hippocampal system, particularly if the damage

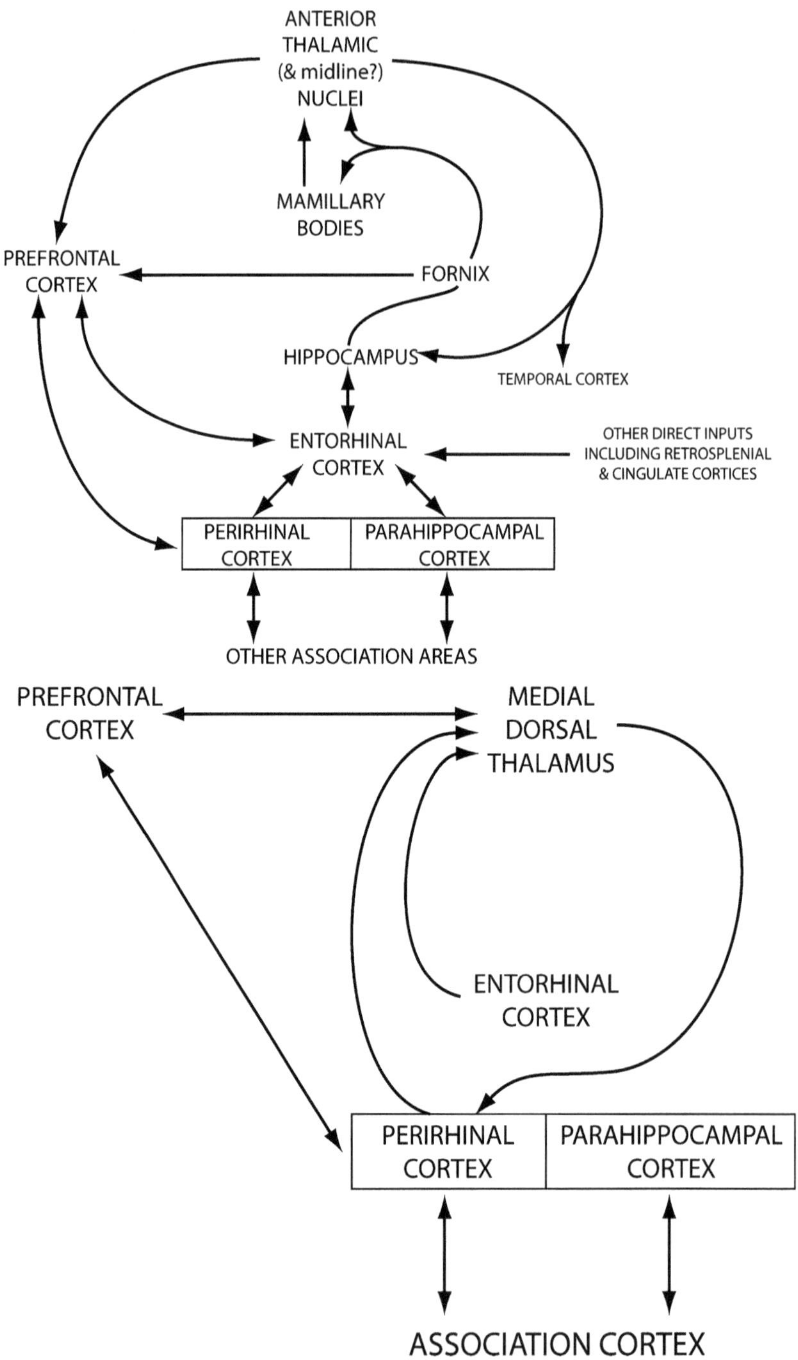

Fig. 9.3 Pathways connecting the medial temporal lobe structures to the thalamus. The *top* schematic represents the connections of the anterior thalamic nuclei. The *bottom* schematic represents the connections of the medio-dorsal nucleus of the thalamus (adapted from [33])

disconnects the thalamus from the hippocampus [39]. In one well-documented case, isolated injury to the left anterior thalamus following a polar thalamic infarct resulted in anterograde amnesia for episodic events, while short-term, semantic, and procedural memory as well as general intelligence remained intact [41]. As with hippocampal amnesia, in cases of thalamic stroke, there is a prominent dissociation on tests of episodic memory that is dependent on recollection and familiarity. For example, patients with thalamic stroke that damages the anterior thalamic nuclei and/or thalamic connections to the hippocampal system often present with deficits in verbal recall (recollection) memory but spared verbal recognition (familiarity) memory as measured, for example, with the California Verbal Learning Test (CVLT) [42] (also see [43–45]). Carlesimo and colleagues [46] tested a thalamic stroke patient with damage that extended to the hippocampal system. Consistent with other reports, this patient was impaired only on the recollective aspects of episodic memory on tests of anterograde and retrograde memory.

These deficits following thalamic stroke may be indistinguishable from those with hippocampal damage, yet the underlying causes of the impairment is different. Hippocampal damage results in impairment in the binding or associative processes of episodic memory at *encoding* and *retrieval*. Thalamic damage that disturbs thalamic-hippocampal pathways appears to cause impairment primarily at the encoding stage [45] in that lesions to the anterior nuclei or their afferent tracts will impair encoding new stimuli. Under ordinary conditions, the thalamus acts as a relay between the hippocampus and cortical regions of the brain where long-term memories are stored. Without these contributions, the details of the encoded memories cannot be appropriately stored, resulting in post-encoding recollection deficits [47].

Thalamic strokes can also result in executive deficits (e.g., attentional control, suppression of irrelevant information) similar to those associated with frontal-lobe damage (see Chap. 8). These deficits are usually a result of damage to the medial dorsal nucleus-perirhinal pathway from paramedian occlusions [48]. In the normal, intact brain, the reciprocal connections between the medial dorsal nucleus and the prefrontal cortex help integrate executive processes with other nonexecutive processes supported by connected brain regions. Damage to this circuit produces widespread impairment that can affect memory and a range of cognitive processes [49]. In terms of memory, patients with this pattern of damage have problems consciously remembering episodic information under conditions of high interference, often exhibiting intrusion errors and false recognitions during various types of testing [50].

The medial dorsal nucleus-perirhinal circuit appears also to be involved in familiarity-based episodic memory decisions. This is based on its connection to perirhinal cortex [33]. Supporting this notion are findings from Carlesimo and colleagues [46] who report that patients with damage to the medial dorsal nucleus-perirhinal circuit present with selective impaired familiarity, but have spared recollection [45] (see [51] for inconsistent results).

The Thalamus: Left Versus Right

Stroke-induced lesions to anterior thalamic nuclei and connections within the extended hippocampal system of the left hemisphere result in episodic memory deficits that are more verbal in nature [39, 52, 53]. Right-sided lesions typically produce the expected pattern of visuospatial, nonverbal memory deficits, but these lesions can result in prominent verbal memory deficits as well [54]. Laterality effects are not as clear-cut following damage to the medial dorsal nucleus-perirhinal circuit, but can still occur [55]. Of note, bilateral lesions from paramedian artery occlusion are more common because a single paramedian artery not infrequently supplies both thalami. These patients present with more severe deficits and limited recovery than patients with unilateral lesions.

The Thalamus: Summary

Memory deficits can arise from thalamic strokes by way of damage to or disconnection of the extended hippocampal system or damage to the medial dorsal nucleus-perirhinal circuit. The memory loss is similar to hippocampal amnesia when there is damage to the extended hippocampal system (worse performance on recollection-recall versus familiarity-based recognition tests). Damage to the medial dorsal nucleus-perirhinal circuit impairs performance on memory tests that require additional *executive processes*, by virtue of the reciprocal connections with the prefrontal cortex, but also impairs familiarity-based memory tests, given the connection to non-hippocampal MTL structures. While these findings suggest a functional specialization for the thalamus in episodic memory, it is important to remember that a stroke does not often result in such clean dissociations: a fusion of the reported impairment is most likely to present after a thalamic stroke.

The Prefrontal Cortex and Episodic Memory

The prefrontal cortex is often referred to as the "executive" of the brain because of its involvement in a number of essential top-down processes, many of which are crucial for episodic memory. Impairment in these regulatory processes negatively impacts memory inasmuch as a given test requires such processes. These deficits manifest most robustly on tests such as free recall memory, tasks that require inhibitory control, and tasks that require recalling the context information.

Failure to organize information or implement strategies when performing a memory task resulting from prefrontal cortex damage will result in free recall impairments [56]. For example, if asked to study a randomly presented, semantically related list of words (e.g., furniture, vegetables), persons with prefrontal damage are less likely to implement a useful strategy to remember the items, such as organizing the related items consecutively in one's mind (a process called *clustering*). Importantly, if a a person with prefrontal damage is aided with a learning strategy at encoding, learning, and recall of information can be improved.

Top-down deficits following prefrontal-lobe damage also manifest as a difficulty in inhibiting irrelevant or competing information. For example, prefrontal damage results in problems with distinguishing between information that was actually learned from information that is similar to what was learned. On a recognition test, this presents as a normal hit rate (correctly indicating that old items are old), but a concomitant high false alarm rate (falsely indicating that new items are old).

As a test of inhibitory control in patients with damage to the prefrontal cortex, Shimamura and colleagues [57] gave patients two lists of word pairs to learn. These pairs were developed so that some pairs from the first list (*lion-hunter*) could interfere with a pair from the next list (*lion-circus*). The prefrontal patients fared worse at learning pairs from the second list that overlapped with words from the first list because of a deficit in inhibiting the first pair (*lion-hunter*) when learning the new pair (*lion-circus*) (also see [58]).

Top-down deficits after prefrontal damage also present as specific impairments in retrieving the source or context of information while sparing memory for factual information (i.e., content). A series of experiments found that prefrontal cortex damage impairs the ability to remember the list order (or temporal source) of words as well as the ability to list factual events in chronological order. The ability to recognize individual words and events was intact [59, 60] (Fig. 9.4).

Patients with damage to the orbitofrontal cortex/basal forebrain, typically as a result of anterior communicating artery aneurysm rupture (Fig. 9.4), provide interesting examples of how executive impairments can produce memory deficits. In most situations, these patients have problems distinguishing between currently relevant and irrelevant (or no longer relevant) memories, and present with context confusions [61–63]. In severe cases, such patients exhibit *confabulation* [64]—the spontaneous production of memories that never occurred. Confabulations can present as simple false memories (e.g., misremembering a canoe trip that took place last week) or more fantastical stories (e.g., misremembering an army operation that took place last week). Either way, these stories are presented without the awareness of their erroneous nature by the confabulators. The cause of these confabulations is thought to arise from deficits in retrieval processes, which appear as an inability to monitor information or to evaluate retrieval cues [64]. Confabulations may indicate an extreme form of disorganization or monitoring of internal information and/or impaired temporal processing, thereby resulting in severe context confusion.

The case of orbitofrontal cortex damage illustrates memory failures resulting from top-down deficits. Other cases of frontal-lobe damage do not produce such clear memory impairment. The inconsistencies in the literature regarding memory problems following prefrontal stroke damage is due, in part, to the large variability in frontal-lobe lesions (given the size of the frontal lobe). Caution should be taken when evaluating memory impairment after frontal-lobe damage.

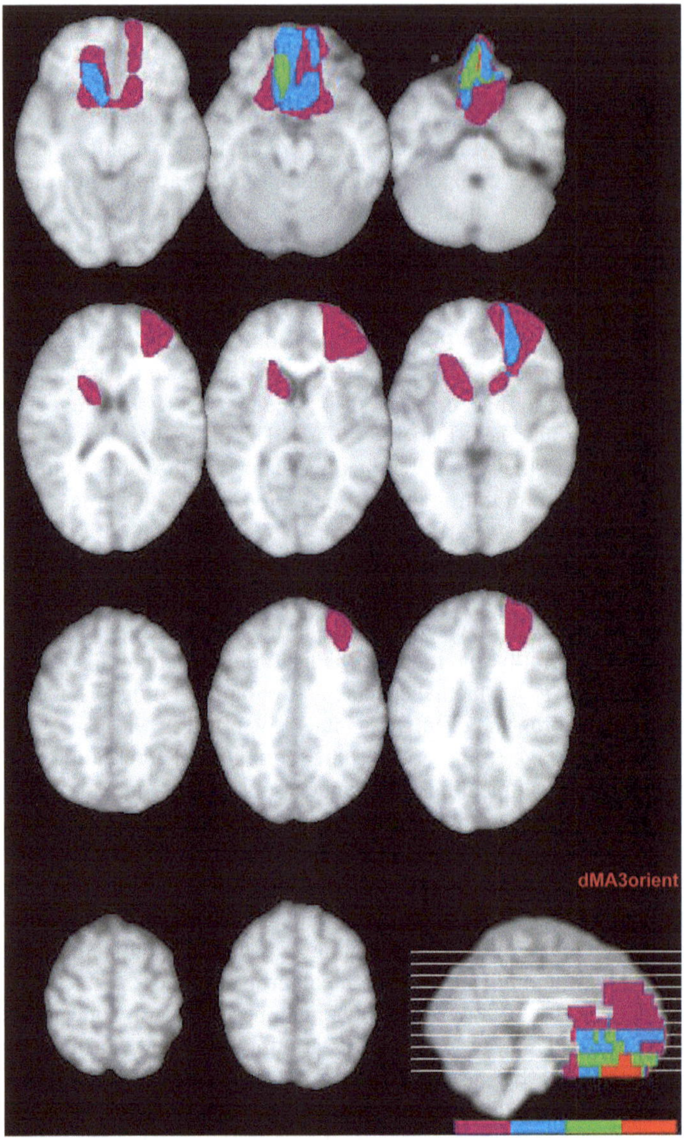

Fig. 9.4 Lesion overlap of four confabulating patients who suffered anterior communicating aneurysm ruptures. Color bars represent the number of patients with lesions that overlap a particular region, with *purple* being one patient and *red* being all patients (Reprinted with permission from Gilboa A, Alain C, Stuss DT, Melo B, Miller S, Moscovitch M. Mechanisms of spontaneous confabulations: a strategic retrieval account. Brain. 2006 Jun;129(Pt 6):1399–414. Oxford University Press)

The Prefrontal Cortex: Left Versus Right

Left lateral regions of the prefrontal cortex are prominently involved in encoding new items and are active when additional processing is required (e.g., contextualization of a memory [65, 66]). An example is found in studies that used the CVLT. This test consists of two different lists of words for participants to study. Each list has four words from four different categories presented in a pseudo-random manner, and delayed recall of the two lists is tested. Left lateral stroke damage results in large deficits on recall of the CVLT lists due to a deficit in encoding and in implementing strategic processes based on the semantic categorization (e.g., clustering) of the lists to optimize memory [67].

The right lateral frontal cortex, on the other hand, is involved in the executive aspects of episodic retrieval. Damage to this region results in an inability to evaluate information appropriately, manifesting behaviorally as an increase in false memories or false alarm rates. For example, Curran and colleagues [68] reported on a patient with right frontal damage who had extremely high false alarms rates on simple recognition memory tests. In another example, right prefrontal damage increased false recognition of faces on a two-alternative forced-choice memory test [69].

The Prefrontal Cortex: Summary

The prefrontal cortex contributes to episodic memory via its control of executive processes. Strokes affecting frontal regions do not produce the classic presentation of amnesia, but rather a range of top-down deficits that affect strategic and organizational processes needed to learn and remember information. These deficits are illustrated in cases of anterior communicating artery aneurysm ruptures that result in orbitofrontal or basal forebrain damage. For these patients, the hallmark of their organizational memory deficits is temporal confusion and confabulation. Finally, there are particular lateralized functions of the prefrontal cortex. The right cortex is involved in executive processes associated with episodic memory retrieval and the left lateral prefrontal cortex, with processes needed for encoding information.

Posterior Association Areas and Episodic Memory

The MTL and frontal lobe are crucial for binding together and coordinating memory, respectively; however, these brain regions do not store the basic elements that comprise memory. Association areas in the temporal, parietal, and occipital cortices primarily involved in sensing and perceiving particular forms of information house the details of long-term memories.

Strokes in association areas can lead to different types of *agnosias*, which are defined as the inability to recognize objects or features that are typically confined to one modality [70]. The precise nature of these deficits will be determined by the location of damage. For example, a medial occipito-temporal stroke can result in a deficit in "seeing" or recognizing common objects that is specific to shape and orientation information, a form of visual agnosia [71]. Right posterior cerebral artery ischemic strokes that damage the fusiform gyrus can result in a select inability to recognize faces (prosopagnosia) [72].

For an object or event to be perceived, sensory information must be sent along neural pathways through primary sensory areas to unimodal and then to multimodal association areas where information is integrated. Adequate functioning of these association areas assures veridical perception and accurate retrieval of information. This has downstream consequences for episodic memory. Consider what happens when you view a new chair in your living room. Regions in the occipital lobe first perceive the visual features of the chair. This information is passed along a ventral pathway dedicated to object properties to higher order association areas in the inferior temporal lobe. At the same time, spatial information related to the chair's location is passed along a dorsal pathway to parietal regions dedicated to spatial processing. To encode this information as a memory, these perceptions must be integrated together. If you want to retrieve these perceptions, the *reverse* pathways that occurred when you perceived or encoded the information must be activated [73–76].

Damage to perceptual systems will impair episodic memory inasmuch as access to those elements at encoding or retrieval will be limited. Going back to the aforementioned example, if there is a stroke along the object-dedicated ventral visual pathway, there will be impairment in perceiving object information. While this deficit is not memory specific, it will make it difficult to encode object information [75]. It will also make it difficult to retrieve object information because these pathways are needed for retrieving specific details from a stored memory.

While this concept applies to perceptual/sensory pathways in general, most of the literature that illustrates this in patients focuses on visual processing. Ogden [77] reported a patient who had bilateral damage to the medial aspects of the occipital lobes. The patient suffered a deficit in visual encoding that affected his performance on visual memory tests. Interestingly, he was also severely impaired in recovering remote autobiographical events that appeared to be due to a failure in integrating elements that comprised old memories. This case highlights that cases of modality/object-specific agnosia typically result in specific memory deficits [70, 71, 78].

An example of stroke-induced visual-perceptual processing loss resulting in memory impairment comes from a study of 61 stroke patients on a test of everyday memory. In this task, participants were shown familiar objects and were asked to remember the location of the objects. While those with strokes to the posterior parietal cortex (the termination point of the dorsal visual stream) recognized the objects, they could not accurately recall the location of the objects. Indeed, they presented with a specific deficit in recalling visual-spatial relations of objects [79] (also see [80]). Following suit are other reports of selective spatial memory deficits (attending to spatial locations) following strokes to the parietal cortex [81, 82].

Posterior Association Areas: Summary

The main contribution of association areas of the parietal, occipital, and temporal lobes is to house corresponding perceptual and sensory representations. While this is illustrated with examples from visual association area damage, deficits in other perceptual/sensory areas are likely to result in similar deficits. This hypothesis has not been thoroughly investigated in stroke patients.

Post-stroke Dementia

Stroke-induced memory loss may progress to more widespread cognitive impairment and even *dementia*. Most of the clinical deficits of post-stroke dementia are the direct result of cerebrovascular disease or vascular lesions, but cases of mixed presentation cerebrovascular disease and neurodegenerative conditions such as Alzheimer's disease (AD) are also reported [83].

In *large vessel dementia*, single or multiple infarcts or hemorrhages precede the development of dementia. The presence of infarcts not only lowers cognition, as described in this chapter, but also increases the risk of dementia. Pathological factors influence the risk of post-stroke dementia. Larger infarcts are associated with higher incidences of dementia, with the location of these infarcts also playing a role: damage to regions of the episodic memory network (e.g., frontal, thalamic, and hippocampal strokes) [84] are associated with an increased risk of dementia.

New research has highlighted the prevalence of microscopic cerebrovascular pathology in post-stroke dementia [85]. In cases of *small vessel dementia*, the underlying pathologies are white matter damage and primary vessel disease [86]. The mechanism by which these pathologies lead to dementia requires investigation, but one suggestion is that these infarcts, particularly white matter damage, represent the disconnection of cortical-cortical or cortical-subcortical pathways. This damage would result in impaired functional neuronal networks and slowed information processing [87].

The diagnosis of *mixed dementia* is the result of cerebrovascular factors, either vascular lesions or small vessel disease [88] acting in synergy with Alzheimer's disease (AD). Strokes could lead to the AD by increasing the deposition of amyloid β (beta) protein, the underlying pathology of AD and assumed pathologic substrate [89]. Vascular insults can also decrease the threshold for clinical manifestation AD pathology: individuals with a pre-dementia brain will exhibit the clinical signs of dementia earlier if they experience a stroke or have stroke-related risk factors [90, 91].

Despite the heterogeneous nature of *vascular dementia*, there are some consistencies in the behavioral manifestations, particularly when vascular dementia (large and small vessel dementia) is compared to AD without the presence of vascular factors. The main complaint of both patient groups is memory, but these difficulties are less associated with episodic memory in cases of vascular dementia compared to AD. Moreover, the pathophysiology of this memory loss is associated with MTL atrophy in AD, which is not always the case for vascular dementia. Vascular dementia is most often associated with impairment in executive functions, processing speed, problem-solving, and reasoning, which have cascading effects on episodic memory [92]. Patients with vascular dementia are generally impaired on fluency tests and measures of executive processing; they are less impaired on free recall and recognition memory compared to patients with AD [93]. From a rehabilitation standpoint, this means that memory problems in vascular dementia are more likely to be improved by providing external support, whereas this is not often the case for patients with AD.

Vascular dementia is often accompanied by psychological symptoms such as personality change, depression, or emotional bluntness. Another distinguishing feature of vascular dementia is that the onset of cognitive impairment is quicker and less gradual than in AD, within the range of days to weeks, and the presentation of cognitive impairment fluctuates from day to day. This pattern of impairment is attributed to the sudden loss of a large number of cells from the vascular incident, which can be contrasted to the hallmark accumulation of protein deposits for AD manifesting as gradual cognitive decline [94, 95].

Progression to vascular dementia may be slowed if dementia-related factors are detected and treated early [96]. Many of the most strongly associated lifestyle factors are ones that can be medically managed: high blood pressure, diabetes, obesity, cigarette smoking, and high cholesterol. Another group of risk factors are vascular in nature. These includes the presence of white matter disease [88], cortical and hippocampal atrophy [97], brain volume loss, and strategic infarcts [84], although the most notable predictor of dementia is the presence of recurrent strokes [98]. It is clear that more research is needed to better understand how these predictors contribute independently and interactively to the conversion of cognitive impairments to post-stroke dementia.

Summary

The bulk of this chapter focused on episodic memory deficits following stroke. We described a framework of memory functioning/processing that emerges from an interaction of different regions of the brain, namely the MTL, thalamus, frontal lobes, and regions of the cortex involved in memory storage. The presentation of mnemonic impairments in stroke survivors is dependent largely on which regions are affected: hippocampal damage affects recollective processes (binding together items in context), thalamic damage impairs encoding processes and possibly executive functions, frontal damage affects organizational and strategic processes, and posterior damage impacts perceptual processing systems that are fundamental to episodic memory.

Viewing post-stroke memory impairments in this framework illustrates two important points. First, comprehensive neuropsychological examination is needed to identify the functional profiles of each stroke patient. Second, an understanding of memory processes in conjunction with these assessments will help clinicians apply the appropriate forms of rehabilitation. Rehabilitative techniques can be tailored to the individual's symptoms, targeting the impaired processes but also making use of intact functions. Encouraging the use of semantic (meaning-based) encoding and retrieval may help improve memory function in people with episodic memory deficits. It may even be possible to address memory loss associated with specific episodic deficits. For individuals with hippocampal damage, interventions that employ reminding cues (e.g., beepers, personal digital assistants, diaries) may reduce the need to recall freely. Teaching compensatory techniques such as the use of visual imagery may also be helpful for this population. In contrast, visual imagery interventions probably would not be helpful for a patient with memory loss resulting from a posterior parietal stroke. These patients may benefit more from interventions that focus on encoding and retrieving information with verbal cues and labels. For those with frontal-lobe damage, interventions that focus on organization and strategy would be the most helpful. At this time, clinical research into the effectiveness of post-stroke tailored rehabilitation techniques is in its infancy. The outcome of future work on this topic may help inform specific guidelines and programs for clinicians to use with their patients.

References

1. Rasquin SM, Verhey FR, Lousberg R, Winkens I, Lodder J. Vascular cognitive disorders: memory, mental speed and cognitive flexibility after stroke. J Neurol Sci. 2002;203–204:115–9. PubMed PMID: 12417368. Epub 2002/11/06. eng.
2. Rasquin SM, Lodder J, Ponds RW, Winkens I, Jolles J, Verhey FR. Cognitive functioning after stroke: a one-year follow-up study. Dement Geriatr Cogn Disord. 2004;18(2):138–44. PubMed PMID: 15211068. Epub 2004/06/24. eng.
3. Stewart FM, Sunderland A, Sluman SM. The nature and prevalence of memory disorder late after stroke. Br J Clin Psychol. 1996;35(Pt 3):369–79. PubMed PMID: 8889078. Epub 1996/09/01. eng.
4. Haacke C, Althaus A, Spottke A, Siebert U, Back T, Dodel R. Long-term outcome after stroke: evaluating health-related quality of life using utility measurements. Stroke. 2006;37(1):193–8. PubMed PMID: 16339458. Epub 2005/12/13. eng.
5. Kotila M, Waltimo O, Niemi ML, Laaksonen R, Lempinen M. The profile of recovery from stroke and factors influencing outcome. Stroke. 1984;15(6):1039–44. PubMed PMID: 6506115. Epub 1984/11/01. eng.
6. Tatemichi TK, Foulkes MA, Mohr JP, Hewitt JR, Hier DB, Price TR, et al. Dementia in stroke survivors in the stroke data bank cohort. Prevalence, incidence, risk factors, and computed tomographic findings. Stroke. 1990;21(6):858–66. PubMed PMID: 2349588. Epub 1990/06/01. eng.
7. Knopman DS. Cerebrovascular disease and dementia. Br J Radiol. 2007;80:S121–7. PubMed PMID: 18445742. Epub 2008/05/01. eng.
8. Kopelman MD. Remote and autobiographical memory, temporal context memory and frontal atrophy in Korsakoff and Alzheimer patients. Neuropsychologia. 1989;27(4):437–60. PubMed PMID: 2733818. Epub 1989/01/01. eng.

9. Jefferies E, Lambon Ralph MA. Semantic impairment in stroke aphasia versus semantic dementia: a case-series comparison. Brain. 2006;129(Pt 8):2132–47. PubMed PMID: 16815878. Epub 2006/07/04. eng.
10. Chertkow H, Bub D, Deaudon C, Whitehead V. On the status of object concepts in aphasia. Brain Lang. 1997;58(2):203–32. PubMed PMID: 9182748. Epub 1997/06/15. eng.
11. Corbett F, Jefferies E, Ehsan S, Lambon Ralph MA. Different impairments of semantic cognition in semantic dementia and semantic aphasia: evidence from the non-verbal domain. Brain. 2009;132(Pt 9):2593–608. PubMed PMID: 19506072. Epub 2009/06/10. eng.
12. Squire LR. Memory systems of the brain: a brief history and current perspective. Neurobiol Learn Mem. 2004;82(3):171–7. PubMed PMID: 15464402. Epub 2004/10/07. eng.
13. Bondi MW, Kaszniak AW. Implicit and explicit memory in Alzheimer's disease and Parkinson's disease. J Clin Exp Neuropsychol. 1991;13(2):339–58. PubMed PMID: 1864919. Epub 1991/03/01. eng.
14. Roncacci S, Troisi E, Carlesimo GA, Nocentini U, Caltagirone C. Implicit memory in parkinsonian patients: evidence for deficient skill learning. Eur Neurol. 1996;36(3):154–9. PubMed PMID: 8738946. Epub 1996/01/01. eng.
15. Laforce Jr R, Doyon J. Distinct contribution of the striatum and cerebellum to motor learning. Brain Cogn. 2001;45(2):189–211. PubMed PMID: 11237366. Epub 2001/03/10. eng.
16. Nikonenko AG, Radenovic L, Andjus PR, Skibo GG. Structural features of ischemic damage in the hippocampus. Anat Rec. 2009;292(12):1914–21. PubMed PMID: 19943345. Epub 2009/11/28. eng.
17. Moscovitch M, Nadel L, Winocur G, Gilboa A, Rosenbaum RS. The cognitive neuroscience of remote episodic, semantic and spatial memory. Curr Opin Neurobiol. 2006;16(2):179–90. PubMed PMID: 16564688. Epub 2006/03/28. eng.
18. Eichenbaum H. A cortical-hippocampal system for declarative memory. Nat Rev Neurosci. 2000;1(1):41–50. PubMed PMID: 11252767. Epub 2001/03/17. eng.
19. Scoville WB, Milner B. Loss of recent memory after bilateral hippocampal lesions. J Neurol Neurosurg Psychiatry. 1957;20(1):11–21. PubMed PMID: 13406589. Epub 1957/02/01. eng.
20. Squire LR. The legacy of patient H.M. For neuroscience. Neuron. 2009;61(1):6–9. PubMed PMID: 19146808. Epub 2009/01/17. eng.
21. Vargha-Khadem F, Gadian DG, Watkins KE, Connelly A, Van Paesschen W, Mishkin M. Differential effects of early hippocampal pathology on episodic and semantic memory. Science. 1997;277(5324):376–80. PubMed PMID: 9219696. Epub 1997/07/18. eng.
22. Viskontas IV, Carr VA, Engel SA, Knowlton BJ. The neural correlates of recollection: hippocampal activation declines as episodic memory fades. Hippocampus. 2009;19(3):265–72. PubMed PMID: 18830998. Epub 2008/10/03. eng.
23. Yonelinas AP, Kroll NE, Quamme JR, Lazzara MM, Sauve MJ, Widaman KF, et al. Effects of extensive temporal lobe damage or mild hypoxia on recollection and familiarity. Nat Neurosci. 2002;5(11):1236–41. PubMed PMID: 12379865. Epub 2002/10/16. eng.
24. Spiers HJ, Maguire EA, Burgess N. Hippocampal amnesia. Neurocase. 2001;7(5):357–82. PubMed PMID: 11744778. Epub 2001/12/18. eng.
25. Rosenbaum RS, Kohler S, Schacter DL, Moscovitch M, Westmacott R, Black SE, et al. The case of K.C.: contributions of a memory-impaired person to memory theory. Neuropsychologia. 2005;43(7):989–1021. PubMed PMID: 15769487. Epub 2005/03/17. eng.
26. Nadel L, Moscovitch M. Memory consolidation, retrograde amnesia and the hippocampal complex. Curr Opin Neurobiol. 1997;7(2):217–27. PubMed PMID: 9142752. Epub 1997/04/01. eng.
27. Ott BR, Saver JL. Unilateral amnesic stroke. Six new cases and a review of the literature. Stroke. 1993;24(7):1033–42. PubMed PMID: 8322379. Epub 1993/07/01. eng.
28. Zola-Morgan S, Squire LR, Amaral DG. Human amnesia and the medial temporal region: enduring memory impairment following a bilateral lesion limited to field CA1 of the hippocampus. J Neurosci. 1986;6(10):2950–67. PubMed PMID: 3760943. Epub 1986/10/01. eng.
29. Szabo K, Forster A, Jager T, Kern R, Griebe M, Hennerici MG, et al. Hippocampal lesion patterns in acute posterior cerebral artery stroke: clinical and MRI findings. Stroke. 2009;40(6):2042–5. PubMed PMID: 19359650. Epub 2009/04/11. eng.

30. Mayes AR, Holdstock JS, Isaac CL, Montaldi D, Grigor J, Gummer A, et al. Associative recognition in a patient with selective hippocampal lesions and relatively normal item recognition. Hippocampus. 2004;14(6):763–84. PubMed PMID: 15318334. Epub 2004/08/20. eng.
31. Kitchener EG, Hodges JR, McCarthy R. Acquisition of post-morbid vocabulary and semantic facts in the absence of episodic memory. Brain. 1998;121(Pt 7):1313–27. PubMed PMID: 9679783. Epub 1998/07/29. eng.
32. Batchelor S, Thompson EO, Miller LA. Retrograde memory after unilateral stroke. Cortex. 2008;44(2):170–8. PubMed PMID: 18387546. Epub 2008/04/05. eng.
33. Aggleton JP, Brown MW. Episodic memory, amnesia, and the hippocampal-anterior thalamic axis. Behav Brain Sci. 1999;22(3):425–44; discussion 444–89. PubMed PMID: 11301518. Epub 2001/04/17. eng.
34. Squire LR, Wixted JT, Clark RE. Recognition memory and the medial temporal lobe: a new perspective. Nat Rev Neurosci. 2007;8(11):872–83. PubMed PMID: 17948032. Epub 2007/10/20. eng.
35. Schouten EA, Schiemanck SK, Brand N, Post MW. Long-term deficits in episodic memory after ischemic stroke: evaluation and prediction of verbal and visual memory performance based on lesion characteristics. J Stroke Cerebrovasc Dis. 2009;18(2):128–38. PubMed PMID: 19251189. Epub 2009/03/03. eng.
36. Aradillas E, Libon DJ, Schwartzman RJ. Acute loss of spatial navigational skills in a case of a right posterior hippocampus stroke. J Neurol Sci. 2011;308(1–2):144–6. PubMed PMID: 21762930. Epub 2011/07/19. eng.
37. Aggleton JP, Saunders RC. The relationships between temporal lobe and diencephalic structures implicated in anterograde amnesia. Memory. 1997;5(1–2):49–71. PubMed PMID: 9156091. Epub 1997/01/01. eng.
38. Graff-Radford NR, Tranel D, Van Hoesen GW, Brandt JP. Diencephalic amnesia. Brain. 1990;113(Pt 1):1–25. PubMed PMID: 2302527. Epub 1990/02/01. eng.
39. Carlesimo GA, Lombardi MG, Caltagirone C. Vascular thalamic amnesia: a reappraisal. Neuropsychologia. 2011;49(5):777–89. PubMed PMID: 21255590. Epub 2011/01/25. eng.
40. Schmahmann JD. Vascular syndromes of the thalamus. Stroke. 2003;34(9):2264–78. PubMed PMID: 12933968. Epub 2003/08/23. eng.
41. Clarke S, Assal G, Bogousslavsky J, Regli F, Townsend DW, Leenders KL, et al. Pure amnesia after unilateral left polar thalamic infarct: topographic and sequential neuropsychological and metabolic (PET) correlations. J Neurol Neurosurg Psychiatry. 1994;57(1):27–34. PubMed PMID: 8301301. Epub 1994/01/01. eng.
42. Schott JM, Crutch SJ, Fox NC, Warrington EK. Development of selective verbal memory impairment secondary to a left thalamic infarct: a longitudinal case study. J Neurol Neurosurg Psychiatry. 2003;74(2):255–7. PubMed PMID: 12531963. Epub 2003/01/18. eng.
43. Kishiyama MM, Yonelinas AP, Kroll NE, Lazzara MM, Nolan EC, Jones EG, et al. Bilateral thalamic lesions affect recollection- and familiarity-based recognition memory judgments. Cortex. 2005;41(6):778–88. PubMed PMID: 16353367. Epub 2005/12/16. eng.
44. Van der Werf YD, Scheltens P, Lindeboom J, Witter MP, Uylings HB, Jolles J. Deficits of memory, executive functioning and attention following infarction in the thalamus; a study of 22 cases with localised lesions. Neuropsychologia. 2003;41(10):1330–44. PubMed PMID: 12757906. Epub 2003/05/22. eng.
45. Van der Werf YD, Jolles J, Witter MP, Uylings HB. Contributions of thalamic nuclei to declarative memory functioning. Cortex. 2003;39(4–5):1047–62. PubMed PMID: 14584566. Epub 2003/10/31. eng.
46. Carlesimo GA, Serra L, Fadda L, Cherubini A, Bozzali M, Caltagirone C. Bilateral damage to the mammillo-thalamic tract impairs recollection but not familiarity in the recognition process: a single case investigation. Neuropsychologia. 2007;45(11):2467–79. PubMed PMID: 17512561. Epub 2007/05/22. eng.
47. Aggleton JP, O'Mara SM, Vann SD, Wright NF, Tsanov M, Erichsen JT. Hippocampal-anterior thalamic pathways for memory: uncovering a network of direct and indirect actions. Eur J Neurosci. 2010;31(12):2292–307. PubMed PMID: 20550571. Epub 2010/06/17. eng.
48. Klein JC, Rushworth MF, Behrens TE, Mackay CE, de Crespigny AJ, D'Arceuil H, et al. Topography of connections between human prefrontal cortex and mediodorsal thalamus stud-

ied with diffusion tractography. Neuroimage. 2010;51(2):555–64. PubMed PMID: 20206702. Epub 2010/03/09. eng.
49. Guberman A, Stuss D. The syndrome of bilateral paramedian thalamic infarction. Neurology. 1983;33(5):540–6. PubMed PMID: 6682494. Epub 1983/05/01. eng.
50. Ghika-Schmid F, Bogousslavsky J. The acute behavioral syndrome of anterior thalamic infarction: a prospective study of 12 cases. Ann Neurol. 2000;48(2):220–7. PubMed PMID: 10939573. Epub 2000/08/12. eng.
51. Cipolotti L, Husain M, Crinion J, Bird CM, Khan SS, Losseff N, et al. The role of the thalamus in amnesia: a tractography, high-resolution MRI and neuropsychological study. Neuropsychologia. 2008;46(11):2745–58. PubMed PMID: 18597798. Epub 2008/07/04. eng.
52. Graff-Radford NR, Damasio H, Yamada T, Eslinger PJ, Damasio AR. Nonhaemorrhagic thalamic infarction. Clinical, neuropsychological and electrophysiological findings in four anatomical groups defined by computerized tomography. Brain. 1985;108(Pt 2):485–516. PubMed PMID: 4005533. Epub 1985/06/01. eng.
53. Parkin AJ, Rees JE, Hunkin NM, Rose PE. Impairment of memory following discrete thalamic infarction. Neuropsychologia. 1994;32(1):39–51. PubMed PMID: 8818153. Epub 1994/01/01. eng.
54. Bogousslavsky J, Regli F, Assal G. The syndrome of unilateral tuberothalamic artery territory infarction. Stroke. 1986;17(3):434–41. PubMed PMID: 2424153. Epub 1986/05/01. eng.
55. Cole M, Winkelman MD, Morris JC, Simon JE, Boyd TA. Thalamic amnesia: Korsakoff syndrome due to left thalamic infarction. J Neurol Sci. 1992;110(1–2):62–7. PubMed PMID: 1506870. Epub 1992/07/01. eng.
56. Baldo JV, Delis D, Kramer J, Shimamura AP. Memory performance on the California Verbal Learning Test-II: findings from patients with focal frontal lesions. J Int Neuropsychol Soc. 2002;8(4):539–46. PubMed PMID: 12030307. Epub 2002/05/28. eng.
57. Shimamura AP. Memory and the prefrontal cortex. Ann N Y Acad Sci. 1995;769:151–9. PubMed PMID: 8595022. Epub 1995/12/15. eng.
58. Verfaellie M, Rapcsak SZ, Keane MM, Alexander MP. Elevated false recognition in patients with frontal lobe damage is neither a general nor a unitary phenomenon. Neuropsychology. 2004;18(1):94–103. PubMed PMID: 14744192. Epub 2004/01/28. eng.
59. Shimamura AP, Janowsky JS, Squire LR. Memory for the temporal order of events in patients with frontal lobe lesions and amnesic patients. Neuropsychologia. 1990;28(8):803–13. PubMed PMID: 2247207. Epub 1990/01/01. eng.
60. Kopelman MD, Stanhope N, Kingsley D. Temporal and spatial context memory in patients with focal frontal, temporal lobe, and diencephalic lesions. Neuropsychologia. 1997;35(12):1533–45. PubMed PMID: 9460723. Epub 1998/02/14. eng.
61. Schnider A, Ptak R. Spontaneous confabulators fail to suppress currently irrelevant memory traces. Nat Neurosci. 1999;2(7):677–81. PubMed PMID: 10404203. Epub 1999/07/15. eng.
62. Hashimoto R, Tanaka Y, Nakano I. Amnesic confabulatory syndrome after focal basal forebrain damage. Neurology. 2000;54(4):978–80. PubMed PMID: 10690998. Epub 2000/02/26. eng.
63. Fischer RS, Alexander MP, D'Esposito M, Otto R. Neuropsychological and neuroanatomical correlates of confabulation. J Clin Exp Neuropsychol. 1995;17(1):20–8. PubMed PMID: 7608298. Epub 1995/02/01. eng.
64. Gilboa A, Alain C, Stuss DT, Melo B, Miller S, Moscovitch M. Mechanisms of spontaneous confabulations: a strategic retrieval account. Brain. 2006;129(Pt 6):1399–414. PubMed PMID: 16638795. Epub 2006/04/28. eng.
65. Wheeler MA, Stuss DT, Tulving E. Frontal lobe damage produces episodic memory impairment. J Int Neuropsychol Soc. 1995;1(6):525–36. PubMed PMID: 9375239. Epub 1995/11/01. eng.
66. Wheeler MA, Stuss DT. Remembering and knowing in patients with frontal lobe injuries. Cortex. 2003;39(4–5):827–46 PMID: 14584555. Epub 2003/10/31. eng.

67. Alexander MP, Stuss DT, Fansabedian N. California Verbal Learning Test: performance by patients with focal frontal and non-frontal lesions. Brain. 2003;126(Pt 6):1493–503. PubMed PMID: 12764068. Epub 2003/05/24. eng.
68. Curran T, Schacter DL, Norman KA, Galluccio L. False recognition after a right frontal lobe infarction: memory for general and specific information. Neuropsychologia. 1997;35(7):1035–49. PubMed PMID: 9226663. Epub 1997/07/01. eng.
69. Rapcsak SZ, Nielsen L, Littrell LD, Glisky EL, Kaszniak AW, Laguna JF. Face memory impairments in patients with frontal lobe damage. Neurology. 2001;57(7):1168–75. PubMed PMID: 11591831. Epub 2001/10/10. eng.
70. Logothetis NK, Sheinberg DL. Visual object recognition. Annu Rev Neurosci. 1996;19:577–621. PubMed PMID: 8833455. Epub 1996/01/01. eng.
71. Karnath HO, Ruter J, Mandler A, Himmelbach M. The anatomy of object recognition–visual form agnosia caused by medial occipitotemporal stroke. J Neurosci. 2009;29(18):5854–62. PubMed PMID: 19420252. Epub 2009/05/08. eng.
72. Rossion B, Caldara R, Seghier M, Schuller AM, Lazeyras F, Mayer E. A network of occipito-temporal face-sensitive areas besides the right middle fusiform gyrus is necessary for normal face processing. Brain. 2003;126(Pt 11):2381–95. PubMed PMID: 12876150. Epub 2003/07/24. eng.
73. Barsalou LW. Simulation, situated conceptualization, and prediction. Philos Trans R Soc Lond B Biol Sci. 2009;364(1521):1281–9. PubMed PMID: 19528009. Epub 2009/06/17. eng.
74. Kable JW, Lease-Spellmeyer J, Chatterjee A. Neural substrates of action event knowledge. J Cogn Neurosci. 2002;14(5):795–805. PubMed PMID: 12167263. Epub 2002/08/09. eng.
75. Greenberg DL, Rubin DC. The neuropsychology of autobiographical memory. Cortex. 2003;39(4–5):687–728. PubMed PMID: 14584549. Epub 2003/10/31. eng.
76. Moscovitch C, Kapur S, Kohler S, Houle S. Distinct neural correlates of visual long-term memory for spatial location and object identity: a positron emission tomography study in humans. Proc Natl Acad Sci U S A. 1995;92(9):3721–5. PubMed PMID: 7731972. Epub 1995/04/25. eng.
77. Ogden JA. Visual object agnosia, prosopagnosia, achromatopsia, loss of visual imagery, and autobiographical amnesia following recovery from cortical blindness: case M.H. Neuropsychologia. 1993;31(6):571–89. PubMed PMID: 8341415. Epub 1993/06/01. eng.
78. Roman GC, Tatemichi TK, Erkinjuntti T, Cummings JL, Masdeu JC, Garcia JH, et al. Vascular dementia: diagnostic criteria for research studies. Report of the NINDS-AIREN International Workshop. Neurology. 1993;43(2):250–60. PubMed PMID: 8094895. Epub 1993/02/01. eng.
79. van Asselen M, Kessels RP, Frijns CJ, Kappelle LJ, Neggers SF, Postma A. Object-location memory: a lesion-behavior mapping study in stroke patients. Brain Cogn. 2009;71(3):287–94. PubMed PMID: 19733953. Epub 2009/09/08. eng.
80. Saito K, Kimura K, Minematsu K, Shiraishi A, Nakajima M. Transient global amnesia associated with an acute infarction in the retrosplenium of the corpus callosum. J Neurol Sci. 2003;210(1–2):95–7. PubMed PMID: 12736096. Epub 2003/05/09. eng.
81. Kessels RP, de Haan EH, Kappelle LJ, Postma A. Selective impairments in spatial memory after ischaemic stroke. J Clin Exp Neuropsychol. 2002;24(1):115–29. PubMed PMID: 11935430. Epub 2002/04/06. eng.
82. Malhotra P, Coulthard EJ, Husain M. Role of right posterior parietal cortex in maintaining attention to spatial locations over time. Brain. 2009;132(Pt 3):645–60. PubMed PMID: 19158107. Pubmed Central PMCID: 2664449. Epub 2009/01/23. eng.
83. Erkinjuntti T. Vascular cognitive deterioration and stroke. Cerebrovasc Dis. 2007;24 Suppl 1:189–94. PubMed PMID: 17971655. Epub 2007/11/29. eng.
84. Gorelick PB. Risk factors for vascular dementia and Alzheimer disease. Stroke. 2004;35 (11 Suppl 1):2620–2. PubMed PMID: 15375299. Epub 2004/09/18. eng.
85. Knopman DS. Cerebrovascular pathology in cognitive impairment: new (in)sights. Neurology. 2012. PubMed PMID: 22377812. Epub 2012/03/02. Eng.
86. Vermeer SE, Longstreth Jr WT, Koudstaal PJ. Silent brain infarcts: a systematic review. Lancet Neurol. 2007;6(7):611–9. PubMed PMID: 17582361. Epub 2007/06/22. eng.

87. Reitz C, Luchsinger JA, Mayeux R. Vascular disease and cognitive impairment. Expert Rev Neurother. 2008;8(8):1171–4. PubMed PMID: 18671658. Epub 2008/08/02. eng.
88. Black S, Gao F, Bilbao J. Understanding white matter disease: imaging-pathological correlations in vascular cognitive impairment. Stroke. 2009;40(3 Suppl):S48–52. PubMed PMID: 19064767. Epub 2008/12/10. eng.
89. Wen Y, Yu WH, Maloney B, Bailey J, Ma J, Marie I, et al. Transcriptional regulation of beta-secretase by p25/cdk5 leads to enhanced amyloidogenic processing. Neuron. 2008;57(5):680–90.
90. Esiri MM, Nagy Z, Smith MZ, Barnetson L, Smith AD. Cerebrovascular disease and threshold for dementia in the early stages of Alzheimer's disease. Lancet. 1999;354(9182):919–20. PubMed PMID: 10489957. Epub 1999/09/18. eng.
91. Snowdon DA, Greiner LH, Mortimer JA, Riley KP, Greiner PA, Markesbery WR. Brain infarction and the clinical expression of Alzheimer disease. The Nun Study. JAMA. 1997;277(10):813–7. PubMed PMID: 9052711. Epub 1997/03/12. eng.
92. Sachdev PS, Brodaty H, Looi JC. Vascular dementia: diagnosis, management and possible prevention. Med J Aust. 1999;170(2):81–5. PubMed PMID: 10026690. Epub 1999/02/23. eng.
93. Looi JC, Sachdev PS. Differentiation of vascular dementia from AD on neuropsychological tests. Neurology. 1999;53(4):670–8. PubMed PMID: 10489025. Epub 1999/09/17. eng.
94. Hayden KM, Zandi PP, Lyketsos CG, Khachaturian AS, Bastian LA, Charoonruk G, et al. Vascular risk factors for incident Alzheimer disease and vascular dementia: the Cache County study. Alzheimer Dis Assoc Disord. 2006;20(2):93–100. PubMed PMID: 16772744. Epub 2006/06/15. eng.
95. Leys D, Henon H, Mackowiak-Cordoliani MA, Pasquier F. Poststroke dementia. Lancet Neurol. 2005;4(11):752–9. PubMed PMID: 16239182. Epub 2005/10/22. eng.
96. Pasi M, Poggesi A, Salvadori E, Pantoni L. Post-stroke dementia and cognitive impairment. Front Neurol Neurosci. 2012;30:65–9. PubMed PMID: 22377866. Epub 2012/03/02. eng.
97. Cordoliani-Mackowiak MA, Henon H, Pruvo JP, Pasquier F, Leys D. Poststroke dementia: influence of hippocampal atrophy. Arch Neurol. 2003;60(4):585–90.
98. Pendlebury ST. Stroke-related dementia: rates, risk factors and implications for future research. Maturitas. 2009;64(3):165–71. PubMed PMID: 19818568. Epub 2009/10/13. eng.

Chapter 10
Behavior After Aneurysmal Subarachnoid Hemorrhage: Cognition and Functional Outcome

Timour Al-Khindi, R. Loch Macdonald, Stephan Mayer, and Tom A. Schweizer

Introduction

Spontaneous subarachnoid hemorrhage (SAH) is a medical emergency characterized by hemorrhage in the subarachnoid space surrounding the brain. In the vast majority of cases (85 %), spontaneous SAH is caused by the rupture of a cerebral aneurysm (aSAH) [1]. The other 15 % are idiopathic and two-thirds of these have a characteristic appearance on computed tomography (CT) and are called benign perimesencephalic SAH (pSAH). The characteristic sign of aSAH is a sudden onset of severe "thunderclap" headache, but patients may also present with vomiting, nausea, photophobia, and nuchal rigidity [2]. A cranial CT scan demonstrates accumulation of blood in the basal cisterns (see Fig. 10.1); patients may also present with hemorrhage into the brain and ventricular system [2]. Hypertension and cigarette smoking

T. Al-Khindi, B.Sc.
Johns Hopkins University School of Medicine, Baltimore, MD, USA

R.L. Macdonald, M.D., Ph.D.
Division of Neurosurgery, St. Michael's Hospital, Toronto, ON, Canada

Department of Surgery, University of Toronto, Toronto, ON, Canada

S. Mayer, M.D., F.C.C.M.
Neurological Intensive Care Unit, Columbia University Medical Center, New York, NY, USA

T.A. Schweizer, Ph.D. (✉)
Keenan Research Centre, Li Ka Shing Knowledge Institute, St. Michael's Hospital, 30 Bond Street, Toronto, ON, Canada M5B 1W8

Institute of Biomedical Engineering, University of Toronto, Toronto, ON, Canada

Heart and Stroke Foundation, Centre for Stroke Recovery, Toronto, ON, Canada

Division of Neurosurgery, Department of Surgery, University of Toronto, St. Michael's Hospital, Toronto, ON, Canada
e-mail: SchweizerT@smh.ca

T.A. Schweizer and R.L. Macdonald (eds.), *The Behavioral Consequences of Stroke*,
DOI 10.1007/978-1-4614-7672-6_10,

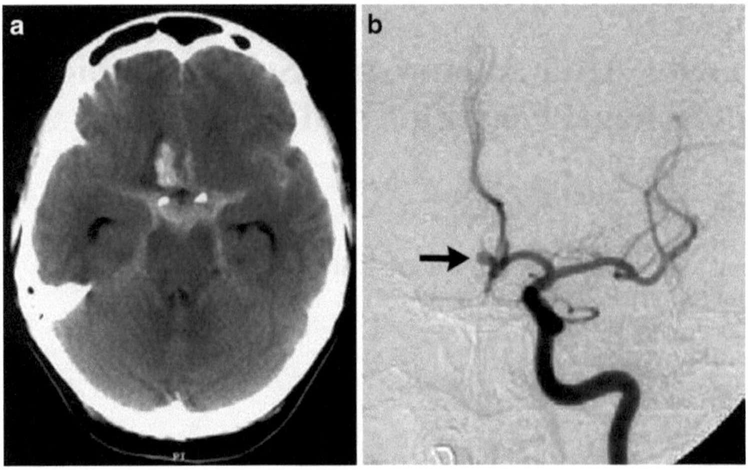

Fig. 10.1 (**a**) aSAH secondary to anterior communicating artery aneurysm rupture. Note the characteristic accumulation of blood in the basal cisterns. (**b**) CT angiography illustrates the anterior communicating artery aneurysm (*arrow*) in the patient in (**a**). Figure adapted with permission from Suarez JI, Tarr RW, Selman WR. Aneurysmal subarachnoid hemorrhage. N Engl J Med. 2006;354:387–396

are risk factors for aSAH [3]. Although uncommon—the incidence of aSAH in North America is approximately 8–11 per 100,000 persons per year [4, 5]—aSAH carries a poor prognosis, with only 35 % of patients surviving [6, 7]. Indeed, although aSAH accounts for only 7 % of all strokes [8], it is responsible for 27 % of all stroke-related years of life lost before age 65 [9]. The high mortality after aSAH may be partially attributed to misdiagnosis. Up to 50 % of cases are misdiagnosed as migraine or tension-type headache due to failure to obtain lumbar puncture or proper neuroimaging [2]. Despite these statistics, advances in the acute management of aSAH over the past 3 decades—namely increased use of vascular imaging and reduced delays to treatment [7]—have substantially reduced mortality after aSAH. In a meta-analysis, Lovelock et al. [7] observed that while the incidence of aSAH has remained stable over the past 30 years, mortality has been reduced by half and the 30-day case fatality rate has decreased by 0.9 % per annum (Fig. 10.2).

Although mortality after aSAH has declined over the past few decades, the extent to which patients' cognitive and functional outcomes have improved over the same time period remains unknown. A large proportion of aSAH patients—up to 76 % [10]—experience persistent cognitive deficits many years after aSAH. These cognitive deficits may lead to difficulties in aspects of functional independence such as return to work, managing finances, and housekeeping. Aneurysmal SAH-associated deficits in cognition and day-to-day functioning are particularly debilitating given that most patients are relatively young (mean age of 55 years [11]), are in their most productive years, and have major responsibilities with respect to work and family.

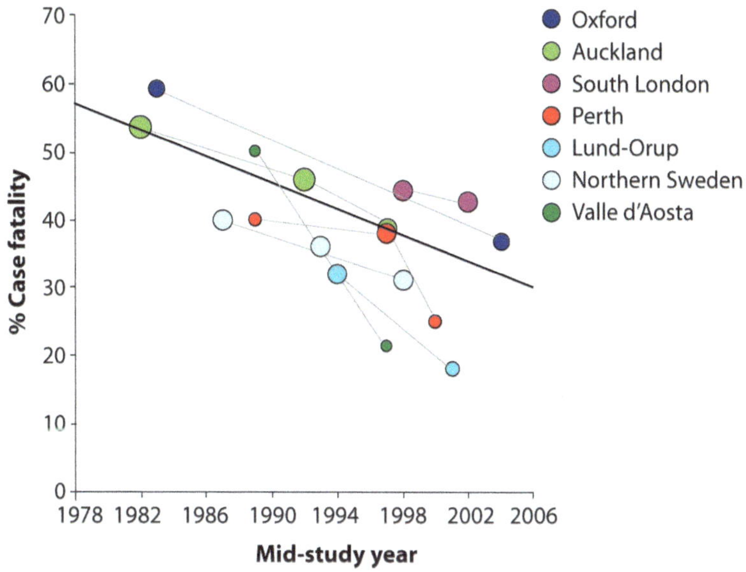

Fig. 10.2 Thirty-day case fatality rate over time. The 30-day case fatality rate after aSAH has decreased by 0.9 % per annum due to increased use of vascular imaging and reduced delays to treatment. Adapted with permission from Lovelock CE, Rinkel GJE, Rothwell PM. Time trends in outcome of subarachnoid hemorrhage. Population-based study and systematic review. Neurology. 2010;74:1494–1501

This chapter describes the cognitive and functional deficits after aSAH and examines how clinical and neuroimaging factors influence cognitive and functional outcome.

Cognitive Outcome

Cognitive impairment is most frequent and severe within the first 3 months following aSAH [12], but residual cognitive dysfunction may persist for a lifetime [13]. The following section reviews the cognitive domains most commonly affected after aSAH: attention, memory, executive function, language, and visuospatial function.

Attention

Attention refers to the ability to allocate cognitive resources to a specific task, thereby producing focused behavior. Attention has many components, including selective attention (i.e., paying attention to a specific task), sustained attention (i.e., paying attention for a prolonged time period), and divided attention (i.e., paying attention to multiple tasks simultaneously). Several studies have found that among all cognitive domains, attentional deficits are the most frequent self-reported complaints by both aSAH patients and their relatives, with estimates as high as 70 % [14–20]. As a result of these attentional problems, patients often report difficulties with maintaining conversations, reading, watching television, and other tasks that involve cognitive processing [21].

The reported frequency and severity of attentional deficits after aSAH has been variable, however. A study by Ravnik et al. [22] found no correlation between subjective attentional impairment and performance on tests of attention; in fact, patients performed better on tests of attention than on tests of other cognitive domains. Other studies report the prevalence of attentional impairment, as measured by cognitive tests, to be modest, ranging from 3 to 19 % [23–28]. In one study, even patients with severe aSAH (i.e., grade 5 Hunt and Hess [29]) showed mild attentional deficits post-aSAH [30]. Hütter et al. [27] found that 70 % of aSAH survivors reported attention problems, but only 13 % were impaired on neuropsychological tests of attention. Discordance between self-reported cognitive deficits and results from objective cognitive tests may be attributed to patients misinterpreting their cognitive deficits as impairments of attention. Additionally, the high self-reported prevalence of attentional dysfunction may result from confounding factors such as depression [17] or excessive fatigue, the latter of which affects up to 75 % of aSAH survivors [14, 25, 31, 32].

Other studies have found that attentional deficits are common and widespread after aSAH, affecting upwards of 76 % of aSAH survivors [26, 33]. The extent of attentional impairment may depend on the timing of testing, with impairment more likely sooner after the hemorrhage, and on the specific type of attention being tested. Benke et al. [13] observed that while divided attention showed impairment after aSAH, sustained and short-term attention did not.

Several clinical variables affect the degree of attentional impairment post-aSAH. Greater aSAH severity, as measured by Hunt and Hess grade [29], the presence of blood in the Sylvian fissure, greater thickness of subarachnoid blood, as measured by Fisher grade [34], older age, intraventricular hemorrhage, hydrocephalus, and temporary clipping have been associated with poorer attentional function [16, 35–38]. Whether attentional function improves over time is controversial; some investigators claim no improvement over the 12-month period after aSAH [12, 33] while others show improvement over the same time period [39, 40].

Several investigators have correlated attentional function with neuroimaging. Mustonen et al. [41] found that cerebral perfusion heterogeneity, as measured by single photon emission tomography (SPET), at 1-week post-aSAH correlated with performance on attention tests at 1-year post-aSAH. Bendel et al. [42] found that hippocampal volumes after aSAH predicted performance on tests of attention 1-year post-aSAH.

Memory

In the aSAH outcome literature, memory is commonly stratified into different components, including visual memory, verbal memory, short-term memory, and long-term memory. Up to 57 % of survivors self-report memory problems after aSAH [14, 15, 19, 20, 43, 44]. Inability to remember new information and maintain short-term (working) memory are particularly common complaints [17, 44]. Verbal memory is commonly impaired in aSAH patients, the prevalence ranging from 14 to 68 % [12, 18, 20, 31, 45–49]. Likewise, deficits in visual memory are also frequent, with impairment prevalence ranging from 14 to 75 % [18, 20, 31, 33, 45–49].

Unlike attention, in which patients tend to overestimate the degree of subjective attentional dysfunction relative to neuropsychological tests, self-reported memory complaints often underestimate the degree of memory dysfunction as detected by neuropsychological impairment. Ljunggren et al. [18] found that 83 % of aSAH survivors had memory deficits on neuropsychological assessment, but only 58 % self-reported memory deficits on interview. Older age, fewer years of education, poorer neurological grade on admission (as measured by World Federations of Neurological Surgeons Grade [50]), hydrocephalus, ischemia, postoperative vasospasm, ruptured aneurysms in the anterior circulation, and thick subarachnoid blood in the anterior interhemispheric fissure and Sylvian fissures have been associated with poorer performance on tests of verbal and visual memory [33, 35, 51, 52]. Temporary clipping and partial resection of the gyrus rectus, meanwhile, have been associated with poorer short-term memory [27]. Some investigators have demonstrated a relationship between memory deficits and cerebral edema [35], but others have observed no correlation [53].

The wide range in the prevalence of memory impairment can be explained by the use of different standardized tests. Commonly used memory tests include the California Verbal Learning Test [54], the Visual Reproduction subtest of the Wechsler Memory Scale [55], and the Rey-Osterrieth Complex Figure—Recall subtest [56, 57]. Differences in difficulty and sensitivity to impairment between different memory tests may contribute to the variable rates of memory impairment. For instance, Mayer et al. [45] found that 31 % of aSAH survivors showed impaired visual memory on the Rey-Osterrieth Complex Figure—Recall subtest, whereas only 22 % showed impaired visual memory on the Visual Reproduction subtest. These findings suggest that the frequency of cognitive impairment after aSAH is determined in part by the particular test used to measure cognitive performance.

Time of testing in relation to the initial insult is also critical in determining aSAH-associated memory deficits. Powell et al. [12, 31] assessed verbal memory at 3, 9, and 18 months post-aSAH. Delayed verbal memory significantly improved from 3 to 9 months and from 9 to 18 months post-ictus, while immediate verbal memory showed no significant improvement in the same time period. Despite improvement in delayed verbal memory over 18 months, 14 % of aSAH patients still had significant delayed verbal memory impairments at 18-month follow-up. Although not all studies differentiate between immediate and delayed verbal memory, the finding that verbal memory improves over time has been replicated by others [39, 53, 58]. Visual memory may also improve over time [33, 39, 53], but results have been inconsistent [58]. Why verbal and visual memory improve over time remains unknown, although reduction of aSAH-associated intracranial pressure [59] and chronic inflammation may play a role [60]. Together, these results show that the prevalence of aSAH-associated memory impairment depends on the type of memory in question and on the length of the follow-up interval.

Although aSAH survivors often report a reduced capacity to learn new information [17], data by D'Esposito et al. [61] suggest that memory impairments are also characterized by an inability to recall information learned prior to aSAH. This form of retrograde amnesia, however, was only observed among patients with medial frontal lobe damage, mainly in arterial territories distal to the anterior communicating artery (ACoA). Given the role of the medial frontal lobe in initiation [62, 63], the inability to recall old information may be due to an inability to initiate memory retrieval. Indeed, Richardson [64] found that, on tests of object naming, aSAH patients had slower retrieval speeds compared to control participants.

The medial temporal lobes, a network of structures that includes the hippocampus together with the entorhinal, perirhinal, and parahippocampal cortices, are critically involved in memory function [65]. A correlation between left hemisphere infarctions and verbal memory impairment in patients with aSAH has been demonstrated [35, 66], but little is known about the specific brain regions in the left hemisphere responsible for verbal memory impairment in aSAH patients. Bendel et al. [42] performed magnetic resonance imaging (MRI) on aSAH survivors and found significantly reduced bilateral hippocampal volumes among aSAH patients relative to healthy controls at 1-year follow-up. Hippocampal volumes correlated with performance on one test of visual memory (the Visual Reproduction subtest of the Wechsler Memory Scale), but not with another visual memory test (the Rey-Osterrieth Complex Figure) or with tests of verbal memory. The investigators did not correlate memory impairment with changes in other brain regions, such as the frontal lobes. This may be important given the role of the frontal lobes in memory function [67] and given that frontal lobe injury is a common sequelae of aSAH [68]. Reduced frontal lobe integrity could account for the unexpected findings reported by Bendel et al. [42]. Indeed, Vilkki et al. [69] found that medial frontal lobe lesions on CT predicted poorer verbal memory in aSAH survivors.

Executive Function

Executive function is predominantly mediated by the frontal lobes [70] and encompasses higher-level cognitive abilities like planning, inhibition, problem-solving, and decision-making. Rather than fractionating executive function into its components, the majority of studies treat executive function as a unitary construct.

Consequently, estimates of the frequency of executive dysfunction in aSAH survivors varies widely, ranging from 3 to 76 % [10, 12, 24, 31, 45–48, 71, 72]. Results from studies using neuropsychological assessments complement patients' self-reported complaints; upwards of 30 % of aSAH survivors self-report a lack of initiative and a reduced capacity for planning and organizing [17, 18]. Executive dysfunction after aSAH is more pronounced in older patients, those with fewer years of education, and those with poorer neurological grade on admission [35, 36]. Similar to memory, a correlation between executive dysfunction and cerebral edema has been documented by some investigators [36], but not by others [53]. Surgical variables also affect outcome; Akyuz et al. [73] found that temporary vessel occlusion exceeding 9 min was a significant predictor of permanent long-term executive deficits.

Kreiter et al. [35] found that executive dysfunction was more severe among patients with ruptured anterior aneurysms than among those with posterior aneurysms. Findings from most other studies, however, suggest no relationship between ruptured aneurysm location and the profile of cognitive impairment [46, 53, 66, 74–76]. Interestingly, Manning et al. [77] found that aSAH patients who had ruptured ACoA aneurysms performed significantly better than patients with aneurysms at other locations on the Tower of London task [78], a test of executive function. Similarly, Papagno et al. [79] observed that patients with ACoA aneurysms performed better than those with aneurysms at other locations on tests of switching and inhibition. The relationship between aneurysm location and profile of cognitive impairment thus remains unclear.

In addition to sometimes causing focal lesions in proximity to the ruptured aneurysm, aSAH often leads to diffuse, global damage to brain tissue in poor-grade patients [10, 80–82]. Global injury is mediated, in part, by elevated intracranial pressure [80], reduction of cerebral blood flow and brain oxygenation, blood–brain barrier breakdown, and global cerebral edema (together these factors contribute to early brain injury; see [83]). Findings from an MRI study by Bendel et al. [84] support the "diffuse damage" hypothesis. The authors found that at 1-year follow-up, aSAH patients showed significantly reduced total grey matter and white matter volumes relative to control subjects. Furthermore, the extent of grey matter volume loss was associated with poorer cognitive test performance, including poorer performance on tests of executive function.

Not all neuroimaging studies, however, support the "diffuse damage" hypothesis. Several MRI studies suggest that executive dysfunction results from focal lesions rather than diffuse damage. Bendel et al. [85] found that 91 % of aSAH patients with executive dysfunction had frontal lobe lesions. Similarly, Martinaud et al. [47] showed that executive dysfunction in aSAH patients was exclusively correlated with lesions to the left frontal lobe, including the left superior frontal gyrus, the left middle frontal gyrus, and the left anterior centrum semiovale. Deficits in behavioral aspects of executive function (e.g., reduction of daily activities, irritability, hyperactivity), meanwhile, were correlated with lesions to the left ventral striatum. An exception to these findings is a study by Vilkki et al. [69] who showed that patients with infarction in the medial frontal lobe had impaired memory, but intact executive function. The relationship between executive dysfunction and frontal lobe injury may thus be qualified by the specific location of the lesion.

Although executive function and memory are often considered distinct cognitive domains, several studies indicate that these domains may interact. As mentioned earlier, D'Esposito et al. [61] observed that patients with medial frontal lobe injury showed deficits in retrieving previously learned information on a memory test. How can injury to the frontal lobes—a region implicated in executive function—affect performance on a memory test? Posterior-inferior midline frontal lesions may damage the septum, which has connection to the hippocampus and plays a role in working memory. Haug et al. [86] suggest that medial frontal lobe injury, which commonly occurs in patients with ACoA aneurysms [47, 85], may lead to deficits in initiation. Indeed, the investigators found that, compared to patients with aneurysms in other locations, these patients tended to perform poorer on the initial portions of memory tests. They also tended to commit more false-positive errors on memory tests, which may arise due to confabulation as opposed to genuine memory impairment. Deficits in executive function may thus "spill over" to affect performance on tests of other cognitive domains, thereby giving a distorted picture of a patient's cognitive functioning.

The use of different neuropsychological tests, each of which probes a different executive function, may partially explain the high variability in executive dysfunction after aSAH. Proust et al. [48] reported that among aSAH survivors treated with endovascular coiling, 29 % showed impairment on the Wisconsin Card Sorting Task (a measure of cognitive flexibility [87]), while only 7 % showed impairment on the Stroop task (a measure of inhibition [88]). Different aspects of executive function may thus have different rates of impairment. In support of this hypothesis, deficits in cognitive flexibility, as measured by the Wisconsin Card Sorting Task, are found in 8–44 % of aSAH patients [24, 45–48, 71], whereas deficits in inhibition, as measured by the Stroop task, are found in 17–56 % of aSAH patients [46–48].

A study by Manning et al. [77] illustrates the importance of differentiating between different aspects of "frontal lobe" function. Manning et al. [77] evaluated executive function 26 weeks post-aSAH in patients who had made a "good recovery" according to the Glasgow Outcome Scale (GOS) [89]. Aneurysmal SAH survivors performed significantly worse than healthy controls on the Wisconsin Card Sorting, Tower of London, and Stroop tasks, while performance on the Cognitive Estimates [90] task was within normal limits. Each of the tests used by Manning et al. [77] captures a slightly different executive "frontal lobe" function: the Tower of London task assesses planning and problem-solving, the Stroop task assesses ability to inhibit impulsive responding, while the Cognitive Estimates task measures ability to make judgments and estimations. These results indicate that some executive functions are impaired in aSAH survivors (e.g., cognitive flexibility, planning, problem-solving, inhibition), while others are intact (e.g., judgment, estimation). Furthermore, the data suggest that patients classified as having made a good recovery on the GOS may still experience profound cognitive deficits more than 6 months after aSAH. This is consistent with earlier studies showing that up to one-third of aSAH survivors who have made a good recovery on the GOS experience persistent neuropsychological impairment [91].

Different aspects of executive function may have different rates of recovery. Haug et al. [53] found that while inhibition improved within 1 year after aSAH, cognitive flexibility and attention did not. Results from longitudinal studies, however, must be interpreted with caution, as tests of executive function may be subject to practice effects (e.g., Stroop test [92], Trail Making Test [92], Wisconsin Card Sorting Test [93]). Studies that found improvement in response inhibition and cognitive flexibility over time [39, 58] used the Stroop test and Trail Making Test, both of which may have practice effects [92]. These findings further illustrate the importance of distinguishing between different executive functions.

Language

Language function involves comprehension and expression of meaningful written and oral information. The frequency of language impairment among aSAH survivors is highly variable, ranging from 0 to 76 % [10, 12, 24, 31, 45–48, 71]. Few studies have examined patients' self-reported language complaints; Passier et al. [17] observed that only 8 % of aSAH survivors self-reported deficits in speaking or writing. Interestingly, language function appears to be critical for predicting patients' perceptions of recovery. Chahal et al. [94] found that 52 % of patients felt they had made a complete recovery from the effects of aSAH 5 years post-ictus. Patients who reported making a full recovery, however, tended to perform better on tests of language and verbal functioning than patients who reported not making a full recovery, suggesting that linguistic function is key for patients' perception of recovery. Older age, fewer years of education, aneurysms in the anterior circulation, and left hemisphere infarction are significant predictors of poorer language function after aSAH [35, 69].

Language function improves significantly within the first 3 months after aSAH [58] and continues to improve in the 18 months after hemorrhage [12, 31, 58]. Longitudinal studies have typically used the verbal and semantic fluency task, which is not subject to practice effects [92, 93]. Mavaddat et al. [10] followed aSAH patients with a favorable neurological outcome (i.e., GOS categories of moderate disability and good recovery) between 6 and 24 months after aSAH. Despite patients having a favorable neurological outcome, 76 % of aSAH survivors were significantly impaired on the verbal and semantic fluency task, indicating that overall patient status, as measured by the GOS, may not be an accurate indicator of language dysfunction after aSAH.

Visuospatial Function

Visuospatial function refers to one's capacity for perception and spatial reasoning. According to several investigators [35, 45], visuospatial function is the least frequently impaired domain after aSAH, with an impairment prevalence of approximately 25 %. Furthermore, any visuospatial impairments that patients do exhibit tend to be mild and close to normative limits [40, 53]. Anterior aneurysms, the presence of blood in the Sylvian fissure, edema, older age, and right hemisphere lesions are significant predictors of poorer visuospatial function after aSAH [35, 69, 95]. Unlike other cognitive functions, deficits in visuospatial function persist over the 1-year period after aSAH [40, 58]. Although less is known about the integrity of visuospatial function after aSAH compared to other cognitive domains, several studies indicate that one test of visuospatial function—the Rey-Osterrieth Complex Figure Test—is the most sensitive test of long-term cognitive functioning in aSAH survivors (i.e., up to 5 years post-ictus) [40, 43].

Functional Outcome

Neuropsychological assessments attempt to characterize performance in individual cognitive domains. Day-to-day behavior, however, involves an interplay between many different cognitive domains rather than single cognitive domains working in isolation. Functional outcome encompasses activities of daily living (ADLs), instrumental activities of daily living (IADLs), and the ability to return to work. Each of these domains will be addressed in turn.

Activities of Daily Living

ADLs are those one performs for self-care [96]. Examples of ADLs include feeding, grooming, dressing, bathing, personal hygiene, and toileting. Aneurysmal SAH-associated ADL deficits are less common compared to cognitive deficits, with impairment ranging from 4 to 12 % [25, 72, 97, 98]. Severity of aSAH, as measured by Hunt and Hess grade [29], and hydrocephalus have been shown to predict ADL impairment after aSAH [99]. Performance on ADLs is key for patients' perceptions of recovery; one recent study showed that ADL performance best differentiated aSAH survivors who reported making a complete recovery from those who reported disability [94]. Common ADL impairments at hospital discharge include incontinence and maintaining personal hygiene [100], but whether these deficits extend to longer follow-up periods remains unclear. Berry et al. [51] showed that the degree of ADL dysfunction correlated with the extent of memory impairment. Similarly, Mayer et al. [45] found deficits of visual memory, visuospatial function, and psychomotor functioning to be significantly associated with impairment in ADL at 3 months post-aSAH, thereby highlighting the link between cognition and functional outcome.

One concern with measuring ADLs after aSAH is that of ceiling effects. In studies that measure ADLs using the Barthel Index [101], over 95 % of assessed patients achieve the highest possible score [102]. Furthermore, many ADL measures (e.g., the Barthel Index, the Katz Index of Independence in ADLs [103]) rely heavily on patient self-report, the accuracy of which is uncertain. For instance, patients may be uncomfortable or embarrassed at disclosing difficulties with toileting or bathing. Thus, the specific profile of ADL impairments after aSAH remains unclear.

Instrumental Activities of Daily Living

IADLs are more complex than ADLs and include tasks like managing finances, shopping, and housekeeping [104]. IADLs are more frequently impaired than ADLs after aSAH, with an estimated prevalence of 44–93 % [12, 31, 105, 106]. Of all IADLs, driving appears to be the one in which patients show the most impairment, with up to 60 % of aSAH survivors requiring assistance [107]. Housekeeping difficulties are also common, as only 51 % of aSAH survivors are capable of managing their households without any limitations [15]. Deficits of visual memory, visuospatial function, language function, and psychomotor function are associated with IADL impairment [45, 94]. Similar to ADLs, the instruments used to measure IADLs may not be sensitive to the cognitive difficulties aSAH survivors may experience. While driving on a busy city street, for instance, aSAH survivors might experience transient lapses of concentration or feelings of being overwhelmed. Despite these subtle difficulties, however, patients would receive the maximum favorable score on the Mode of Transportation subscale of the Lawton IADL scale [104], as their performance fulfills the criteria of traveling independently. Furthermore, commonly used measures of IADL performance, like the Lawton IADL scale, often employ a binary scoring system that does not allow for different gradients of IADL performance. IADL measures also rely heavily on self-report, which may not accurately reflect patients' true performance capabilities. Patients may be reluctant to disclose difficulties with driving, for instance, because they fear having their driver's license suspended. Additional research is needed to characterize the specific IADL profile in aSAH survivors.

Return to Work

One of the most important components of day-to-day functioning is the ability to return to one's previous occupation. This is especially important in the case of aSAH, as many aSAH survivors are young and have financial responsibilities and families to support. Among patients who were employed prior to aSAH, up to 40 % are unable to return to their previous occupation [14, 66, 80, 97, 108–111]. Other studies have found less favorable results, demonstrating that only 6–17 % of patients return to their previous occupation [32, 106, 112]. Even among those who do return to work, however, up to 59 % report reduced working capacity [15]. Return to work is an important issue even among patients with mild cognitive deficits; Haug et al. [30] found that only 21 % of previously employed aSAH patients with mild cognitive dysfunction were able to resume full-time employment. Patients frequently return to jobs with less responsibility, and they must often work fewer hours [14, 19, 32, 110, 111]. The impact of fatigue is profound, for up to 34 % of patients require daytime naps in order to function normally [20]. Research by Cedzich and Roth [15] indicates that, among patients who return to work, most do so between 6 and 12 months post-aSAH. Interestingly, patients' employment status prior to aSAH has been shown to predict the extent of cognitive impairment 1-year post-ictus. Haug et al. [30] observed that the fraction of patients working full-time prior to hemorrhage was less among patients with severe cognitive impairment than among those with mild cognitive impairment 1 year after aSAH. Although only correlational, these results suggest that one's resources before aSAH influence the course of post-stroke cognitive recovery.

Return to work and working ability are also influenced by aneurysm location and focal brain lesions. Unlike cognition, in which aneurysm location is less predictive of the profile of cognitive impairment [46, 53, 66, 74–76], return to work is greater among patients with middle cerebral artery aneurysms than among those with ACoA aneurysms [86]. Additionally, Vilkki et al. [66] found that left hemisphere lesions from aSAH were associated with failure to return to work and significantly reduced working ability. Self-reported planning and reasoning impairments and poorer performance on tests of executive function were also associated with failure to return to work. Unfortunately, although the investigators observed a correlation between left hemisphere lesions and failure to return to work, they did not examine language function in their sample, a cognitive domain predominantly mediated by the left hemisphere [113].

Additional Factors Affecting Outcome

Clipping vs. Coiling

There has been debate over the effects of microsurgical clipping and endovascular coiling of ruptured aneurysms on cognitive and functional outcome (Fig. 10.3). A randomized clinical trial, the International Subarachnoid Aneurysm Trial (ISAT), suggested that clinical outcome, as measured by the modified Rankin scale [114], was better after coiling than clipping [115].

With respect to cognitive function, most studies indicate that verbal memory, visual memory, visuospatial function, attention, executive function, information processing speed, psychomotor function, language, and return to work are similar in patients treated by clipping or coiling [19, 39, 48, 58, 80, 116–118]. Several studies, however, have demonstrated that clipping is associated with poorer executive function, visual memory, and verbal memory compared to coiling [48, 80, 119, 120].

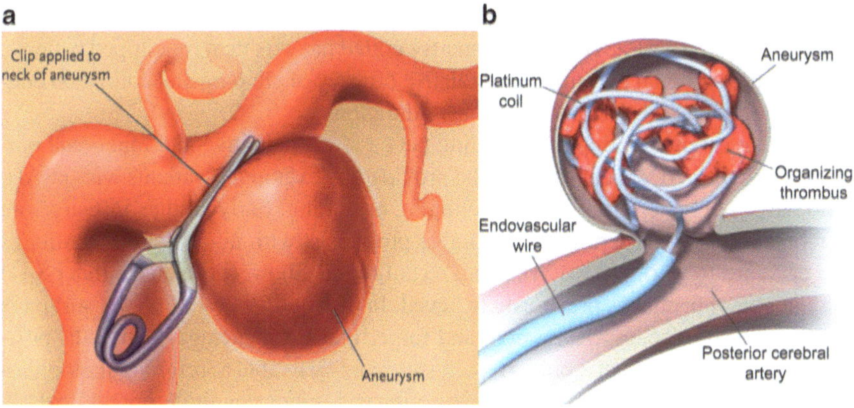

Fig. 10.3 (**a**) Surgical clipping of an aneurysm. Figure adapted from the University of Illinois College of Medicine at Chicago. Figure adapted with permission from Brisman JL, Song JK, Newell DW. Cerebral aneurysms. N Engl J Med. 2006;355(9):928–939. (**b**) Endovascular coiling of an aneurysm. Figure adapted from Suarez JI, Tarr RW, Selman WR. Aneurysmal subarachnoid hemorrhage. N Engl J Med. 2006;354:387–396

Bendel et al. [85], for instance, observed executive dysfunction in 73 % of clipped patients but in only 27 % of coiled patients. Why clipping is associated with greater cognitive impairment in these studies is unclear. Patients were not randomly assigned to clipping or coiling treatment, so differences in baseline risk for cognitive impairment cannot be excluded. One possibility is that clipped patients may have more brain injury. Hadjivassiliou et al. [120] found that 48 % of clipped patients suffered from encephalomalacia, compared to 0 % of coiled patients. Moreover, 87 % of clipped patients had infarcts compared to 57 % of coiled patients. Similarly, Bendel et al. [121] observed more frequent lesions after clipping than after coiling, particularly in the frontal lobes (70 % and 51 % had lesions after clipping and coiling, respectively) and the temporal lobes (48 % and 22 % had lesions after clipping and coiling, respectively).

Not all studies, however, have found an association between clipping and greater cognitive impairment. Santiago-Ramajo et al. [117, 122] observed poorer verbal and visual memory among coiled patients compared to clipped patients 4 months after aSAH. Likewise, Frazer et al. [39] found that coiled patients performed significantly poorer on tests of visuospatial function, executive function, and information processing speed 6 months after aSAH compared to clipped patients. Studies that observed poorer outcome among coiled patients examined cognitive performance within 6 months of SAH. Of the four studies that observed poorer outcome among clipped patients [48, 80, 119, 120], three examined cognitive function at least 1 year after SAH. Clipping and coiling may thus have different effects on cognitive outcome depending on time since treatment, with clipped patients having superior cognitive outcome in the short term and coiled patients having superior cognitive outcome in the long term. This hypothesis is consistent with data from

Santiago-Ramajo et al. [122], who found that the rate of cognitive recovery between 4 and 12 months post-aSAH was greater for coiled patients than for clipped patients. Future studies should investigate the relative efficacy of clipping and coiling on cognitive and functional outcome over longer durations.

To what extent do clipping or coiling per se affect cognitive and functional outcome? Insight into this question can be provided by examining patients with pSAH, who have SAH but do not undergo clipping or coiling. Early studies investigating outcome after pSAH reported a favorable prognosis, with no change in quality of life or return to work [123, 124]. More recent reports, however, indicate that pSAH may not be as benign as previously believed. In one study, 62 % of pSAH survivors experienced headaches, depression, and forgetfulness on average 23 months post-pSAH [125]. Additionally, only 41 % of survivors were able to return to their previous occupations. Madureira et al. [126] observed minor cognitive impairments in 72 % of pSAH survivors and depression in 33 % at 39 months post-pSAH. These findings suggest that poor cognitive and functional outcome cannot be explained by surgical or endovascular procedures alone. Indeed, data by Hillis and colleagues [75] indicate that patients with aSAH who have undergone aneurysm clipping have poorer verbal and visual memory than patients who have undergone clipping of unruptured aneurysms, suggesting that extravasation of blood per se has detrimental effects on cognition. Surgical and endovascular procedures may not be innocuous, however; Hütter et al. [38] showed that, compared to SAH patients who were treated conservatively, patients who were treated surgically tended to exhibit poorer concentration. Patients in this study, however, were only examined in the acute phase (i.e., 6 days post-ictus) and it is unclear whether this difference would have persisted in the long term.

Neuroimaging

Many studies have found neuroanatomical changes accompanying SAH. Kivisaari et al. [68] found that CT, though useful as a diagnostic tool, was less sensitive than MRI in detecting lesions after aSAH. CT detected lesions in 57 % of patients, whereas MRI detected lesions in 81 % of patients. Most of this difference in sensitivity can be attributed to detecting frontal lobe lesions; CT only revealed frontal lobe injury in 16 % of cases, whereas MRI revealed frontal lobe injury in 48 % of cases. Other studies using SPECT have observed reduced cerebral perfusion after aSAH, particularly in subcortical regions and, in the case of ACoA aneurysm patients, anterior cortical areas [127]. Egge et al. [74] found that cerebral perfusion as measured by SPECT correlated with the degree of cognitive deficit post-aSAH, but the investigators only used composite scores of cognitive impairment and did not stratify by cognitive domain.

MRI studies have largely been limited to gross measures of atrophy such as ratios of grey matter volume to intracranial volume. Consistent with the hypothesis that aSAH causes diffuse brain injury, Bendel et al. [84, 85] observed gross brain atrophy and ventriculomegaly. The degree of gross brain atrophy predicted impairment in several domains, including memory, executive function, and language. Among a cohort of patients with anterior cerebral artery aneurysms, the degree of atrophy was greatest in frontal regions, namely the orbitofrontal cortex, the inferior frontal gyrus, the gyrus rectus, and thalamic and hypothalamic regions [85]. Increased risk of frontal lobe damage in patients with aneurysms of the anterior circulation is consistent with data by Rabinstein et al. [128], who showed that infarct location predicted aneurysm location in 77 % of cases. Future studies should extend these findings by examining how lesions to specific brain areas contribute to cognitive dysfunction.

Summary

Studies show that a high percentage of patients with aSAH have long-standing or permanent deficits in attention, memory, executive function, language, visuospatial function, IADLs, and returning to work. The variability in reported rates is likely due to multiple factors such as varying patient populations tested, different times of testing after aSAH, and different tests and definitions of abnormal performance on the tests. Subtle cognitive and real-world deficits that accompany aSAH often go undetected by gross outcome measures like the GOS. As a result, patients who are classified as having made a "good recovery" continue to experience deficits particularly in memory, executive function, and language many years after aSAH. These patients frequently cannot go back to work and often experience reduced life satisfaction. By raising clinical awareness about residual impairments in cognition and day-to-day functioning, it is our hope that clinicians and researchers will move toward developing comprehensive rehabilitation strategies that target these subtle yet debilitating symptoms.

References

1. van Gijn J, Rinkel GJE. Subarachnoid hemorrhage: diagnosis, causes and management. Brain. 2001;124:249–78.
2. Suarez JI, Tarr RW, Selman WR. Aneurysmal subarachnoid hemorrhage. N Engl J Med. 2006;354:387–96.
3. Feigin VL, Rinkel GJE, Lawes CMM, Algra A, Bennett DA, van Gijn J, et al. Risk factors for subarachnoid hemorrhage: an updated systematic review of epidemiological studies. Stroke. 2005;36:2773–80.
4. King Jr JT. Epidemiology of aneurysmal subarachnoid hemorrhage. Neuroimaging Clin N Am. 1997;7:659–68.
5. Østbye T, Levy AR, Mayo NE. Hospitalization and case-fatality rates for subarachnoid hemorrhage in Canada from 1982 through 1991. Stroke. 1997;28:793–8.
6. Feigin VL, Lawes CM, Bennett DA, Barker-Collo SL, Parag V. Worldwide stroke incidence and early case fatality reported in 56 population-based studies: a systematic review. Lancet Neurol. 2009;8:355–69.

7. Lovelock CE, Rinkel GJE, Rothwell PM. Time trends in outcome of subarachnoid hemorrhage. Population-based study and systematic review. Neurology. 2010;74:1494–501.
8. Feigin VL, Lawes CMM, Bennett DA, Anderson CS. Stroke epidemiology: a review of population-based studies of incidence, prevalence, and case-fatality in the late 20th century. Lancet Neurol. 2003;2:43–53.
9. Johnston SC, Selvin S, Gress DR. The burden, trends, mortality from and demographics of subarachnoid hemorrhage. Neurology. 1998;50:1413–8.
10. Mavaddat N, Sahakian BJ, Hutchinson PJA, Kirkpatrick PJ. Cognition following subarachnoid hemorrhage from anterior communicating artery aneurysm: relation to timing of surgery. J Neurosurg. 1999;91:402–7.
11. Rinkel GJ, Algra A. Long-term outcomes of patients with aneurysmal subarachnoid haemorrhage. Lancet Neurol. 2011;10:349–56.
12. Powell J, Kitchen N, Heslin J, Greenwood R. Psychosocial outcomes at three and nine months after good neurological recovery from aneurysmal subarachnoid haemorrhage: predictors and prognosis. J Neurol Neurosurg Psychiatry. 2002;72:772–81.
13. Benke T, Koylu B, Delazer M, Trinka E, Kemmler G. Cholinergic treatment of amnesia following basal forebrain lesion due to aneurysm rupture—an open-label pilot study. Eur J Neurol. 2005;12:791–6.
14. Wermer MJH, Kool H, Albrecht KW, Rinkel GJE. Subarachnoid hemorrhage treated with clipping: long-term effects on employment, relationships, personality and mood. Neurosurgery. 2007;60:91–8.
15. Cedzich C, Roth A. Neurological and psychosocial outcome after subarachnoid haemorrhage, and the hunt and hess scale as a predictor of clinical outcome. Zentralbl Neurochir. 2005;66:112–8.
16. Hütter BO, Gilsbach JM. Which neuropsychological deficits are hidden behind a good outcome (Glasgow = I) after aneurysmal subarachnoid hemorrhage? Neurosurgery. 1993;33:999–1005.
17. Passier PECA, Visser-Meily JMA, van Zandvoort MJE, Post MWM, Rinkel GJE, van Heugten C. Prevalence and determinants of cognitive complaints after aneurysmal subarachnoid hemorrhage. Cerebrovasc Dis. 2010;29:557–63.
18. Ljunggren B, Sonesson B, Säveland H, Brandt L. Cognitive impairment and adjustment in patients without neurological deficits after aneurysmal SAH and early operation. J Neurosurg. 1985;62:673–9.
19. Schuiling WJ, Rinkel GJE, Walchenbach R, de Weerd AW. Disorders of sleep and wake in patients after subarachnoid hemorrhage. Stroke. 2005;36:578–82.
20. Sonesson B, Ljunggren B, Säveland H, Brandt L. Cognition and adjustment after late and early operation for ruptured aneurysm. Neurosurgery. 1987;21:279–87.
21. McKenna P, Willison JR, Lowe D, Neil-Dwyer G. Recovery after subarachnoid haemorrhage. Br Med J. 1989;299:485–7.
22. Ravnik J, Starovasnik B, Šešok S, Pirtosek Z, Svigelj V, Bunc G, et al. Long-term cognitive deficits in patients with good outcomes after aneurysmal subarachnoid hemorrhage from anterior communicating artery. Croat Med J. 2006;47:253–63.
23. Cheng H, Shi J. Neuropsychological evaluations and cognitive deficits in patients with aneurysmal SAH. Chin J Clin Psychol. 2005;13:86–7.
24. Cheng H, Shi J, Zhou M. Cognitive assessment in Chinese patients with aneurysmal subarachnoid hemorrhage. Behav Neurol. 2006;17:117–20.
25. Noble AJ, Baisch S, Mendelow AD, Allen L, Kane P, Schenk T. Posttraumatic stress disorder explains reduced quality of life in subarachnoid hemorrhage patients in both short and long term. Neurosurgery. 2008;63:1095–105.
26. Hütter BO, Gilsbach JM. Introspective capacities in patients with cognitive deficits after subarachnoid hemorrhage. J Clin Exp Neuropsychol. 1995;17:499–517.
27. Hütter BO, Gilsbach JM, Kreitschmann I. Quality of life and cognitive deficits after subarachnoid haemorrhage. Br J Neurosurg. 1995;9:465–75.

28. Thompson JN, Sheldrick R, Berry E. Cognitive and mental health difficulties following subarachnoid hemorrhage. Neuropsychol Rehabil. 2011;21:92–102.
29. Hunt WE, Hess RM. Surgical risk as related to time of intervention in the repair of intracranial aneurysms. J Neurosurg. 1968;28:14–20.
30. Haug T, Sorteberg A, Finset A, Lindegaard KF, Lundar T, Sorteberg W. Cognitive functioning and health-related quality of life 1 year after aneurysmal subarachnoid hemorrhage in preoperative comatose patients (Hunt and Hess Grade V patients). Neurosurgery. 2010;66:474–84.
31. Powell J, Kitchen N, Heslin J, Greenwood R. Psychosocial outcomes at 18 months after good neurological recovery from aneurysmal subarachnoid haemorrhage. J Neurol Neurosurg Psychiatry. 2004;75:1119–24.
32. Buchanan KM, Elias LJ, Goplen GB. Differing perspectives on outcome after subarachnoid haemorrhage: the patient, the relative, the neurosurgeon. Neurosurgery. 2000;46:831–8.
33. Ogden JA, Mee EW, Henning M. A prospective study of impairment of cognition and memory and recovery after subarachnoid hemorrhage. Neurosurgery. 1993;33:572–86.
34. Fisher C, Kistler J, Davis J. Relation of cerebral vasospasm to subarachnoid hemorrhage visualized by computerized tomographic scanning. Neurosurgery. 1980;6:1–9.
35. Kreiter KT, Copeland D, Bernardini GL, Bates JE, Peery S, Claassen J, et al. Predictors of cognitive dysfunction after subarachnoid hemorrhage. Stroke. 2002;33:200–9.
36. Egge A, Waterloo K, Sjøholm H, Solberg T, Ingebrigtsen T, Romner B. Prophylactic hyperdynamic postoperative fluid therapy after aneurysmal subarachnoid hemorrhage: a clinical, prospective, randomized, controlled study. Neurosurgery. 2001;49:593–606.
37. Hütter BO, Gilsbach JM. Early neuropsychological sequelae of aneurysm surgery and subarachnoid hemorrhage. Acta Neurochir (Wien). 1996;138:1370–8.
38. Hütter BO, Kreitschmann-Andermahr I, Gilsbach JM. Cognitive deficits in the acute stage after subarachnoid hemorrhage. Neurosurgery. 1998;43:1054–65.
39. Frazer D, Ahuja A, Watkins L, Cipolotti L. Coiling versus clipping for the treatment of aneurysmal subarachnoid hemorrhage: a longitudinal investigation into cognitive outcome. Neurosurgery. 2007;60:434–42.
40. Penninx JF, Visser-Meily JMA, Passier PECA, Rinkel GJE, Post MW, Van Zandvoort MJE. The prognostic value of neuropsychological examination after SAH. Behav Neurol. 2010;23:173–5.
41. Mustonen T, Koivisto T, Vanninen R, Hänninen T, Vapalahti M, Hernesniemi J, et al. Heterogeneity of cerebral perfusion 1 week after haemorrhage is an independent predictor of clinical outcome in patients with aneurysmal subarachnoid haemorrhage. J Neurol Neurosurg Psychiatry. 2008;79:1128–33.
42. Bendel P, Koivisto T, Hänninen T, Kolehmainen A, Könönen M, Hurskainen H, et al. Subarachnoid hemorrhage is followed by temporomesial volume loss. Neurology. 2006;67:575–82.
43. Ogden JA, Utley T, Mee EW. Neurological and psychosocial outcome 4 to 7 years after subarachnoid hemorrhage. Neurosurgery. 1997;41:25–34.
44. Barth M, Thomé C, Schmiedek P, Weiss C, Kasuya H, Vajcokzy P. Characterization of functional outcome and quality of life following subarachnoid hemorrhage in patients treated with and without nicardipine prolonged-release implants. J Neurosurg. 2009;110:955–60.
45. Mayer SA, Kreiter KT, Copeland D, Bernardini GL, Bates JE, Peery S, et al. Global and domain-specific cognitive impairment and outcome after subarachnoid hemorrhage. Neurology. 2002;59:1750–8.
46. Ørbo M, Waterloo K, Egge A, Isaksen J, Ingebrigtsen T, Romner B. Predictors for cognitive impairment one year after surgery for aneurysmal subarachnoid hemorrhage. J Neurol. 2008;255:1770–6.
47. Martinaud O, Perin B, Gérardin E, Proust F, Bioux S, Gars DL, et al. Anatomy of executive deficit following ruptured anterior communicating artery aneurysm. Eur J Neurol. 2009;16:595–601.

48. Proust F, Martinaud O, Gérardin E, Derrey S, Levèque S, Bioux S, et al. Quality of life and brain damage after microsurgical clip occlusion or endovascular coil embolization for ruptured anterior communicating artery aneurysms: neuropsychological assessment. J Neurosurg. 2009;110:19–29.
49. Romner B, Sonesson B, Ljunggren B, Brandt L, Säveland H, Holtås S. Late magnetic resonance imaging related to neurobehavioral functioning after aneurysmal subarachnoid hemorrhage. Neurosurgery. 1989;25:390–7.
50. Drake CG. Report of World Federation of Neurological Surgeons Committee on a universal subarachnoid hemorrhage grading scale. J Neurosurg. 1988;68:985–6.
51. Berry E, Jones RA, West CG, Brown JD. Outcome of subarachnoid haemorrhage. An analysis of surgical variables, cognitive and emotional sequelae related to SPECT scanning. Br J Neurosurg. 1997;11:378–87.
52. Richardson JTE. Performance in free recall following rupture and repair of intracranial aneurysm. Brain Cogn. 1989;9:210–26.
53. Haug T, Sorteberg A, Sorteberg W, Lindegaard K, Lundar T, Finset A. Cognitive outcome after subarachnoid hemorrhage: time course of recovery and relationship to clinical, radiological and management parameters. Neurosurgery. 2007;60:649–57.
54. Elwood RE. The California Verbal Learning Test: psychometric characteristics and clinical application. Neuropsychol Rev. 1995;5:173–7.
55. Wechsler D. Wechsler memory scale-revised. San Antonio, TX: The Psychological Corporation; 1987.
56. Osterrieth PA. Le test de copie d'une figure complexe. Arch Psychol. 1944;30:206–356.
57. Rey A. L'examen psychologique dans les cas d'encéphalopathie traumatique. Arch Psychol. 1941;28:286–340.
58. Koivisto T, Vanninen R, Hurskainen H, Saari T, Hernesniemi J, Vapalahti M. Outcomes of early endovascular versus surgical treatment of ruptured cerebral aneurysms: a prospective randomized study. Stroke. 2000;31:2369–77.
59. Graetz D, Nagel A, Schlenk F, Sakowitz O, Vajkoczy P, Sarrafzadeh A. High ICP as trigger of proinflammatory IL-6 cytokine activation in aneurysmal subarachnoid hemorrhage. Neurol Res. 2010;32:728–35.
60. Chaichana KL, Pradilla G, Huang J, Tamargo RJ. Role of inflammation (leukocyte-endothelial cell interactions) in vasospasm after subarachnoid hemorrhage. World Neurosurg. 2010;73:22–41.
61. D'Esposito M, Alexander MP, Fischer R, McGlinchey-Berroth R, O'Connor M. Recovery of memory and executive function following anterior communicating artery aneurysm rupture. J Int Neuropsychol Soc. 1996;2:565–70.
62. Rushworth MF, Kennerley SW, Walton ME. Cognitive neuroscience: resolving conflict in and over the medial frontal cortex. Curr Biol. 2005;15:R54–6.
63. Stuss DT, Alexander MP, Hamer L, Palumbo C, Dempster R, Binns M, et al. The effects of focal anterior and posterior brain lesions on verbal fluency. J Int Neuropsychol Soc. 1998;4:265–78.
64. Richardson JTE. Cognitive performance following rupture and repair of intracranial aneurysm. Acta Neurol Scand. 1991;83:110–22.
65. Squire LR, Stark CEL, Clark RE. The medial temporal lobe. Annu Rev Neurosci. 2004;27:279–306.
66. Vilkki JS, Juvela S, Siironen J, Ilvonen T, Varis J, Porras M. Relationship of local infarctions to cognitive and psychosocial impairments after aneurysmal subarachnoid hemorrhage. Neurosurgery. 2004;55:790–803.
67. Fletcher PC, Henson RNA. Frontal lobes and human memory: insights from functional neuroimaging. Brain. 2001;124:849–81.
68. Kivisaari RP, Salonen O, Servo A, Autti T, Hernesniemi J, Öhman J. MR imaging after aneurysmal subarachnoid hemorrhage and surgery: a long-term follow-up study. AJNR Am J Neuroradiol. 2001;22:1143–8.

69. Vilkki J, Holst P, Öhman J, Servo A, Heiskanen O. Cognitive deficits related to computed tomographic findings after surgery for a ruptured intracranial aneurysm. Neurosurgery. 1989;25:166–72.
70. Stuss DT, Levine B. Adult clinical neuropsychology: lessons from studies of the frontal lobes. Annu Rev Psychol. 2002;53:401–33.
71. Anderson SW, Todd MM, Hindman BJ, Clarke WR, Torner JC, Tranel D, IHAST Investigators, et al. Effects of intraoperative hypothermia on neuropsychological outcomes after intracranial aneurysm surgery. Ann Neurol. 2006;60:518–27.
72. Noble AJ, Schenk T. Posttraumatic stress disorder in the family and friends of patients who have suffered spontaneous subarachnoid hemorrhage. J Neurosurg. 2008;109:1027–33.
73. Akyuz M, Erylmaz M, Ozdemir C, Goksu E, Ucar T, Tuncer R. Effect of temporary clipping on frontal lobe functions in patients with ruptured aneurysm of the anterior communicating artery. Acta Neurol Scand. 2005;112:293–7.
74. Egge A, Waterloo K, Sjøholm H, Ingebrigtsen T, Forsdahl S, Jacobsen EA, et al. Outcome 1 year after aneurysmal subarachnoid hemorrhage: relation between cognitive performance and neuroimaging. Acta Neurol Scand. 2005;112:76–80.
75. Hillis AE, Anderson N, Sampath P, Rigamonti D. Cognitive impairments after surgical repair of ruptured and unruptured aneurysms. J Neurol Neurosurg Psychiatry. 2000;69:608–15.
76. Samra SK, Giordani B, Caveney AF, Clarke WR, Scott PA, Anderson S, et al. Recovery of cognitive function after surgery for aneurysmal subarachnoid hemorrhage. Stroke. 2007;38:1864–72.
77. Manning L, Pierot L, Dufour A. Anterior and non-anterior ruptured aneurysms: memory and frontal lobe function performance following coiling. Eur J Neurol. 2005;12:466–74.
78. Shallice T. Specific impairments of planning. Philos Trans R Soc Lond B Biol Sci. 1982;298:199–209.
79. Papagno C, Rizzo S, Ligori L, Lima J, Riggio A. Memory and executive functions in aneurysms of the anterior communicating artery. J Clin Exp Neuropsychol. 2003;25:24–35.
80. Bellebaum C, Schäfers L, Schoch B, Wanke I, Stolke D, Forsting M, et al. Clipping versus coiling: neuropsychological follow up after aneurysmal subarachnoid haemorrhage (SAH). J Clin Exp Neuropsychol. 2004;26:1081–92.
81. Salmond CH, DeVito EE, Clark L, Menon DK, Chatfield DA, Pickard JD, et al. Impulsivity, reward sensitivity, and decision-making in subarachnoid hemorrhage survivors. J Int Neuropsychol Soc. 2006;12:697–706.
82. Uski TK, Lilja Å, Säveland H, Ekman R, Sonesson B, Brandt L. Cognitive functioning and cerebrospinal fluid concentrations of neuropeptides for patients with good neurological outcomes after aneurysmal subarachnoid hemorrhage. Neurosurgery. 2000;47:812–8.
83. Cahill J, Zhang JH. Subarachnoid hemorrhage: is it time for a new direction? Stroke. 2009;40:S86–7.
84. Bendel P, Koivisto T, Aikiä M, Niskanen E, Könönen M, Hänninen T, et al. Atrophic enlargement of CSF volume after subarachnoid hemorrhage: correlation with neuropsychological outcome. AJNR Am J Neuroradiol. 2010;31:370–6.
85. Bendel P, Koivisto T, Niskanen E, Könönen M, Aikiä M, Hänninen T, et al. Brain atrophy and neuropsychological outcome after treatment of ruptured anterior cerebral artery aneurysms: a voxel-based morphometric study. Neuroradiology. 2009;51:711–22.
86. Haug T, Sorteberg A, Sorteberg W, Lindegaard K, Lundar T, Finset A. Cognitive functioning and health related quality of life after rupture of an aneurysm on the anterior communicating artery versus middle cerebral artery. Br J Neurosurg. 2009;23:507–15.
87. Nelson HE. A modified card sorting test sensitive to frontal lobe defects. Cortex. 1976;12:313–24.
88. Stroop JR. Studies of interference in serial verbal reactions. J Exp Psychol. 1935;18:643–62.
89. Jennett B, Bond M. Assessment of outcome after severe brain damage. Lancet. 1975;1:480–4.

90. Shallice T, Evans ME. The involvement of the frontal lobes in cognitive estimation. Cortex. 1978;14:294–303.
91. Desantis A, Laiacona M, Barbarotto R, Basso A, Villani R, Spagnoli D, et al. Neuropsychological outcome of patients operated upon for an intracranial aneurysm: analysis of general prognostic factors and of the effects of the location of the aneurysm. J Neurol Neurosurg Psychiatry. 1989;52:1135–40.
92. Dikmen SS, Heaton RK, Grant I, Temkin NR. Test-retest reliability and practice effects of expanded Halstead-Reitan Neuropsychological Test Battery. J Int Neuropsychol Soc. 1999;5:346–56.
93. Basso MR, Bornstein RA, Lang JM. Practice effects on commonly used measures of executive function across twelve months. Clin Neuropsychol. 1999;13:283–92.
94. Chahal N, Barker-Collo S, Feigin V. Cognitive and functional outcomes of 5-year subarachnoid haemorrhage survivors: comparison to matched healthy controls. Neuroepidemiology. 2011;37:31–8.
95. Satzger W, Niedermeier N, Schönberger J, Engel RR, Beck OJ. Timing of operation for ruptured cerebral aneurysm and long-term recovery of cognitive functions. Acta Neurochir (Wien). 1995;136:168–74.
96. Christiansen C, Baum C. Occupational therapy. Enabling function and well-being. Thorofare, NJ: Slack; 1997.
97. Hackett ML, Anderson CS. Health outcomes 1 year after subarachnoid hemorrhage: an international population-based study. Neurology. 2000;55:658–62.
98. Pasternak JJ, McGregor DG, Schroeder DR, Lanier WL, Shi Q, Hindman BJ, et al. Hyperglycemia in patients undergoing cerebral aneurysm surgery: its association with long-term gross neurologic and neuropsychological function. Mayo Clin Proc. 2008;83:406–17.
99. Dombovy ML, Drew-Cates J, Serdans R. Recovery and rehabilitation following subarachnoid haemorrhage. Part I: outcome after inpatient rehabilitation. Brain Inj. 1998;12:443–54.
100. Ślusarz R, Beuth W, Książkiewicz B. Postsurgical examination of functional outcome of patients having undergone surgical treatment of intracranial aneurysm. Scand J Caring Sci. 2009;23:130–9.
101. Mahoney F, Barthel D. Functional evaluation: the Barthel Index. Md State Med J. 1965;14:61–5.
102. Kim DH, Haney CL, van Ginhoven G. Utility of outcome measures after treatment for intracranial aneurysms: a prospective trial involving 520 patients. Stroke. 2005;36:792–6.
103. Katz S, Down TD, Cash HR, Grotz RC. Progress in development of the index of ADL. Gerontologist. 1970;10:20–30.
104. Lawton MP, Brody EM. Assessment of older people: self-maintaining and instrumental activities of daily living. Gerontologist. 1969;9:179–86.
105. Fernandez A, Schmidt JM, Claassen J, Pavlicova M, Huddleston D, Kreiter KT, et al. Fever after subarachnoid hemorrhage: risk factors and impact on outcome. Neurology. 2007;68:1013–9.
106. Kirkness CJ, Thompson JM, Ricker BA, Buzaitis A, Newell DW, Dikmen S, et al. The impact of aneurysmal subarachnoid hemorrhage on functional outcome. J Neurosci Nurs. 2002;34:134–41.
107. Dombovy ML, Drew-Cates J, Serdans R. Recovery and rehabilitation following subarachnoid haemorrhage. Part II: long-term follow-up. Brain Inj. 1998;12:887–94.
108. Carter BS, Buckley D, Ferraro R, Rordorf G, Ogilvy CS. Factors associated with reintegration to normal living after subarachnoid hemorrhage. Neurosurgery. 2000;46:1326–33.
109. Haug T, Sorteberg A, Sorteberg W, Lindegaard KF, Lundar T, Finset A. Surgical repair of unruptured and ruptured middle cerebral artery aneurysms: impact on cognitive functioning and health-related quality of life. Neurosurgery. 2009;64:412–20.
110. Morris PG, Wilson JTL, Dunn L. Anxiety and depression after spontaneous subarachnoid hemorrhage. Neurosurgery. 2004;54:47–54.
111. Tomberg T, Orasson A, Linnamägi Ü, Toomela A, Pulver A, Asser T. Coping strategies in patients following subarachnoid haemorrhage. Acta Neurol Scand. 2001;104:148–55.

112. Hop JW, Rinkel GJE, Algra A, van Gijn J. Changes in functional outcome and quality of life in patients and caregivers after aneurysmal subarachnoid hemorrhage. J Neurosurg. 2001;95:957–63.
113. Witelson SF, Pallie W. Left hemisphere specialization for language in the newborn: neuroanatomical evidence of asymmetry. Brain. 1973;96:641–6.
114. van Swieten JC, Koudstaal PJ, Visser MC, Schouten HJA, van Gijn J. Interobserver agreement for the assessment of handicap in stroke patients. Stroke. 1988;19:604–7.
115. Molyneux A, Kerr R, Stratton I, Sandercock P, Clarke M, Shrimpton J, International Subarachnoid Aneurysm Trial (ISAT) Collaborative Group, et al. International Subarachnoid Aneurysm Trial (ISAT) of neurosurgical clipping versus endovascular coiling in 2143 patients with ruptured intracranial aneurysms: a randomised trial. Lancet. 2002;360:1267–74.
116. Mukerji N, Holliman D, Baisch S, Noble A, Schenk T, Nath F. Neuropsychologic impact of treatment modalities in subarachnoid hemorrhage: clipping is no different from coiling. World Neurosurg. 2010;74:129–38.
117. Santiago-Ramajo S, Katati MJ, Pérez-García M, Coín-Mejias MA, Vilar-Lopez R, Caracuel-Romero A, et al. Neuropsychological evaluation of the treatments applied to intracranial aneurysms in a Spanish sample. J Clin Exp Neuropsychol. 2007;29:634–41.
118. Wong GKC, Wong R, Mok VCT, Fan DS, Leung G, Wong A, et al. Clinical study on cognitive dysfunction after spontaneous subarachnoid haemorrhage: patient profiles and relationship to cholinergic dysfunction. Acta Neurochir. 2009;151:1601–7.
119. Fontanella M, Perozzo P, Ursone R, Garbossa D, Bergui M. Neuropsychological assessment after microsurgical clipping or endovascular treatment for anterior communicating artery aneurysm. Acta Neurochir. 2003;145:867–72.
120. Hadjivassiliou M, Tooth CL, Romanowski CAJ, Byrne J, Battersby RD, Oxbury S, et al. Aneurysmal SAH: cognitive outcome and structural damage after clipping or coiling. Neurology. 2001;56:1672–7.
121. Bendel P, Koivisto T, Könönen M, Hänninen T, Hurskainen H, Saari T, et al. MR imaging of the brain 1 year after aneurysmal subarachnoid hemorrhage: randomize study comparing surgical with endovascular treatment. Radiology. 2008;246:543–52.
122. Santiago-Ramajo S, Katati MJ, Pérez-García M, Arjona-Moron V. Evaluating the recovery of cognitive impairment in subarachnoid hemorrhage taking into consideration the practice effects. Neurosurgery. 2010;67:1497–504.
123. Brilstra EH, Hop JW, Rinkel GJE. Quality of life after perimesencephalic haemorrhage. J Neurol Neurosurg Psychiatry. 1997;63:382–4.
124. Rinkel GJ, Wijdicks EF, Vermeulen M, Hageman LM, Tans JT, van Gijn J. Outcome in perimesencephalic (nonaneurysmal) subarachnoid hemorrhage: a follow-up study in 37 patients. Neurology. 1990;40:1130–2.
125. Marquardt G, Niebauer T, Schick U, Lorenz R. Long term follow up after perimesencephalic subarachnoid haemorrhage. J Neurol Neurosurg Psychiatry. 2000;69:127–30.
126. Madureira S, Canhão P, Guerreiro M, Ferro JM. Cognitive and emotional consequences of perimesencephalic subarachnoid hemorrhage. J Neurol. 2000;415:862–7.
127. Tooth CL, Tindale WB, Hadjivassiliou M, Romanowski CA, Hunt E, Pantke R, et al. Subcortical hypoperfusion following surgery for aneurysmal subarachnoid haemorrhage: implications for cognitive performance? Behav Neurol. 2000;12:39–51.
128. Rabinstein AA, Friedman JA, Weigand SD, McClelland RL, Fulgham JR, Manno EM, et al. Predictors of cerebral infarction in aneurysmal subarachnoid hemorrhage. Stroke. 2004;35:1862–6.

Chapter 11
Cognitive Dysfunction After Intracerebral Hemorrhage, Vasculitis, and Other Stroke Syndromes

Eric E. Smith and José Andrés Venegas-Torres

Introduction

Cognitive impairment following stroke is predominantly related to the region of destroyed brain tissue, the capacity for accommodating the loss of function that results, and the degree of recovery. However, some of the variance in cognitive function after stroke is related to the mechanism of stroke. Most strokes result from brain ischemia, but up to 15 % may be due to subarachnoid (see Chap. 10) or intracerebral hemorrhage (ICH). Among the strokes resulting from brain ischemia, most are caused by hypertensive arterial disease, atherosclerotic disease, or cardiac disease. However, a minority may be caused by a wide range of less common arterial or hematologic diseases, including vasculitis, hypercoagulable states, and others. In some cases, these diseases may have their own effects on cognitive function independent of the presence of stroke, with their own tempo and characteristics. Here, we review the associations between cognitive function and ICH, and cognitive function and less common causes of ischemic stroke.

E.E. Smith, M.D., M.P.H. (✉)
Department of Clinical Neurosciences, Foothills Medical Centre,
Room C1212, 1403 29 Street NW, Calgary, AB, Canada T2N 2T9
e-mail: Eric.Smith@albertahealthservices.ca

J.A. Venegas-Torres, M.D.
Department of Clinical Neurosciences, Calgary Stroke Program,
University of Calgary, Foothills Medical Centre, Calgary, AB, Canada

T.A. Schweizer and R.L. Macdonald (eds.), *The Behavioral Consequences of Stroke*,
DOI 10.1007/978-1-4614-7672-6_11, © Springer Science+Business Media New York 2014

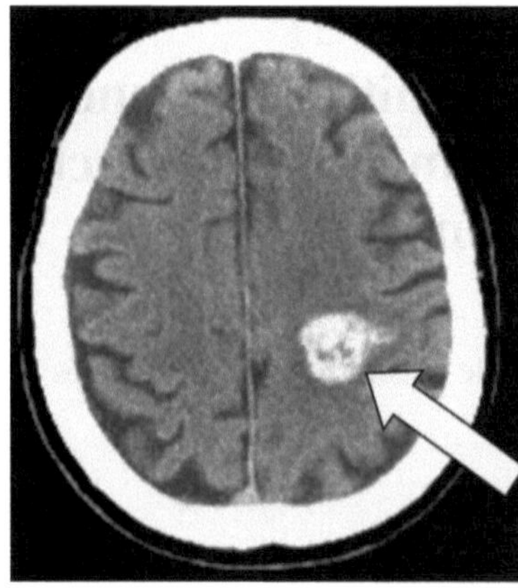

Fig. 11.1 Acute intracerebral hemorrhage (ICH), appearing as a hyperdense (*bright*) lesion on computed tomography (CT) scan

Intracerebral Hemorrhage

Etiology and Epidemiology

ICH results from rupture of blood vessels in the brain (Fig. 11.1) [1]. In some cases the blood may extend into the ventricles or, less commonly, the subarachnoid space. The annual incidence is 10–30 per 100,000 population, accounting for two million (10–15 %) of about 15 million strokes that occur worldwide each year. Although the age-specific incidence of ICH is stable or decreasing, probably due to better population control of hypertension, the absolute number of ICHs is expected to rise in the next decades because of aging of the population. Compared to ischemic stroke, the case fatality rate of ICH is much higher, with only 38 % of affected patients surviving the first year. Many survivors are left with significant residual disability.

Depending on the underlying cause of bleeding, ICH is classified as either primary or secondary [1]. Primary ICH, accounting for 78–88 % of cases, originates from the spontaneous rupture of small vessels damaged by chronic hypertension or cerebral amyloid angiopathy (CAA). Secondary ICH occurs in a minority of patients in association with vascular abnormalities (such as arteriovenous malformations and aneurysms), tumors, or impaired coagulation. However, ICH associated with anticoagulant medication is typically considered a form of primary ICH; for example, evidence suggests that warfarin-related ICH is associated with the same small vessel diseases, such as CAA, that cause primary ICH [2]. There are many potential causes of secondary ICH (see Table 11.1). Except for vascular malformations as a cause of ICH in younger patients, these secondary causes are rare. Younger patients with ICH should have angiography to rule out secondary causes.

The most common locations of ICH are the cerebral lobes (35 %), the basal ganglia (30 %), the thalamus (20 %), the brainstem (5 %), and the cerebellum (10 %) (Fig. 11.2) [3]. It is useful to discriminate between lobar and non-lobar locations of

Table 11.1 Primary and secondary causes of intracerebral hemorrhage (ICH)

Primary ICH
Hypertensive arteriolosclerosis
Age-related arteriolosclerosis
Cerebral amyloid angiopathy
Secondary causes of ICH
Vascular malformations
Arteriovenous malformation
AV dural fistula
Cavernous hemangioma
Hemorrhagic transformation of ischemic stroke
Related to arterial infarction
Related to venous infarction
Vasculitis
Moyamoya disease
Coagulopathy
Related to anticoagulant use
Related to thrombolytic use
Thrombocytopenia
Decreased synthesis of clotting factors (e.g., hemophilia, liver disease)
Increased consumption of clotting factors (e.g., disseminated intrasvacular coagulation)
Brain tumor
Aneurysm
Ruptured saccular aneurysm
Ruptured mycotic aneurysm
Related to sympathomimetic drug use
Amphetamines
Cocaine
Phenylpropanolamine
Ephedrine
Trauma

primary ICH, because the associated causes and risk factors differ by location. Lobar ICH may be caused by either CAA or hypertensive arteriosclerosis, while non-lobar primary ICH is almost exclusively caused by hypertensive arteriosclerosis. This is because CAA affects the cortical and leptomeningeal arteries but not the arteries penetrating into the substance of the brain, while most hypertension-related ICHs occur at or near the bifurcation of small penetrating arteries that originate from the major branches of the circle of Willis. Accordingly, hypertension is a stronger risk factor for non-lobar ICH than lobar ICH, while conversely the presence of one or more apolipoprotein E epsilon 4 (apoE ε4) alleles, associated with the presence of CAA, is a stronger risk factor for lobar ICH than non-lobar ICH [4, 5]. Other risk factors for ICH include age, male sex, cigarette smoking, low (not high) serum cholesterol levels, and Hispanic, African, or Asian racial origin [1, 6].

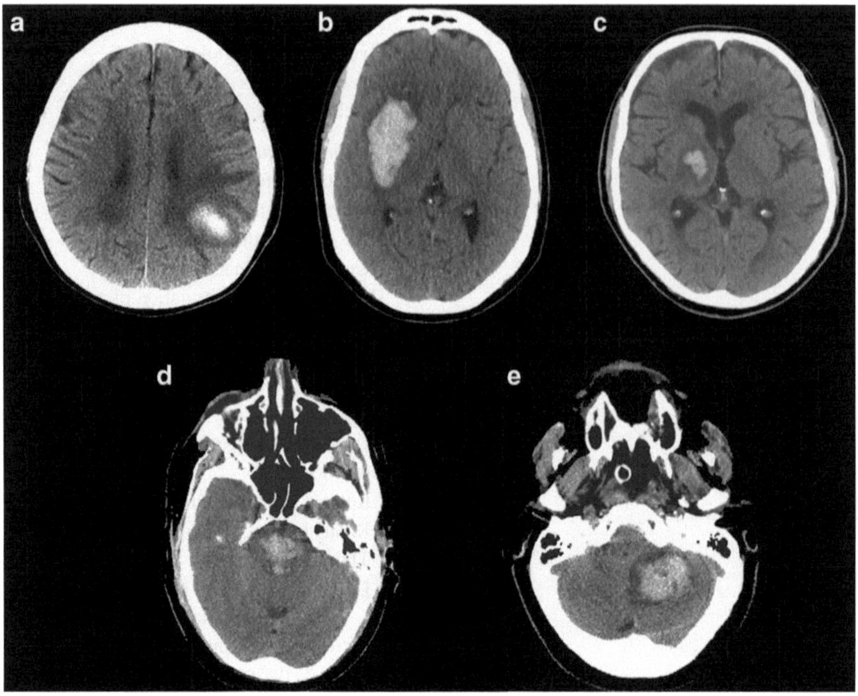

Fig. 11.2 Common locations of ICH: (**a**) cerebral lobe (lobar), (**b**) basal ganglia (putamen), (**c**) thalamus, (**d**) brainstem, (**e**) cerebellum

Cognitive Symptoms and Signs in Acute ICH

Cognitive symptoms in acute ICH depend on the location of bleeding. ICH may cause neurological dysfunction by local destruction of tissue, local mass effect with compression of tissue, potential toxic effects of iron or other molecules found in coagulating blood, or increased intracranial pressure caused by mass effect or obstruction of the ventricular system.

Small- to medium-size ICHs that do not increase intracranial pressure typically present with focal neurological symptoms reflecting the site of origin of the ICH. These may include hemiparesis, hemiplegia, aphasia (resulting from ICH in the left perisylvian cortex), or neglect or visual–spatial difficulty (most commonly resulting from ICH in the parietal cortex). ICH in the thalamus may also present with these "cortical" signs of aphasia (left thalamus) or neglect (right thalamus). ICH in the basal ganglia is usually associated with more prominent motor signs, such as hemiparesis, than cognitive signs; however, survivors of basal ganglia ICH often have cognitive deficits as well (discussed in the subsequent section: "Cognitive Consequences of ICH").

Larger ICHs that cause increased intracranial pressure, or ICHs in the brainstem or thalamus that affect the reticular activating system or its connections in the thalamus, often present with depressed level of consciousness in addition to other signs. In many cases the level of consciousness is significantly depressed and confounds the ability to assess other aspects of cognition. Depressed level of consciousness at presentation is more common in ICH than ischemic stroke; however, there are no signs or symptoms that reliably distinguish between ICH and ischemic stroke.

Delirium is a frequent complication in the first several weeks following ICH. Many factors may contribute to the risk of delirium; among them are the size of the ICH and the presence of intraventricular hemorrhage. Neurological toxic effects of blood break-down products may also contribute. In animal studies, experimental ICH is followed by extensive brain inflammation that worsens cognitive and motor function. The significance of post-ICH secondary injury in humans remains unclear, although post-ICH edema can be detected using neuroimaging. Patients with ICH have as much as a fivefold increased odds of post-stroke cognitive impairment and delirium in the first few weeks compared to patients with ischemic stroke [7].

Cognitive Consequences of ICH

ICH is the most severe form of stroke. Most survivors will have residual disabilities, and cognitive symptoms are common.

Dementia frequently accompanies stroke, both ischemic and hemorrhagic. According to a recent systematic review, about one in ten patients have dementia before their first stroke and one in ten develop new dementia after their first stroke. The risk of post-stroke dementia is higher in ICH than in ischemic stroke, reflecting the greater severity of ICH [8]. Risk factors for dementia following ICH specifically, as opposed to stroke in general, have not been well defined. Based on studies done in consecutive patients with either ischemic stroke or ICH, the expected risk factors for dementia post-ICH might be higher initial stroke severity, advanced age, left hemisphere origin, aphasia, and prior history of stroke or dementia [8].

A history of dementia preceding ICH is seen in 15–23 % of patients [9, 10]. Preexisting dementia is associated with increasing age, female gender, low educational level, severity of cerebral or temporal lobe atrophy, previous stroke or transient ischemic attack, severity of leukoaraiosis, diabetes, atrial fibrillation, and arterial hypertension. The high prevalence of pre-ICH dementia suggests that dementia may be more common in persons who develop ICH than other similar-aged persons, although this has not been proven in an epidemiologic study. Patients who develop ICH could be at risk for dementia, independent of the ICH, because of shared risk factors. Hypertensive arteriolosclerosis and CAA, the most common causes of ICH, also cause brain microscopic infarction and ischemic white matter demyelination (also known as leukoaraiosis) that contribute to the risk of dementia. Leukoaraiosis is frequently seen in the brains of persons with ICH and is associated with pre-ICH dementia (Fig. 11.3) [10, 11]. CAA may be accompanied by

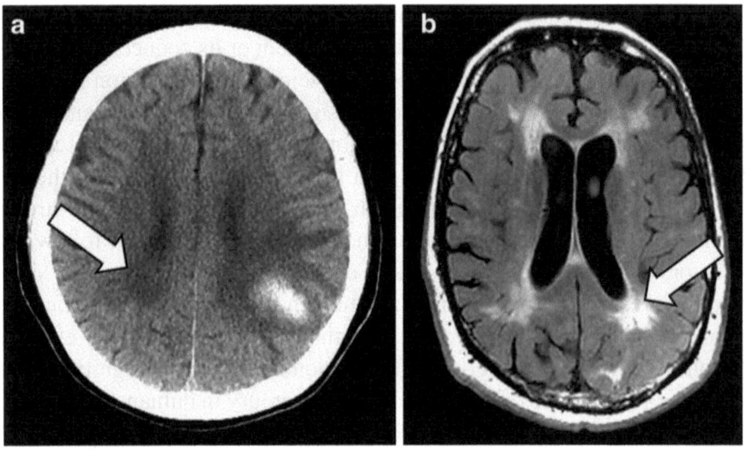

Fig. 11.3 Leukoaraiosis appears on CT as periventricular white matter hypodensities in this patient with a left parietal lobar ICH (**a**, *arrow*), and on T2-weighted MRI as hyperintensities (different patient) (**b**, fluid attenuated inversion recovery [FLAIR] sequence, *arrow*)

Alzheimer's pathology, although fewer than half meet pathological criteria for Alzheimer's disease [12].

Milder forms of cognitive impairment are even more common than dementia following stroke [13]. As a screening test for cognitive impairment, the Montreal Cognitive Assessment tool (MoCA) is more sensitive at detecting milder forms of cognitive impairment than the Mini-Mental State Examination (MMSE) [14]. However, there are very limited data on the prevalence of mild cognitive impairment following ICH specifically.

The pattern of deficits on neuropsychological testing in ICH patients depends on the location and severity of the hemorrhage, the ability of the brain to accommodate the cerebral damage (which in turn depends on the overall health of the brain, including the degree to which the brain is damaged by other effects of small vessel disease such as leukoaraiosis), and the degree of neurological recovery. Therefore, the resulting neuropsychological syndrome may reflect a combination of focal deficits related to the site of origin of the ICH, as well as global deficits as a consequence of more widespread neuronal injury caused by globally raised intracranial pressure, for example. Although it is useful to consider the neuropsychological deficits from ICH in terms of site of origin, it also must be recognized that the ICH may have extended from the site of origin to involve neighboring white matter or other brain structures.

Patients with lobar brain hemorrhages are expected to display abnormalities on neuropsychological testing that reflect a decline in function in the areas affected by the hemorrhage. For example, a patient with a left posterior superior temporal ICH may show symptoms of Wernicke's aphasia.

Patients with the more common non-lobar brain hemorrhages may display a variety of deficits. Non-lobar hemorrhages most commonly affect subcortical brain regions—the putamen, head of the caudate nucleus, and thalamus—that communicate widely with the frontal lobes via feedback loops to regulate thought and mood.

The striatum (caudate nucleus and putamen) is the main input structure of the basal ganglia. Given the connections between the striatum and the convergent loops related to the major cortical areas of the brain, damage to the striatum may result in a multitude of deficits pertaining to motor, oculomotor, cognitive, associative, and limbic functions. Indeed, ICH in the basal ganglia has been associated with poor performance in multiple cognitive domains including language, attention, memory, visuospatial processing, and executive function abilities, with many patients showing impairment in multiple domains [15].

Patients with ICH in the putamen may have impairments in executive function, visuospatial function (including neglect), apraxia, and aphasia. On the Wisconsin Card Sorting Test (WCST), patients with putaminal ICH have been shown to have impaired mental set shifting with disrupted concept formation or problem solving (the inability to comprehend the possible sorting rules) and perseveration (the inability to flexibly shift attention and response preparation from one set of rules to another) [16]. Unilateral spatial neglect can be detected in up to half of patients [17, 18]. Motor apraxias may occur in 20–40 % [19]. Aphasia may occur, and is related, at least in part, to extension of the hemorrhage into adjacent white matter pathways [20]. Extension into the deep frontal and anterior periventricular white matter causes decreased fluency, while extension to the deep temporal white matter causes impaired word comprehension. Hemorrhage extension into both the anterior and posterior periventricular white matter may cause a global aphasia. Extension more laterally into the region of the external capsule may cause conduction aphasia [21, 22]. Buccofacial oral apraxia often accompanies the aphasia [23].

Patients with ICH in the head of the caudate nucleus may have impairments in executive function, short-term and long-term memory, apathy, or aphasia [24–26]. Impaired mental set shifting and perseveration may be seen on the WCST [24]. One study suggests that the most prominent impairments are in short-term memory, long-term memory, and verbal fluency [24]. Aphasia following hemorrhage restricted to the caudate is rare, but is not infrequent when the hemorrhage also involves the neighboring white matter tracts [26].

Patients with ICH in the thalamus may have impairments in executive function, short-term and long-term memory, apraxia, aphasia, or neglect. Thalamic aphasia may be seen in lesions located in the dominant hemisphere and is characterized by fluent speech with paraphasias, perseveration, and lack of spontaneous speech but relatively preserved comprehension and repetition. Unilateral spatial neglect is more common with right-sided lesions, and often improves clinically over time [27].

In sum, a wide variety of neuropsychological deficits may be detected in survivors of ICH. Patients with small to moderate-sized lobar ICHs may have focal neuropsychological deficits related to the site of injury, while patients with larger ICHs or ICHs originating in non-lobar subcortical structures typically have deficits in multiple domains, including aspects of executive function, visuospatial function, aphasia (for dominant hemisphere lesions), and neglect (for nondominant hemisphere lesions). In patients with subcortical ICHs, particularly those originating in the putamen and caudate nucleus, the cognitive deficits may predominantly reflect injury to adjacent structures including white matter pathways, rather than damage to the putamen or caudate themselves. The cognitive profiles of ICH are not distinct enough to discriminate ICH from ischemic stroke based on neuropsychological testing alone, although patients with ICH have been described to have somewhat more prominent problems with higher-level perceptual functions [25].

Diagnosis and Medical Management

There are no signs or symptoms that reliably distinguish between ICH and ischemic stroke; therefore, neuroimaging is needed to distinguish between the two [28]. Computed tomography (CT) shows similar sensitivity to magnetic resonance imaging (MRI) for acute hemorrhage and is often the initial investigation of choice. Medical management includes supportive care, control of excessively high blood pressure, lowering intracranial pressure if necessary, provision of hydration and nutrition, and prevention and treatment of medical complications such as seizure, deep venous thrombosis, and infection [28]. Severely affected patients may require intubation and mechanical ventilation. Limitation of care or palliation may be appropriate in severely affected patients with little hope of survival or meaningful quality of life [29, 30]. Surgical removal of the hematoma is often performed for patients with secondary ICH, cerebellar ICH, or with lobar ICH and clinical deterioration, and less commonly for patients with non-lobar-hypertensive ICH [31]. Expansion of the ICH occurs in up to 40 % of patients within 24 h of hospital presentation and may be lessened by good blood pressure control [32]. Trials of hemostatic therapies have not shown a clinical benefit despite demonstration of a modest reduction in ICH growth [33].

The risk of recurrent hemorrhage is approximately 2 % per year for non-lobar ICH and 5–10 % per year for lobar ICH. Recurrences are more frequent when blood pressure is poorly controlled. Among lobar ICH patients, risk factors for recurrence include prior symptomatic ICH, a greater number of MRI-defined microbleeds, the presence of one or more apoE ε(epsilon)2 or ε(epsilon)4 alleles, and the probable presence of underlying CAA. Risk factors for recurrence in patients with proven CAA are the same but also include the use of aspirin and the presence of leukoaraiosis [10, 34, 35].

Rational steps to prevent recurrence include blood pressure reduction and cessation of smoking and alcohol use [28]. However, these secondary prevention strategies are of unproven benefit with the exception that data from a randomized trial support an effect of blood pressure reduction [36, 37].

Whether better supportive care leads to improved cognitive outcomes is currently not known. There are no proven therapies for cognitive rehabilitation that are

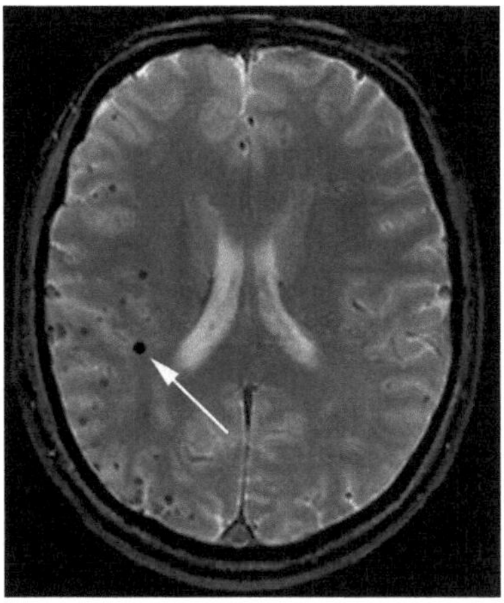

Fig. 11.4 Microbleeds, evident as small *black dots* on MRI T2*-weighted gradient-recalled echo (GRE) sequence. The *white arrow* points to the largest microbleed

specific to ICH. Therefore, cognitive rehabilitation in ICH should follow the same principles as for ischemic stroke. This may include pharmacotherapy for depression or apathy. In our experience, stimulants such as methylphenidate may be used in ICH patients provided that blood pressure is monitored and controlled to avoid the risk of hypertension.

Brain Microbleeds

Microbleeds represent small (usually less than 5 mm) prior asymptomatic brain hemorrhages. Histopathologically, they appear as areas of perivascular hemosiderin deposition [38]. The hemosiderin molecule contains iron atoms derived from the breakdown of hemoglobin from red blood cells that previously leaked into the tissue. After deposition, the hemosiderin molecules remain in location indefinitely. Therefore the hemosiderin deposits across the brain represent the cumulative distribution of all the small hemorrhages experienced during the patient's life. These hemosiderin deposits are visible pathologically but may be overlooked because of their small size (usually 5 mm or less, and often as small as 1–2 mm). The deposits cannot be visualized on CT or on conventional MRI sequences but can be seen as "microbleeds" on specialized MRI sequences—T2*-weighted gradient-recalled echo (GRE) or susceptibility-weighted imaging (SWI)—where they appear as small dark round (i.e., hypointense) foci [39] (Fig. 11.4).

Microbleeds are a common finding in asymptomatic elderly, being present in at least 5–10 % of persons over the age of 60 [40]. However, the prevalence is much higher in persons with ischemic stroke (20–30 %) or ICHs (50–70 %). Microbleeds are a sign of cerebral small vessel disease, and when seen in large numbers, are most commonly associated with either poorly controlled hypertension or CAA. When multiple and restricted to the cerebral cortex or subcortical junction, without involvement of subcortical structures such as the basal ganglia, they are highly suggestive of the pathological presence of CAA (see also the subsequent section: "Cerebral Amyloid Angiopathy") [41].

Previously, microbleeds were considered too small to directly affect brain function. More recently, this assumption has been questioned because of emerging evidence that microbleeds may be associated with cognitive impairment. For example, a case–control study done in patients seen in a neurovascular clinic found that executive dysfunction was more frequent in 25 patients with microbleeds compared to 30 controls without microbleeds, matched for stroke type, stroke location, and the extent of MRI white matter lesions [42]. In this study, executive dysfunction was defined as poor performance on two or more of the following tests: word fluency, Stroop, trail making, WCST, or Weigl Sorting Task. The stroke patients with executive dysfunction were more likely to have microbleeds in the frontal lobes and basal ganglia [42]. Microbleeds are frequently seen in persons with vascular dementia, but are also seen in up to 20 % of patients with mild cognitive impairment or clinically probable Alzheimer's disease [43–50]. Microbleeds have been associated with worse cognitive performance in adults without diagnosed neurological diseases [51–53].

However, caution is warranted before ascribing cognitive symptoms to the direct effects of microbleeds. Microbleeds may be merely a sign of an underlying cerebral small vessel disease causing cognitive dysfunction by mechanisms other than bleeding. For example, both hypertensive arteriopathy and CAA are associated with microbleeds but also cause microscopic infarctions that, in contrast to microbleeds, cannot be detected on MRI. Further studies are needed to clarify whether microbleeds affect brain function directly, or are merely an associated finding.

Cerebral Amyloid Angiopathy

CAA is caused by deposition of amyloid, a protein aggregate, in the media and adventitia of small arteries of the brain [54]. In most cases the amyloid is beta-amyloid, derived from the abeta protein. In some rare autosomal dominant hereditary diseases, the amyloid is derived from other proteins including transthyretin (in transthyretin amyloidosis), cystatin C (Icelandic familial dementia), or BRI (British familial dementia) [55]. CAA is most commonly clinically recognized as a cause of approximately half of lobar ICHs. However, the spectrum of clinical presentations of CAA also includes small sulcal subarachnoid hemorrhages, vasculitis, and transient neurological symptoms, and there is increasing evidence that CAA is a frequent contributor to the risk of dementia.

Arteries affected by amyloid deposition have thickened walls, fibrin deposition within the wall, and loss of wall integrity including circumferential cracking (a "vessel within a vessel" appearance) [54]. There may be perivascular hemosiderin deposits indicating leakage of red blood cells through the arterial wall [56].

Pathologically, the presence of amyloid may be diagnosed by the characteristic apple-green colored birefringence seen when affected arteries are stained with Congo red and then viewed under polarized light. An alternate means of pathological diagnosis is to identify the amyloid protein using an immunohistochemical stain.

Sporadic non-inherited CAA caused by beta-amyloid is a common age-associated pathology [57]. As with Alzheimer's disease, the prevalence of CAA is highly age-dependent and increases exponentially with increasing age. Some degree of CAA is seen in more than 40 % of persons over 80 [58]. In many cases, affected persons appear asymptomatic, although increasing evidence suggests that CAA is an important under-recognized contributor to the risk of dementia.

The common sporadic CAA is caused by aggregation of abeta into beta-amyloid with deposition in the vascular media. Abeta is derived from proteolytic processing of the amyloid precursor protein by beta-secretase and gamma-secretase. After formation, abeta can have several fates including aggregation into beta-amyloid with deposition in the arterial wall or brain parenchyma, proteolytic cleavage by enzymes such as neprilysin or insulin-degrading enzyme, or clearance from the brain via a perivascular interstitial fluid drainage pathway [59]. Aggregation of abeta with deposition as beta-amyloid may result from an imbalance between production and clearance of abeta [60]. Vascular beta-amyloid deposits are almost exclusively restricted to the leptomeningeal and cortical arteries.

The beta-amyloid seen in CAA is the same beta-amyloid that is the main constituent of the neuritic plaques seen in Alzheimer's disease. Indeed, the frequency and severity of Alzheimer's pathology is higher in the brains of persons with CAA than persons without [12]. CAA and Alzheimer's pathology may be conceived as being situated on the opposite ends of a spectrum, with some patients predominantly exhibiting vascular amyloid deposition (i.e., CAA) with increased risk of lobar ICH, and other patients predominantly exhibiting parenchymal amyloid deposition (i.e., Alzheimer's pathology) with increased risk of Alzheimer's type dementia.

The cause of nonhereditary beta-amyloid CAA is unknown. Other than age, the only other known risk factor is the presence of one or more apoE ε(epsilon)4 alleles [12]. Patients with one or more apoE ε(epsilon)2 alleles have more severe vasculopathic changes and may be at increased risk for ICH [61, 62]. Lowering of blood pressure and avoidance of antithrombotics and anticoagulants are reasonable steps to reduce the risk of recurrence.

Clinically, CAA is best recognized as a cause of approximately half of lobar ICHs in developed countries where there is good treatment of chronic hypertension. The hemorrhages may occur in any brain lobe, but have a predilection for posterior brain regions including the occipital lobe and posterior temporal and parietal lobes [63]. In about half of the cases of pathologically proven CAA-associated ICH, the lobar ICH is accompanied by one or more lobar microbleeds (Fig. 11.5) [41]. The presence of microbleeds in other locations not involved with vascular amyloid, such as the thalamus, basal ganglia, or brainstem, is not consistent with CAA and should prompt consideration of other causes such as hypertensive arteriolosclerosis. Diagnostic criteria for clinical research, termed the Boston criteria, discriminate

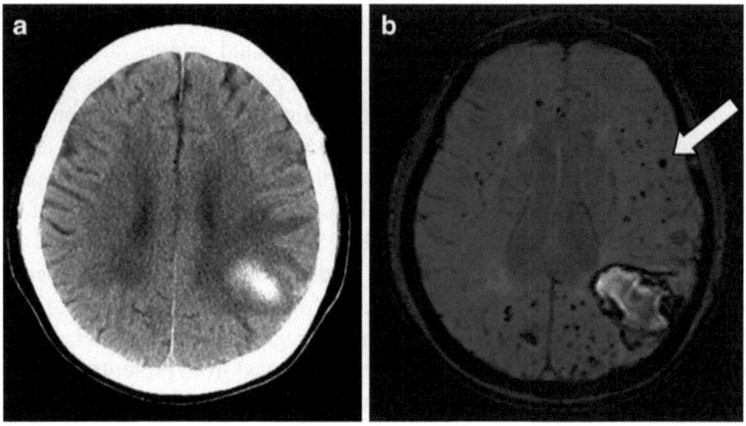

Fig. 11.5 Cerebral amyloid angiopathy (CAA). In this 70-year-old woman with left parietal lobar ICH (**a**, CT scan), MRI susceptibility-weighted imaging reveals dozens of small hypointensities indicating lobar microbleeds (**b**, *arrow*) not apparent on CT scan, consistent with probable CAA

Table 11.2 Boston criteria for diagnosis of cerebral amyloid angiopathy (CAA)-related ICH

Definite CAA	Full postmortem examination showing severe CAA
Probable CAA with supporting pathology	Pathologic tissue (evacuated hematoma or cortical biopsy) showing • Lobar ICH with some degree of CAA • Absence of other diagnostic lesion
Probable CAA	Clinical data and MRI or CT demonstrating • Multiple hemorrhages restricted to lobar regions or a single lobar hemorrhage with additional evidence of superficial siderosis • Age ≥55 years • Absence of other cause of hemorrhage or superficial siderosis
Possible CAA	Clinical data and MRI or CT demonstrating • Single lobar hemorrhage, or focal or disseminated superficial siderosis • Age ≥55 years • Absence of other cause of hemorrhage or superficial siderosis

These criteria were validated against autopsy data [41] and were recently modified to include incorporate evidence of superficial siderosis on MRI. Superficial siderosis is an MRI finding indicative of past bleeding in the subarachnoid space adjacent to the cortex or in the superficial cerebral cortex [65]

between *probable* and *possible* CAA based on presentation with lobar ICH and the presence or absence of accompanying lobar hemorrhages or microbleeds, without identifiable alternate causes (Table 11.2) [41]. Positron emission tomography (PET) with Pittsburgh Compound B or other beta-amyloid ligands will identify vascular beta-amyloid as well as parenchymal beta-amyloid, and could therefore be useful to confirm beta-amyloid deposition [64].

There are several less common presentations of CAA. Some patients present with vasculitis or perivascular inflammation and signs of patchy focal or multifocal edema in the periventricular white matter or juxta-cortical white matter, often involving the temporal or parietal lobes [66]. These patients may have a subacute cognitive decline, headaches, seizures, or focal neurological deficits such as hemiparesis. In some cases, the clinical course may be that of a rapidly progressive dementia. Neuroimaging shows brain edema, usually without infarction. In contrast to chronic white matter lesions, which are also common in CAA, the white matter lesions seen in CAA-associated vasculitis are not associated with atrophy but rather exhibit swelling and mass effect. When the edema is very focal, it may be mistaken for a low-grade tumor. There are usually multiple lobar microbleeds, although we have also seen pathologically proven cases without microbleeds. The cerebrospinal fluid often exhibits increased protein and sometimes a lymphocytic pleocytosis, but may also be normal. The diagnosis of CAA with vasculitis may be suspected on clinical and radiological grounds, but can only be definitively proven by biopsy. Immunosuppressive therapy, for example with steroids, can produce a dramatic clinical and radiological improvement in up to two-thirds of patients. Responders may stay in remission for years without need for chronic immunosuppression [66].

CAA is an important contributor to the risk of dementia, even in the absence of lobar ICH and accounting for the effects of accompanying Alzheimer's pathology. In the population-based Medical Research Council-Cognitive Function and Aging Study (MRC-CFAS) autopsy study, CAA accounted for 7 % of the risk of dementia, controlling for Alzheimer's disease pathology [67]. Similarly, another population-based study showed that for a given severity of Alzheimer's pathology, a higher degree of CAA was associated with worse performance on a global cognitive test [68].

The cognitive impairment seen in cerebral amyloid-angiopathy that is independent of hemorrhagic stroke is probably caused by brain ischemia. CAA is associated with a high burden of periventricular white matter demyelination [10], presumably on an ischemic basis, and small cerebral infarctions. Disturbed blood flow regulation with impaired vasodilation might play a role in generating these small infarctions [57].

There is limited information on the pattern of cognitive deficits that can be attributed to CAA, independent of the effects of lobar ICHs and coexisting Alzheimer's pathology. Based on the association between CAA and subcortical white matter lesions, deficits may be expected to include those associated with white matter disease, including poor performance on timed tests. However, cortical microscopic infarctions and microbleeds could produce signs of cortical dysfunction dependent on the location and severity of the pathological changes. A prospective cohort study with brain autopsy found that more severe CAA was correlated with worse performance on the symbol digit modalities test and a test of episodic memory, controlling for the amount of Alzheimer's pathology and in the absence of ICHs [69]. This poor performance on a timed test influenced by visual perception may reflect a combination of the subcortical ischemic white matter demyelination and predilection for the posterior parietal lobe and occipital lobe that are seen in cerebral amyoid angiopathy.

In summary, three potential contributing factors must be considered when evaluating cognition in a patient with probable CAA: (1) the effects of lobar ICH, if one has occurred; (2) the effects of accompanying Alzheimer's pathology, which may or may not be present; and (3) the effects of CAA independent of either lobar ICH or accompanying Alzheimer's pathology, which are presumably mediated by brain ischemia. The neuropsychological evaluation should therefore include tests sensitive to the cortical deficits expected based on the lobar ICH location, tests sensitive to the pattern of impaired memory retrieval seen in Alzheimer's disease, and tests sensitive to the impairments in psychomotor processing speed and executive dysfunction that may be seen with subcortical ischemic white matter disease.

Other Less Common Causes of Ischemic Stroke

The common causes of ischemic stroke are arterial disease related to hypertension and atherosclerosis, or cardiac disease related to coronary heart disease or atrial fibrillation. A substantial number of ischemic strokes are cryptogenic, which is defined as having no diagnosable cause. Other determined causes of stroke are rare. A comprehensive discussion of all uncommon causes of stroke is outside the scope of this chapter. Instead, we will highlight selected less common causes of stroke where cognitive impairment may be a prominent feature.

Cerebral Autosomal Dominant Arteriopathy with Subcortical Ischemic Leukoencephalopathy

Cerebral autosomal dominant arteriopathy with subcortical ischemic leukoencephalopathy (CADASIL) is a rare genetic cause of stroke, migraine, and vascular dementia [70]. The disease is caused by mutations of the NOTCH3 gene, which encodes the Notch3 receptor. Small arteries become thickened and fibrosed. Electron microscopy shows granular osmiophilic material adjacent to the vascular basement membrane, which is pathognomonic for the disease. Symptoms are limited to the central nervous system (CNS), even though arteries outside the brain are affected, in addition to those within the brain. Neuroimaging reveals multiple lacunar infarcts in the basal ganglia, thalamus, and white matter, along with extensive white matter demyelination (Fig. 11.6). Typically, the disease begins with migraine with aura in the victim's late 20s, lacunar strokes occur in their 40s, and vascular cognitive impairment and dementia occur in their 40s and 50s [71]. The diagnosis can be made by genetic analysis, or by skin or muscle biopsy with electron microscopy to identify the characteristic vascular changes with granular osmiophilic material.

CADASIL is an instructive disease because it is the prototypical example of a pure subcortical ischemic dementia. Because CADASIL patients develop dementia

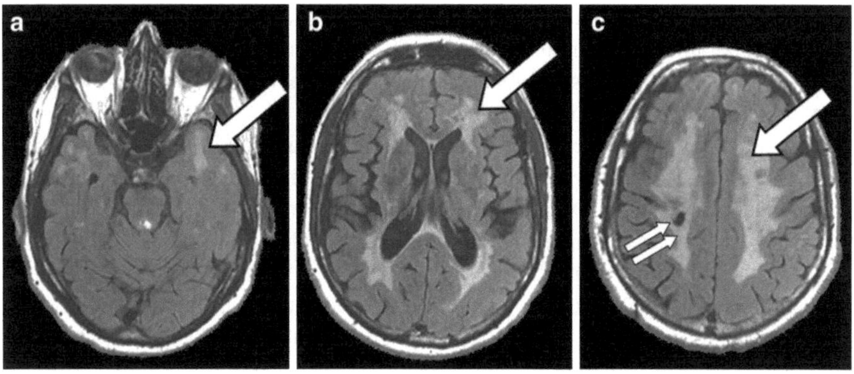

Fig. 11.6 Cerebral autosomal dominant arteriopathy with subcortical ischemic leukoencephalopathy (CADASIL): a 41-year-old man with CADASIL confirmed by genetic testing. The MRI FLAIR sequence shows extensive white matter T2-hyperintensities (**a–c**, *arrows*), including hyperintensities in the anterior temporal white matter (**a**) that are characteristic of CADASIL compared to other white matter diseases, and a lacunar infarct in the right corona radiate (**c**, *double arrow*)

at an early age, the cognitive profile of CADASIL is uncomplicated by other late-life pathologies of dementia such as Alzheimer's disease. In the earlier stages, CADASIL patients may display poor performance on neuropsychological tests sensitive to executive dysfunction including the Trail Making Test and tests of verbal fluency [72, 73]. As the disease progresses, more global cognitive impairments are seen, including memory. Apathy is a prominent neuropsychiatric symptom [74]. A randomized placebo-controlled trial of the acetylcholinesterase inhibitor donepezil showed that treated patients had modest improvements on the Trail Making Test but no improvement in the global cognitive endpoint compared to the placebo group [75].

Vasculitis

CNS vasculitis is caused by vascular or perivascular inflammation. Pathologically, lymphocytic infiltration of the vascular wall or perivascular tissue is seen, with microglial activation. The inflammation may be idiopathic and isolated to the CNS, or may occur in the context of a systemic inflammatory disorder. CNS inflammation, of either proven or suspected vascular origin, has been reported in polyarteritis nodosa, Churg–Strauss vasculitis, systemic lupus erythematosis (SLE), sarcoidosis, Wegener's granulomatosis, rheumatoid arthritis, Sjögren's syndrome, Behcet's disease, temporal arteritis, thromboangiitis obliterans (Buerger's disease), Kawasaki's syndrome, infections (particularly fungal infections causing basal meningitis), and other conditions. Generally, CNS vasculitis is a rare complication of these systemic diseases. Diagnosing CNS vasculitis is challenging because it is rare and has many nonspecific manifestations. The literature on CNS vasculitis is often difficult to interpret because of changing perceptions and nomenclatures, with lack of pathological confirmation in many suspected cases.

The most common type of CNS vasculitis is primary angiitis of the CNS, also called isolated CNS angiitis [76]. Because giant cells are sometimes seen pathologically, this form of vasculitis has also been termed granulomatous angiitis of the CNS. In some cases vascular beta-amyloid is also identified and is probably the target of the autoimmune attack [66, 77].

Symptoms of CNS vasculitis may include headache, neurological deficits, and confusion [78]. In contrast to acute stroke, the symptoms are subacute and progressive or waxing and waning, and evolve over days to weeks. Impaired cognition is common. There is minimal information on the pattern of cognitive deficits; a few small case series suggest involvement of attention, processing speed, and visual–spatial functions [79, 80]. Age at onset is usually younger than for typical ischemic stroke, but there is wide variation. Neuroimaging may show edema, small infarcts or hemorrhages, or may be normal. Lumbar puncture may show increased protein and lymphocytic pleocytosis or may be normal. Peripheral markers of inflammation, such as erythrocyte sedimentation rate or C-reactive protein, are usually normal in vasculitis isolated to the CNS but may be elevated when CNS vasculitis complicates a systemic inflammatory disorder. Cerebral angiography shows vascular narrowing when medium-sized arteries are involved, but is usually normal in small vessel vasculitis such as that seen in idiopathic CNS vasculitis. Brain biopsy is needed to make a definitive diagnosis of CNS vasculitis. However, because the disease tends to have a patchy distribution across the brain, even brain biopsy may be falsely negative if an unaffected area was sampled. Brain tissue at the margin of an affected brain region may provide the highest diagnostic yield; however, care should be taken not to biopsy the center of an infarct because it would show only non-diagnostic tissue destruction. There are no proven therapies for CNS vasculitis. Intravenous steroids and immunosuppression are frequently tried, and the disease may go into remission.

Intravascular Large B Cell Lymphoma

Intravascular large B cell lymphoma (also called intravascular lymphomatosis or malignant angioendotheliomatosis) is a rare manifestation of lymphoma caused by occlusion of capillaries and small venules by lymphoma cells. Affected organs include the brain, skin, bone marrow, spleen, and liver. Common neurological manifestations include stroke and progressive cognitive impairment, including rapid-onset dementia [78]. Brain MRI may show small infarcts that enhance after gadolinium contrast injection. Many of the presenting symptoms and signs mimic CNS vasculitis. Biopsy is required to make the diagnosis.

Systemic Lupus Erythematosis

SLE is an autoimmune disease of unknown cause that affects multiple organs including the brain. Non-neurological manifestations include malar rash, arthritis, and anemia. Neuropsychiatric symptoms are relatively common and are part of the diagnostic criteria for the disease [81, 82]. Manifestations of the disease may include stroke, seizures, CNS inflammation, psychosis, and mood disorders. Several different pathophysiologies may lead to CNS complications in lupus, including vasculitis, noninflammatory vasculopathy, hypercoagulability, and autoantibodies reacting against various epitopes including neurons [83].

Stroke may complicate SLE in up to 20 % of persons [84]. Stroke in SLE may be caused by CNS vasculitis, a CNS vasculopathy without frank inflammation, cardioembolism due to nonbacterial valvular vegetations (Libman Sacks endocarditis), or a hypercoagulable state due to the presence of antiphospholipid antibodies [85].

The neuropsychological syndrome in patients with stroke and SLE will depend on the brain regions affected by the stroke. However, cognitive dysfunction in SLE also occurs independent of stroke. Cognitive deficits may also be identified in the context of concurrent psychiatric illness. Therefore, a combination of both stroke-related and non-stroke-related neuropsychological deficits may be seen. Some degree of cognitive dysfunction may be seen in 20–50 % of stroke-free SLE patients [86–90] and is frequently transitory, with subsequent improvement [88]. Impaired immediate memory and concentration may be seen with acute disease activity, while impaired delayed recognition memory may represent the sequelae of past CNS involvement [91]. Even patients without overt neurological or psychiatric symptoms may exhibit impairments on tests of verbal memory, logical reasoning, visual attention, and working memory [92]. Cognitive outcomes have been proposed as secondary endpoints in trials of disease-modifying therapies [93].

Moyamoya Syndrome

The moyamoya syndrome is marked by gradual tapering occlusion of the terminal carotid artery or proximal middle cerebral artery with associated development of small, fragile arterial collaterals that supply the anterior circulation of the brain [94, 95]. Most commonly the syndrome is idiopathic, but can also occur as a complication of sickle cell disease, autoimmune diseases such as SLE, connective tissue diseases such as neurofibromatosis type I, Down's syndrome, and other diseases. Some cases are familial but the causative genes are unknown. The term *moyamoya* is Japanese and refers to the characteristic appearance of the collateral arteries as a "puff of smoke" on conventional angiography (see Fig. 11.7) [94]. These arteries are often insufficient to maintain cerebral perfusion, leaving the brain in a tenuous, chronically hypoperfused state. Fluctuations in brain perfusion may lead to

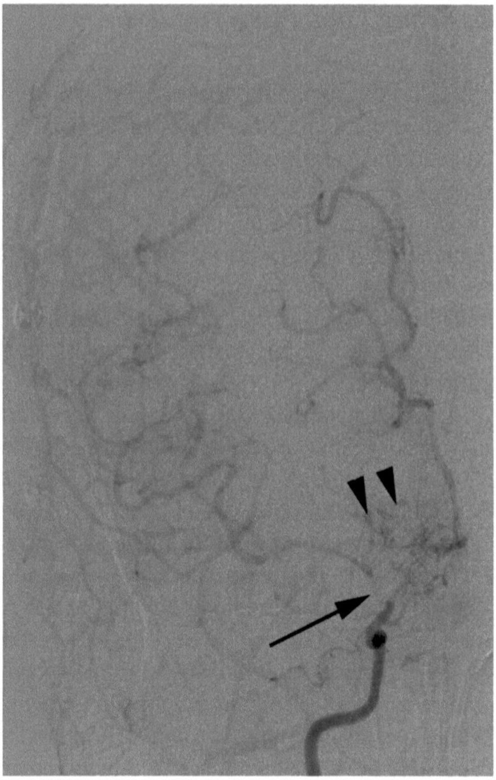

Fig. 11.7 Moyamoya disease. Angiography (right internal carotid artery injection, anterior–posterior view) in a 17-year-old woman with familial moyamoya disease shows occlusion of the terminal portion of the right internal carotid artery (*arrow*) with multiple fine collateral vessels (*arrowheads*) supplying blood flow to the right middle cerebral artery

transient ischemia with focal neurological symptoms. Progressive stepwise brain infarction over weeks and months may ensue. Alternatively, the fragile collaterals may rupture and cause ICH.

Moyamoya syndrome usually presents in childhood before the age of 10 years. Transient focal neurological symptoms are the most common presenting symptoms but sometimes cognitive impairment is the dominant symptom [96]. Poor school performance is frequent [97]. Low IQ scores may be seen and may decline over time in some children [98]. Mental retardation with very low IQ is uncommon at presentation, however [99]. More severe global cognitive impairments are usually the result of multiple ischemic strokes. Neuropsychological studies in adults with moyamoya demonstrate impairment in up to a third of patients. Some studies point to impaired executive function as the most prominent finding [100–102], while others have found broader impairments affecting memory as well [103].

Revascularization procedures that restore normal cerebral blood flow may improve IQ in affected children [104, 105]. Long-term outcomes after revascularization are uncertain, with some patients demonstrating stable improvement but others still suffering from periodic transient ischemic attacks and with borderline low IQ [99, 106]. In some children, IQ worsens over time in the absence of stroke, which may be considered a possible indication for surgical revascularization [107, 108].

Sickle Cell Disease

Sickle cell disease is caused by a mutation in the HBB gene encoding beta-globin, one of the constituents of the hemoglobin molecule. The red blood cells in affected patients can assume a characteristic sickle shape, particularly under conditions of low oxygen, leading to obstruction of capillaries. Patients may have intermittent sickle cell crises, marked by fever, chest pain and diffuse pain, dyspnea, and confusion. Stroke is a common complication and is a leading cause of death [109]. Patients with sickle cell disease are at risk for both ischemic stroke and hemorrhagic stroke. Most strokes are ischemic (67 %) and are caused by occlusion of a large intracranial artery. Progressive stenosis of intracranial arteries may occur, in some cases leading to the moyamoya syndrome. High intracranial arterial blood flow velocities seen on transcranial Doppler ultrasound identify children at higher risk of stroke, and this risk can be reduced by blood transfusions [110]. "Silent" or minimally symptomatic small infarcts can occur in the cerebral white matter and are visible on MRI in 20–30 % of children [111, 112]. Risk factors for ischemic stroke include prior transient ischemic attack, prior episodes of chest pain, lower hemoglobin, and higher blood pressure [113]. Hemorrhagic stroke in sickle cell patients may result from bleeding in the subarachnoid space or the brain and usually occurs in young adults, not children [113]. Sickle cell disease may promote aneurysm formation by causing endothelial dysfunction with secondary changes in the vessel wall. The fragile collateral arteries seen in the moyamoya syndrome can be another source of bleeding.

Cognitive dysfunction in sickle cell disease may be the consequence of symptomatic stroke, or the consequence of "silent" cerebrovascular disease in the absence of symptomatic stroke. Up to one-third of children with sickle cell disease have low IQ [114]. Neuropsychological testing reveals executive dysfunction [114] and lower performance on tests of arithmetic, vocabulary, and visual motor speed and coordination [112]. Neuropsychological test performance is worse in children with silent brain infarcts [112]. However, neuropsychological deficits may also be present in the absence of silent brain infarction, and are correlated with the severity of anemia, microstructural alterations in the cerebral white matter, and lower cerebral blood flow [115–117]. As with children, about one-third of adults have low IQ (more than one standard deviation below the age- and sex-adjusted means) [118]. Executive dysfunction, slower psychomotor processing speed and poor working memory may be seen in adults, even in the absence of symptomatic stroke. Cognitive deficits are worse in patients with more severe anemia [118].

Exchange transfusion lowers the risk of stroke in susceptible patients with sickle cell disease [110]. However, it is not known whether exchange transfusions prevent silent brain infarction or improve cognition. Hydroxyurea is not as effective at preventing strokes as exchange transfusion [119].

Hypercoagulability and Hyperviscosity States

Ischemic stroke caused by hypercoagulability or hyperviscosity is not common. The most frequently encountered cause of hypercoagulability in ischemic stroke is active cancer. Otherwise, ischemic stroke due to hypercoagulability is rare. Autoimmune coagulopathies such as the antiphospholipid antibody syndrome may cause ischemic stroke with secondary vascular dementia [120]. Sneddon's syndrome is caused by antiphospholipid antibodies with thrombosis of small arteries in the skin and brain. Patients present with livedo reticularis, a lacy purple rash usually on the thighs, in conjunction with ischemic stroke. Rarely, patients may present with a progressive dementia caused by multiple "silent" brain infarcts [121, 122]. Thrombotic thrombocytopenic purpura presents with some combination of microangiopathic hemolytic anemia, thrombocytopenia, acute delirium, acute renal failure, and fever [123]. The simultaneous presence of all of these manifestations is uncommon, however, and the diagnosis can be made based on the presence of microangiopathic hemolytic anemia and thrombocytopenia without other apparent cause. Neurological manifestations are caused by small arterial occlusions due to platelet dysfunction. Headache and confusion are common; infarction, coma, and seizures may also occur. A hyperviscosity syndrome may occur due to the very high serum protein levels encountered in Waldenstrom's macroglobulinemia or multiple myeloma [124]. The high blood viscosity impairs cerebral blood flow and can cause symptoms including headache, dizziness, confusion, vision loss, and coma [124].

Cerebral Venous Thrombosis

Thrombosis of cerebral sinuses or veins usually presents subacutely with progressive headache, seizures, or neurological deficits. The cause may be idiopathic, or may be secondary to infection in the mastoid or middle ear, or as a complication of neurosurgery. Risk factors include female sex, oral contraceptive use, factor V Leiden mutation or prothrombin gene mutation, and smoking [125]. Occlusion of cerebral veins or sinuses prevents outflow of blood from the brain that may lead to increased intracranial pressure, edema, and venous infarction. Cognitive symptoms may include decreased arousal and awareness due to globally increased intracranial pressure or focal symptoms due to venous infarction.

Dural Arteriovenous Fistula

Dural arteriovenous fistulas are direct connections between a dural artery and a meningeal vein or dural sinus. Exposure of the venous sinus to arterial pressures can impair brain venous outflow. Rarely, this can lead to an encephalopathy due to venous hypertension that mimics a progressive dementia [126]. In some cases the symptoms may be improved after surgical obliteration of the fistula [127, 128].

Summary

ICH comprises 10–15 % of stroke but is associated with high mortality and morbidity. There are few specific studies of cognitive dysfunction in these patients. Deficits are related to the location and volume of the hemorrhage and will be similar to those occurring in patients with ischemic strokes affecting the same brain regions. In addition, however, patients with non-lobar ICH usually have chronic hypertension that can itself produce neurocognitive deficits. Patients with primary lobar ICH usually have CAA. This may also be associated with preexisting, progressive neurocognitive deficits, and in some cases with Alzheimer's disease. Treatment of cognitive deficits after ICH has seldom been investigated.

References

1. Qureshi AI, Mendelow AD, Hanley DF. Intracerebral haemorrhage. Lancet. 2009;373(9675):1632–44. PubMed PMID: 19427958. English.
2. Rosand J, Hylek EM, O'Donnell HC, Greenberg SM. Warfarin-associated hemorrhage and cerebral amyloid angiopathy: a genetic and pathologic study. Neurology. 2000;55(7):947–51. PubMed PMID: 11061249.
3. Smith EE, Rosand J, Greenberg SM. Imaging of hemorrhagic stroke. Magn Reson Imaging Clin N Am. 2006;14(2):127–40. v. PubMed PMID: 16873007.
4. Biffi A, Sonni A, Anderson CD, Kissela B, Jagiella JM, Schmidt H, et al. Variants at APOE influence risk of deep and lobar intracerebral hemorrhage. Ann Neurol. 2010;68(6):934–43. PubMed PMID: 21061402. Epub 2010/11/10. Eng.
5. Woo D, Sauerbeck LR, Kissela BM, Khoury JC, Szaflarski JP, Gebel J, et al. Genetic and environmental risk factors for intracerebral hemorrhage: preliminary results of a population-based study. Stroke. 2002;33(5):1190–5. PubMed PMID: 11988589.
6. O'Donnell MJ, Xavier D, Liu L, Zhang H, Chin SL, Rao-Melacini P, et al. Risk factors for ischaemic and intracerebral haemorrhagic stroke in 22 countries (the INTERSTROKE study): a case-control study. Lancet. 2010;376(9735):112–23. PubMed PMID: 20561675. Epub 2010/06/22. eng.
7. Nys GM, van Zandvoort MJ, de Kort PL, Jansen BP, de Haan EH, Kappelle LJ. Cognitive disorders in acute stroke: prevalence and clinical determinants. Cerebrovasc Dis. 2007;23 (5–6):408–16. PubMed PMID: 17406110. Epub 2007/04/05. eng.
8. Pendlebury ST, Rothwell PM. Prevalence, incidence, and factors associated with pre-stroke and post-stroke dementia: a systematic review and meta-analysis. Lancet Neurol. 2009;8(11):1006–18. PubMed PMID: 19782001. Epub 2009/09/29. eng.

9. Cordonnier C, Leys D, Dumont F, Deramecourt V, Bordet R, Pasquier F, et al. What are the causes of pre-existing dementia in patients with intracerebral haemorrhages? Brain. 2010;133(11):3281–9. PubMed PMID: 20852266. Epub 2010/09/21. eng.
10. Smith EE, Gurol ME, Eng JA, Engel CR, Nguyen TN, Rosand J, et al. White matter lesions, cognition, and recurrent hemorrhage in lobar intracerebral hemorrhage. Neurology. 2004;63(9):1606–12. PubMed PMID: 15534243.
11. Smith EE, Nandigam KR, Chen YW, Jeng J, Salat D, Halpin A, et al. MRI markers of small vessel disease in lobar and deep hemispheric intracerebral hemorrhage. Stroke. 2010;41(9):1933–8. PubMed PMID: 20689084. Epub 2010/08/07. eng.
12. Viswanathan A, Greenberg SM. Cerebral amyloid angiopathy in the elderly. Ann Neurol. 2011;70(6):871–80. PubMed PMID: 22190361. Epub 2011/12/23. eng.
13. Sachdev PS, Chen X, Brodaty H, Thompson C, Altendorf A, Wen W. The determinants and longitudinal course of post-stroke mild cognitive impairment. J Int Neuropsychol Soc. 2009;15(6):915–23. PubMed PMID: 19891821. Epub 2009/11/07. eng.
14. Pendlebury ST, Cuthbertson FC, Welch SJ, Mehta Z, Rothwell PM. Underestimation of cognitive impairment by Mini-Mental State Examination versus the Montreal Cognitive Assessment in patients with transient ischemic attack and stroke: a population-based study. Stroke. 2010;41(6):1290–3. PubMed PMID: 20378863. Epub 2010/04/10. eng.
15. Su CY, Chen HM, Kwan AL, Lin YH, Guo NW. Neuropsychological impairment after hemorrhagic stroke in basal ganglia. Arch Clin Neuropsychol. 2007;22(4):465–74. PubMed PMID: 17336034. English.
16. Su CY, Wuang YP, Chang JK, Guo NW, Kwan AL. Wisconsin Card Sorting Test performance after putaminal hemorrhagic stroke. Kaohsiung J Med Sci. 2006;22(2):75–84. PubMed PMID: 16568724. English.
17. Kawahara N, Sato K, Muraki M, Tanaka K, Kaneko M, Uemura K. CT classification of small thalamic hemorrhages and their clinical implications. Neurology. 1986;36(2):165–72. PubMed PMID: 3945386. Epub 1986/02/01. eng.
18. Mijovic-Prelec D, Bentley P, Caviness Jr VS. Selective rotation of egocentric spatial representation following right putaminal hemorrhage. Neuropsychologia. 2004;42(13):1827–37. PubMed PMID: 15351631. Epub 2004/09/08. eng.
19. Maeshima S, Truman G, Smith DS, Dohi N, Nakai K, Itakura T, et al. Apraxia and cerebral haemorrhage: the relationship between haematoma volume and prognosis. J Clin Neurosci. 2000;7(4):309–11. PubMed PMID: 10938607. Epub 2000/08/12. eng.
20. Kawanishi M, Kajikawa H, Yamamura K, Nomura E, Kajikawa M, Hihara R, et al. Aphasia following left putaminal hemorrhage. Statistical analysis of factors affecting prognosis. Neurol Res. 2002;24(8):817–21. PubMed PMID: 12500706. Epub 2002/12/26. eng.
21. Damasio H, Damasio AR. The anatomical basis of conduction aphasia. Brain. 1980;103(2):337–50. PubMed PMID: 7397481. Epub 1980/06/01. eng.
22. Naeser MA, Hayward RW. The resolving stroke and aphasia. A case study with computerized tomography. Arch Neurol. 1979;36(4):233–5. PubMed PMID: 426669. Epub 1979/04/01. eng.
23. Maeshima S, Truman G, Smith DS, Dohi N, Itakura T, Komai N. Buccofacial apraxia and left cerebral haemorrhage. Brain Inj. 1997;11(11):777–82. PubMed PMID: 9354254. Epub 1997/11/14. eng.
24. Fuh JL, Wang SJ. Caudate hemorrhage: clinical features, neuropsychological assessments and radiological findings. Clin Neurol Neurosurg. 1995;97(4):296–9. PubMed PMID: 8599895. Epub 1995/11/01. eng.
25. Su CY, Chang JJ, Chen HM, Su CJ, Chien TH, Huang MH. Perceptual differences between stroke patients with cerebral infarction and intracerebral hemorrhage. Arch Phys Med Rehabil. 2000;81(6):706–14. PubMed PMID: 10857510. Epub 2000/06/17. eng.
26. Bhatia KP, Marsden CD. The behavioural and motor consequences of focal lesions of the basal ganglia in man. Brain. 1994;117(Pt 4):859–76. PubMed PMID: 7922471. Epub 1994/08/01. eng.

27. Maeshima S, Ueyoshi A, Matsumoto T, Boh-oka S, Yoshida M, Itakura T, et al. Unilateral spatial neglect in patients with cerebral hemorrhage: the relationship between hematoma volume and prognosis. J Clin Neurosci. 2002;9(5):544–8. PubMed PMID: 12383412. Epub 2002/10/18. eng.
28. Morgenstern LB, Hemphill III JC, Anderson C, Becker K, Broderick JP, Connolly Jr ES, et al. Guidelines for the management of spontaneous intracerebral hemorrhage: a guideline for healthcare professionals from the American Heart Association/American Stroke Association. Stroke. 2010;41(9):2108–29. PubMed PMID: 20651276. Epub 2010/07/24. eng.
29. Hemphill III JC, Bonovich DC, Besmertis L, Manley GT, Johnston SC. The ICH score: a simple, reliable grading scale for intracerebral hemorrhage. Stroke. 2001;32(4):891–7. PubMed PMID: 11283388.
30. Rost NS, Smith EE, Chang Y, Snider RW, Chanderraj R, Schwab K, et al. Prediction of functional outcome in patients with primary intracerebral hemorrhage: the FUNC score. Stroke. 2008;39(8):2304–9. PubMed PMID: 18556582.
31. Mendelow AD, Gregson BA, Fernandes HM, Murray GD, Teasdale GM, Hope DT, et al. Early surgery versus initial conservative treatment in patients with spontaneous supratentorial intracerebral haematomas in the International Surgical Trial in Intracerebral Haemorrhage (STICH): a randomised trial. Lancet. 2005;365(9457):387–97. PubMed PMID: 15680453.
32. Anderson CS, Huang Y, Wang JG, Arima H, Neal B, Peng B, et al. Intensive blood pressure reduction in acute cerebral haemorrhage trial (INTERACT): a randomised pilot trial. Lancet Neurol. 2008;7(5):391–9. PubMed PMID: 18396107.
33. Mayer SA, Brun NC, Begtrup K, Broderick J, Davis S, Diringer MN, et al. Efficacy and safety of recombinant activated factor VII for acute intracerebral hemorrhage. N Engl J Med. 2008;358(20):2127–37. PubMed PMID: 18480205.
34. O'Donnell HC, Rosand J, Knudsen KA, Furie KL, Segal AZ, Chiu RI, et al. Apolipoprotein E genotype and the risk of recurrent lobar intracerebral hemorrhage. N Engl J Med. 2000;342(4):240–5. PubMed PMID: 10648765.
35. Biffi A, Halpin A, Towfighi A, Gilson A, Busl K, Rost NS, et al. Antiplatelet agents and recurrent intracerebral hemorrhage in cerebral amyloid angiopathy. Neurology. 2010;75(8):693–8.
36. Chapman N, Huxley R, Anderson C, Bousser MG, Chalmers J, Colman S, et al. Effects of a perindopril-based blood pressure-lowering regimen on the risk of recurrent stroke according to stroke subtype and medical history: the PROGRESS Trial. Stroke. 2004;35(1):116–21. PubMed PMID: 14671247.
37. Arima H, Tzourio C, Anderson C, Woodward M, Bousser MG, MacMahon S, et al. Effects of perindopril-based lowering of blood pressure on intracerebral hemorrhage related to amyloid angiopathy: the PROGRESS trial. Stroke. 2010;41(2):394–6. PubMed PMID: 20044530. Epub 2010/01/02. eng.
38. Fazekas F, Kleinert R, Roob G, Kleinert G, Kapeller P, Schmidt R, et al. Histopathologic analysis of foci of signal loss on gradient-echo T2*-weighted MR images in patients with spontaneous intracerebral hemorrhage: evidence of microangiopathy-related microbleeds. AJNR Am J Neuroradiol. 1999;20(4):637–42.
39. Greenberg SM, Vernooij MW, Cordonnier C, Viswanathan A, Salman RA, Warach S, et al. Cerebral microbleeds: a guide to detection and interpretation. Lancet Neurol. 2009;8(2):165–74. PubMed PMID: 19161908. English.
40. Jeerakathil T, Wolf PA, Beiser A, Hald JK, Au R, Kase CS, et al. Cerebral microbleeds: prevalence and associations with cardiovascular risk factors in the Framingham Study. Stroke. 2004;35(8):1831–5. PubMed PMID: 15155954.
41. Knudsen KA, Rosand J, Karluk D, Greenberg SM. Clinical diagnosis of cerebral amyloid angiopathy: validation of the Boston criteria. Neurology. 2001;56(4):537–9. PubMed PMID: 11222803.

42. Werring DJ, Frazer DW, Coward LJ, Losseff NA, Watt H, Cipolotti L, et al. Cognitive dysfunction in patients with cerebral microbleeds on T2*-weighted gradient-echo MRI. Brain. 2004;127(Pt 10):2265–75. PubMed PMID: 15282216.
43. Cordonnier C, Potter GM, Jackson CA, Doubal F, Keir S, Sudlow CL, et al. Improving interrater agreement about brain microbleeds: development of the Brain Observer MicroBleed Scale (BOMBS). Stroke. 2009;40(1):94–9. PubMed PMID: 19008468.
44. Pettersen JA, Sathiyamoorthy G, Gao FQ, Szilagyi G, Nadkarni NK, St George-Hyslop P, et al. Microbleed topography, leukoaraiosis, and cognition in probable Alzheimer disease from the Sunnybrook dementia study. Arch Neurol. 2008;65(6):790–5. PubMed PMID: 18541799. English.
45. Atri A, Locascio JJ, Lin JM, Yap L, Dickerson BC, Grodstein F, et al. Prevalence and effects of lobar microhemorrhages in early-stage dementia. Neurodegener Dis. 2005;2(6):305–12. PubMed PMID: 16909013. English.
46. Goos JD, Kester MI, Barkhof F, Klein M, Blankenstein MA, Scheltens P, et al. Patients with Alzheimer disease with multiple microbleeds: relation with cerebrospinal fluid biomarkers and cognition. Stroke. 2009;40(11):3455–60. PubMed PMID: 19762705. Epub 2009/09/19. eng.
47. Henneman WJ, Sluimer JD, Cordonnier C, Baak MM, Scheltens P, Barkhof F, et al. MRI biomarkers of vascular damage and atrophy predicting mortality in a memory clinic population. Stroke. 2009;40(2):492–8. PubMed PMID: 19109551. Epub 2008/12/26. eng.
48. Kirsch W, McAuley G, Holshouser B, Petersen F, Ayaz M, Vinters HV, et al. Serial susceptibility weighted MRI measures brain iron and microbleeds in dementia. J Alzheimers Dis. 2009;17(3):599–609. PubMed PMID: 19433895. Pubmed Central PMCID: 2788087. Epub 2009/05/13. eng.
49. Staekenborg SS, Koedam EL, Henneman WJ, Stokman P, Barkhof F, Scheltens P, et al. Progression of mild cognitive impairment to dementia: contribution of cerebrovascular disease compared with medial temporal lobe atrophy. Stroke. 2009;40(4):1269–74. PubMed PMID: 19228848. Epub 2009/02/21. eng.
50. Seo SW, Hwa Lee B, Kim EJ, Chin J, Sun Cho Y, Yoon U, et al. Clinical significance of microbleeds in subcortical vascular dementia.[see comment]. Stroke. 2007;38(6):1949–51. PubMed PMID: 17510457. English.
51. Yakushiji Y, Nishiyama M, Yakushiji S, Hirotsu T, Uchino A, Nakajima J, et al. Brain microbleeds and global cognitive function in adults without neurological disorder. Stroke. 2008;39(12):3323–8. PubMed PMID: 18688000. English.
52. Van Es AC, van der Grond J, de Craen AJ, Westendorp RG, Bollen EL, Blauw GJ, et al. Cerebral microbleeds and cognitive functioning in the PROSPER study. Neurology. 2011;77(15):1446–52.
53. Takashima Y, Mori T, Hashimoto M, Kinukawa N, Uchino A, Yuzuriha T, et al. Clinical correlating factors and cognitive function in community-dwelling healthy subjects with cerebral microbleeds. J Stroke Cerebrovasc Dis. 2011;20(2):105–10.
54. Vinters HV. Cerebral amyloid angiopathy. A critical review. Stroke. 1987;18(2):311–24.
55. Revesz T, Holton JL, Lashley T, Plant G, Frangione B, Rostagno A, et al. Genetics and molecular pathogenesis of sporadic and hereditary cerebral amyloid angiopathies. Acta Neuropathol. 2009;118(1):115–30. PubMed PMID: 19225789. Epub 2009/02/20. eng.
56. Vonsattel JP, Myers RH, Hedley-Whyte ET, Ropper AH, Bird ED, Richardson Jr EP. Cerebral amyloid angiopathy without and with cerebral hemorrhages: a comparative histological study. Ann Neurol. 1991;30(5):637–49. PubMed PMID: 1763890.
57. Keage HA, Carare RO, Friedland RP, Ince PG, Love S, Nicoll JA, et al. Population studies of sporadic cerebral amyloid angiopathy and dementia: a systematic review. BMC Neurol. 2009;9(1):3. PubMed PMID: 19144113.
58. Neuropathology Group of the Medical Research Council Cognitive Function and Ageing Study (MRC CFAS). Pathological correlates of late-onset dementia in a multicentre, community-based population in England and Wales Lancet. 2001;357(9251):169–175. PubMed PMID: 11213093.

59. Smith EE, Greenberg SM. Beta-amyloid, blood vessels, and brain function. Stroke. 2009;40(7):2601–6. PubMed PMID: 19443808. Pubmed Central PMCID: 2704252. Epub 2009/05/16. eng.
60. Weller RO, Boche D, Nicoll JA. Microvasculature changes and cerebral amyloid angiopathy in Alzheimer's disease and their potential impact on therapy. Acta Neuropathol. 2009;118(1):87–102. PubMed PMID: 19234858. English.
61. Greenberg SM, Rebeck GW, Vonsattel JP, Gomez-Isla T, Hyman BT. Apolipoprotein E epsilon 4 and cerebral hemorrhage associated with amyloid angiopathy. Ann Neurol. 1995;38(2):254–9.
62. Greenberg SM, Vonsattel JP, Segal AZ, Chiu RI, Clatworthy AE, Liao A, et al. Association of apolipoprotein E epsilon2 and vasculopathy in cerebral amyloid angiopathy. Neurology. 1998;50(4):961–5.
63. Rosand J, Muzikansky A, Kumar A, Wisco JJ, Smith EE, Betensky RA, et al. Spatial clustering of hemorrhages in probable cerebral amyloid angiopathy. Ann Neurol. 2005;58(3):459–62. PubMed PMID: 16130107.
64. Johnson KA, Gregas M, Becker JA, Kinnecom C, Salat DH, Moran EK, et al. Imaging of amyloid burden and distribution in cerebral amyloid angiopathy. Ann Neurol. 2007;62:229–34.
65. Linn J, Halpin A, Demaerel P, Ruhland J, Giese AD, Dichgans M, et al. Prevalence of superficial siderosis in patients with cerebral amyloid angiopathy. Neurology. 2010;74(17):1346–50.
66. Kinnecom C, Lev MH, Wendell L, Smith EE, Rosand J, Frosch MP, et al. Course of cerebral amyloid angiopathy-related inflammation. Neurology. 2007;68(17):1411–6. PubMed PMID: 17452586.
67. Matthews FE, Brayne C, Lowe J, McKeith I, Wharton SB, Ince P. Epidemiological pathology of dementia: attributable-risks at death in the Medical Research Council Cognitive Function and Ageing Study. PLoS Med. 2009;6(11):e1000180. PubMed PMID: 19901977. Epub 2009/11/11. eng.
68. Pfeifer LA, White LR, Ross GW, Petrovitch H, Launer LJ. Cerebral amyloid angiopathy and cognitive function: the HAAS autopsy study. Neurology. 2002;58(11):1629–34.
69. Arvanitakis Z, Leurgans SE, Wang Z, Wilson RS, Bennett DA, Schneider JA. Cerebral amyloid angiopathy pathology and cognitive domains in older persons. Ann Neurol. 2011;69(2):320–7. PubMed PMID: 21387377. Epub 2011/03/10. eng.
70. Chabriat H, Joutel A, Dichgans M, Tournier-Lasserve E, Bousser MG. Cadasil. Lancet Neurol. 2009;8(7):643–53. PubMed PMID: 19539236. Epub 2009/06/23. eng.
71. Dichgans M, Mayer M, Uttner I, Bruning R, Muller-Hocker J, Rungger G, et al. The phenotypic spectrum of CADASIL: clinical findings in 102 cases. Ann Neurol. 1998;44(5):731–9.
72. Charlton RA, Morris RG, Nitkunan A, Markus HS. The cognitive profiles of CADASIL and sporadic small vessel disease. Neurology. 2006;66(10):1523–6. PubMed PMID: 16717212. English.
73. Dichgans M. Cognition in CADASIL. Stroke. 2009;40(3 Suppl):S45–7. PubMed PMID: 19064778. English.
74. Reyes S, Viswanathan A, Godin O, Dufouil C, Benisty S, Hernandez K, et al. Apathy: a major symptom in CADASIL. Neurology. 2009;72(10):905–10. PubMed PMID: 19273824. Epub 2009/03/11. eng.
75. Dichgans M, Markus HS, Salloway S, Verkkoniemi A, Moline M, Wang Q, et al. Donepezil in patients with subcortical vascular cognitive impairment: a randomised double-blind trial in CADASIL. Lancet Neurol. 2008;7(4):310–8. PubMed PMID: 18296124.
76. Hajj-Ali RA, Singhal AB, Benseler S, Molloy E, Calabrese LH. Primary angiitis of the CNS. Lancet Neurol. 2011;10(6):561–72. PubMed PMID: 21601163. Epub 2011/05/24. eng.
77. Scolding NJ, Joseph F, Kirby PA, Mazanti I, Gray F, Mikol J, et al. Abeta-related angiitis: primary angiitis of the central nervous system associated with cerebral amyloid angiopathy. Brain. 2005;128(Pt 3):500–15. PubMed PMID: 15659428.

78. Glass J, Hochberg F, Miller DC. Intravascular lymphomatosis: a systemic disease with neurologic manifestations. Cancer. 2006;71(10):3145–64.
79. Casato M, Saadoun D, Marchetti A, Limal N, Picq C, Pantano P, et al. Central nervous system involvement in hepatitis C virus cryoglobulinemia vasculitis: a multicenter case-control study using magnetic resonance imaging and neuropsychological tests. J Rheumatol. 2005;32(3):484–8.
80. Mattioli F, Capra R, Rovaris M, Chiari S, Codella M, Miozzo A, et al. Frequency and patterns of subclinical cognitive impairment in patients with ANCA-associated small vessel vasculitides. J Neurol Sci. 2002;195(2):161–6.
81. Hochberg MC. Updating the American College of Rheumatology revised criteria for the classification of systemic lupus erythematosus. Arthritis Rheum. 1997;40(9):1725. PubMed PMID: 9324032. Epub 1997/10/27. eng.
82. American College of Rheumatology Ad Hoc Committee on Neuropsychiatric Lupus Nomenclature. The American College of Rheumatology nomenclature and case definitions for neuropsychiatric lupus syndromes. Arthritis Rheum. 1999;42(4):599–608. PubMed PMID: 10211873.
83. Colasanti T, Delunardo F, Margutti P, Vacirca D, Piro E, Siracusano A, et al. Autoantibodies involved in neuropsychiatric manifestations associated with systemic lupus erythematosus. J Neuroimmunol. 2009;212(1–2):3–9. PubMed PMID: 19500858. Epub 2009/06/09. eng.
84. Mikdashi J, Handwerger B, Langenberg P, Miller M, Kittner S. Baseline disease activity, hyperlipidemia, and hypertension are predictive factors for ischemic stroke and stroke severity in systemic lupus erythematosus. Stroke. 2007;38(2):281–5. PubMed PMID: 17218611. Epub 2007/01/16. eng.
85. Futrell N, Millikan C. Frequency, etiology, and prevention of stroke in patients with systemic lupus erythematosus. Stroke. 1989;20(5):583–91. PubMed PMID: 2718197. Epub 1989/05/01. eng.
86. Monastero R, Bettini P, Del Zotto E, Cottini E, Tincani A, Balestrieri G, et al. Prevalence and pattern of cognitive impairment in systemic lupus erythematosus patients with and without overt neuropsychiatric manifestations. J Neurol Sci. 2001;184(1):33–9. PubMed PMID: 11231030. Epub 2001/03/07. eng.
87. Loukkola J, Laine M, Ainiala H, Peltola J, Metsanoja R, Auvinen A, et al. Cognitive impairment in systemic lupus erythematosus and neuropsychiatric systemic lupus erythematosus: a population-based neuropsychological study. J Clin Exp Neuropsychol. 2003;25(1):145–51. PubMed PMID: 12607179. Epub 2003/02/28. eng.
88. Hanly JG, Fisk JD, Sherwood G, Eastwood B. Clinical course of cognitive dysfunction in systemic lupus erythematosus. J Rheumatol. 1994;21(10):1825–31. PubMed PMID: 7837145. Epub 1994/10/01. eng.
89. Denburg SD, Denburg JA, Carbotte RM, Fisk JD, Hanly JG. Cognitive deficits in systemic lupus erythematosus. Rheum Dis Clin North Am. 1993;19(4):815–31. PubMed PMID: 8265824. Epub 1993/11/01. eng.
90. Nowicka-Sauer K, Czuszynska Z, Smolenska Z, Siebert J. Neuropsychological assessment in systemic lupus erythematosus patients: clinical usefulness of first-choice diagnostic tests in detecting cognitive impairment and preliminary diagnosis of neuropsychiatric lupus. Clin Exp Rheumatol. 2011;29(2):299–306. PubMed PMID: 21418780. Epub 2011/03/23. eng.
91. Fisk JD, Eastwood B, Sherwood G, Hanly JG. Patterns of cognitive impairment in patients with systemic lupus erythematosus. Br J Rheumatol. 1993;32(6):458–62. PubMed PMID: 8508281. Epub 1993/06/01. eng.
92. Kozora E, Arciniegas DB, Filley CM, West SG, Brown M, Miller D, et al. Cognitive and neurologic status in patients with systemic lupus erythematosus without major neuropsychiatric syndromes. Arthritis Rheum. 2008;59(11):1639–46. PubMed PMID: 18975359. Epub 2008/11/01. eng.
93. Mikdashi JA, Esdaile JM, Alarcon GS, Crofford L, Fessler BJ, Shanberg L, et al. Proposed response criteria for neurocognitive impairment in systemic lupus erythematosus clinical trials. Lupus. 2007;16(6):418–25. PubMed PMID: 17664232. Epub 2007/08/01. eng.

94. Suzuki J, Takaku A. Cerebrovascular "moyamoya" disease. Disease showing abnormal net-like vessels in base of brain. Arch Neurol. 1969;20(3):288–99. PubMed PMID: 5775283.
95. Scott RM, Smith ER. Moyamoya disease and moyamoya syndrome. N Engl J Med. 2009;360(12):1226–37. PubMed PMID: 19297575. English.
96. Levin SD, Hoare RD, Robinson RO. Childhood moyamoya presenting as dementia: report of a case. Dev Med Child Neurol. 1983;25(6):794–7. PubMed PMID: 6653912.
97. Sato H, Sato N, Tamaki N, Matsumoto S. Chronic low-perfusion state in children with moyamoya disease following revascularization. Childs Nerv Syst. 1990;6(3):166–71. PubMed PMID: 2357714.
98. Hogan AM, Kirkham FJ, Isaacs EB, Wade AM, Vargha-Khadem F. Intellectual decline in children with moyamoya and sickle cell anaemia. Dev Med Child Neurol. 2005;47(12):824–9. PubMed PMID: 16288673. Epub 2005/11/18. eng.
99. Karasawa J, Touho H, Ohnishi H, Miyamoto S, Kikuchi H. Long-term follow-up study after extracranial-intracranial bypass surgery for anterior circulation ischemia in childhood moyamoya disease. J Neurosurg. 1992;77(1):84–9. PubMed PMID: 1607976.
100. Weinberg DG, Rahme RJ, Aoun SG, Batjer HH, Bendok BR. Moyamoya disease: functional and neurocognitive outcomes in the pediatric and adult populations. Neurosurg Focus. 2011;30(6):E21. PubMed PMID: 21631223. Epub 2011/06/03. eng.
101. Calviere L, Catalaa I, Marlats F, Viguier A, Bonneville F, Cognard C, et al. Correlation between cognitive impairment and cerebral hemodynamic disturbances on perfusion magnetic resonance imaging in European adults with moyamoya disease. Clinical article. J Neurosurg. 2010;113(4):753–9. PubMed PMID: 20469988. Epub 2010/05/18. eng.
102. Karzmark P, Zeifert PD, Tan S, Dorfman LJ, Bell-Stephens TE, Steinberg GK. Effect of moyamoya disease on neuropsychological functioning in adults. Neurosurgery. 2008;62(5):1048–51; discussion 51–2. PubMed PMID: 18580802. Epub 2008/06/27. eng.
103. Festa JR, Schwarz LR, Pliskin N, Cullum CM, Lacritz L, Charbel FT, et al. Neurocognitive dysfunction in adult moyamoya disease. J Neurol. 2010;257(5):806–15. PubMed PMID: 20033200. Epub 2009/12/25. eng.
104. Ishii R, Takeuchi S, Ibayashi K, Tanaka R. Intelligence in children with moyamoya disease: evaluation after surgical treatments with special reference to changes in cerebral blood flow. Stroke. 1984;15(5):873–7. PubMed PMID: 6474539.
105. Lee JY, Phi JH, Wang KC, Cho BK, Shin MS, Kim SK. Neurocognitive profiles of children with moyamoya disease before and after surgical intervention. Cerebrovasc Dis. 2011;31(3):230–7. PubMed PMID: 21178347. Epub 2010/12/24. eng.
106. Matsushima Y, Aoyagi M, Nariai T, Takada Y, Hirakawa K. Long-term intelligence outcome of post-encephalo-duro-arterio-synangiosis childhood moyamoya patients. Clin Neurol Neurosurg. 1997;99 Suppl 2:S147–50. PubMed PMID: 9409426.
107. Imaizumi C, Imaizumi T, Osawa M, Fukuyama Y, Takeshita M. Serial intelligence test scores in pediatric moyamoya disease. Neuropediatrics. 1999;30(6):294–9. PubMed PMID: 10706023.
108. Matsushima Y, Aoyagi M, Niimi Y, Masaoka H, Ohno K. Symptoms and their pattern of progression in childhood moyamoya disease. Brain Dev. 1990;12(6):784–9. PubMed PMID: 2092589.
109. Platt OS, Brambilla DJ, Rosse WF, Milner PF, Castro O, Steinberg MH, et al. Mortality in sickle cell disease. Life expectancy and risk factors for early death. N Engl J Med. 1994;330(23):1639–44. PubMed PMID: 7993409. Epub 1994/06/09. eng.
110. Adams RJ, McKie VC, Hsu L, Files B, Vichinsky E, Pegelow C, et al. Prevention of a first stroke by transfusions in children with sickle cell anemia and abnormal results on transcranial Doppler ultrasonography. N Engl J Med. 1998;339(1):5–11. PubMed PMID: 9647873.
111. Debaun MR, Sarnaik SA, Rodeghier MJ, Minniti CP, Howard TH, Iyer RV, et al. Associated risk factors for silent cerebral infarcts in sickle cell anemia: low baseline hemoglobin, gender and relative high systolic blood pressure. Blood. 2011;119(16):3684–90. PubMed PMID: 22096242. Epub 2011/11/19. Eng.

112. Armstrong FD, Thompson Jr RJ, Wang W, Zimmerman R, Pegelow CH, Miller S, et al. Cognitive functioning and brain magnetic resonance imaging in children with sickle cell disease. Neuropsychology Committee of the Cooperative Study of Sickle Cell Disease. Pediatrics. 1996;97(6 Pt 1):864–70. PubMed PMID: 8657528. Epub 1996/06/01. eng.
113. Ohene-Frempong K, Weiner SJ, Sleeper LA, Miller ST, Embury S, Moohr JW, et al. Cerebrovascular accidents in sickle cell disease: rates and risk factors. Blood. 1998;91(1):288–94. PubMed PMID: 9414296. Epub 1998/02/07. eng.
114. Hijmans CT, Fijnvandraat K, Grootenhuis MA, van Geloven N, Heijboer H, Peters M, et al. Neurocognitive deficits in children with sickle cell disease: a comprehensive profile. Pediatr Blood Cancer. 2011;56(5):783–8. PubMed PMID: 21370411. Epub 2011/03/04. eng.
115. Scantlebury N, Mabbott D, Janzen L, Rockel C, Widjaja E, Jones G, et al. White matter integrity and core cognitive function in children diagnosed with sickle cell disease. J Pediatr Hematol Oncol. 2011;33(3):163–71. PubMed PMID: 21325970. Epub 2011/02/18. eng.
116. Hijmans CT, Grootenhuis MA, Oosterlaan J, Heijboer H, Peters M, Fijnvandraat K. Neurocognitive deficits in children with sickle cell disease are associated with the severity of anemia. Pediatr Blood Cancer. 2011;57(2):297–302. PubMed PMID: 21671366. Epub 2011/06/15. eng.
117. Strouse JJ, Cox CS, Melhem ER, Lu H, Kraut MA, Razumovsky A, et al. Inverse correlation between cerebral blood flow measured by continuous arterial spin-labeling (CASL) MRI and neurocognitive function in children with sickle cell anemia (SCA). Blood. 2006;108(1):379–81. PubMed PMID: 16537809. Pubmed Central PMCID: 1482738. Epub 2006/03/16. eng.
118. Vichinsky EP, Neumayr LD, Gold JI, Weiner MW, Rule RR, Truran D, et al. Neuropsychological dysfunction and neuroimaging abnormalities in neurologically intact adults with sickle cell anemia. JAMA. 2010;303(18):1823–31. PubMed PMID: 20460621. Pubmed Central PMCID: 2892214. Epub 2010/05/13. eng.
119. Ware RE, Helms RW. Stroke with transfusions changing to hydroxyurea (SWiTCH). Blood. 2012;119(17):3925–32. PubMed PMID: 22318199. Epub 2012/02/10. Eng.
120. Gomez-Puerta JA, Cervera R, Calvo LM, Gomez-Anson B, Espinosa G, Claver G, et al. Dementia associated with the antiphospholipid syndrome: clinical and radiological characteristics of 30 patients. Rheumatology (Oxford). 2005;44(1):95–9. PubMed PMID: 15367749. Epub 2004/09/16. eng.
121. Wright RA, Kokmen E. Gradually progressive dementia without discrete cerebrovascular events in a patient with Sneddon's syndrome. Mayo Clin Proc. 1999;74(1):57–61. PubMed PMID: 9987534. Epub 1999/02/13. eng.
122. Wilson WB, Smith DB, Wright RR, Seibert CE. Sneddon's disease presenting with visual loss and dementia. J Clin Neuroophthalmol. 1988;8(4):255–61. PubMed PMID: 2977141. Epub 1988/12/01. eng.
123. George JN. Clinical practice. Thrombotic thrombocytopenic purpura. N Engl J Med. 2006;354(18):1927–35. PubMed PMID: 16672704. Epub 2006/05/05. eng.
124. Mehta J, Singhal S. Hyperviscosity syndrome in plasma cell dyscrasias. Semin Thromb Hemost. 2003;29(5):467–71. PubMed PMID: 14631546. Epub 2003/11/25. eng.
125. Bousser MG, Ferro JM. Cerebral venous thrombosis: an update. Lancet Neurol. 2007;6(2):162–70. PubMed PMID: 17239803. Epub 2007/01/24. eng.
126. Hurst RW, Bagley LJ, Galetta S, Glosser G, Lieberman AP, Trojanowski J, et al. Dementia resulting from dural arteriovenous fistulas: the pathologic findings of venous hypertensive encephalopathy. AJNR Am J Neuroradiol. 1998;19(7):1267–73. PubMed PMID: 9726465. Epub 1998/09/03. eng.
127. Datta NN, Rehman SU, Kwok JC, Chan KY, Poon CY. Reversible dementia due to dural arteriovenous fistula: a simple surgical option. Neurosurg Rev. 1998;21(2–3):174–6. PubMed PMID: 9795956. Epub 1998/10/31. eng.
128. Zeidman SM, Monsein LH, Arosarena O, Aletich V, Biafore JA, Dawson RC, et al. Reversibility of white matter changes and dementia after treatment of dural fistulas. AJNR Am J Neuroradiol. 1995;16(5):1080–3. PubMed PMID: 7543726. Epub 1995/05/01. eng.

Chapter 12
Post-stroke Depression

Bradleigh D. Hayhow, Simone Brockman, and Sergio E. Starkstein

Introduction

Post-stroke depression (PSD) is one of the most common neuropsychiatric sequelae of stroke. Cross-sectional studies of stroke survivors have demonstrated that about one-third of patients develop acute PSD and more than half suffer depression at some later point in their lives [1, 2]. PSD is strongly associated with a range of adverse clinical outcomes including increased length of hospital stay, higher risk of dependency, increased degree of neurological impairment, and increased patient mortality [3–6]. Therefore, there is significant potential benefit for effective prevention, diagnosis, and treatment of PSD.

Despite the acknowledged clinical and social impacts of PSD, the definitive management of this condition remains unresolved [7–10]. In this chapter, we will address the main consequences of PSD and highlight the best practice diagnosis, prevention, and treatment options based on current evidence.

Diagnosis of Post-stroke Depression

PSD is defined within the *Diagnostic and Statistical Manual of Mental Disorders, fourth edition, text revision (DSM-IV-TR)* as a "mood disorder due to a general medical condition" [11]. To satisfy the diagnostic criteria, there must be evidence from the history, physical examination, or laboratory findings that the disturbance is

B.D. Hayhow, B.A.(Hons.), B.M.B.S. • S. Brockman, B.Sc.(Hons.)
S.E. Starkstein, M.D., Ph.D. (✉)
Department of Psychiatry, Fremantle Hospital, Alma Street, Fremantle, WA 6959, Australia
e-mail: Sergio.starkstein@uwa.edu.au

T.A. Schweizer and R.L. Macdonald (eds.), *The Behavioral Consequences of Stroke*, DOI 10.1007/978-1-4614-7672-6_12, © Springer Science+Business Media New York 2014 a direct physiological consequence of medical illness. A competing explanation for the genesis of PSD identifies it as a complex psychosocial reaction to critical illness, rather than a direct physiological consequence of stroke. However, this objection is partly overcome by the finding that patients suffering equivalent levels of disability due to orthopedic illness or traumatic brain injury have a lower incidence of depression than stroke patients [12, 13]. Nevertheless, psychosocial factors have been identified to play a role in PSD, and the most parsimonious explanation may be that PSD results from a combination of bio-psychosocial factors [14].

The DSM-IV-TR criteria for the diagnosis of "mood disorder due to a general medical condition" include that symptoms must occur independently of delirium and must cause clinically significant distress or impairment in social, occupational, or other important areas of functioning. The diagnosis is further refined by the specifiers "with depressive features," "with major depressive-like episode," "with manic features," and "with mixed features." The DSM-IV-TR also outlines research criteria for "minor depression," which is a concept that some investigators have utilized in relation to their study of PSD [15]. While variations in diagnostic criteria may account for some of the inconsistency in the research of PSD, major and minor depression as diagnosed by DSM-IV criteria remain the two most useful clinical diagnostic categories [16].

A major depressive episode is diagnosed based on the presence of either a depressed mood or a loss of interest or pleasure, in addition to at least four other depressive symptoms with at least 2 weeks duration. These include an increase or decrease in weight or appetite, insomnia or hypersomnia, psychomotor agitation or retardation, fatigue or loss of energy, feelings of worthlessness or excessive/inappropriate guilt, diminished ability to think or concentrate, and recurrent thoughts of death or suicidal ideation. The criteria for a minor depressive episode are identical to those for a major depressive episode, but require only two depressive symptoms in addition to the core symptom of either a depressed mood or a loss of interest or pleasure. When minor depression is taken into account, the prevalence of PSD rises to capture approximately half of all stroke survivors [2].

A number of tools are available to aid in the measurement of depressive features in clinical settings. These can be found in the form of structured diagnostic interviews that provide a categorical diagnosis, or severity rating scales that indicate a diagnosis on the basis of a cutoff score. It is important to note that depression should only be diagnosed with the standardized diagnostic criteria using information obtained from structured psychiatric interviews. Rating scales should be used to measure the severity of depression only.

A confounder in diagnosing depression in stroke is the potential for overlap between symptoms of depression and symptoms of stroke. Whereas some studies have indicated that this may be particularly problematic for the motor symptoms of depression [17, 18], more recent research has demonstrated that stroke patients with depression are still significantly more likely to endorse overlapping symptoms, such as insomnia, poor appetite, loss of energy, and poor concentration, than stroke patients who are not depressed [19].

Patients suffering aphasic deficits (see Chap. 6) also present inherent assessment difficulties, and this population has typically been excluded from PSD research. Given that 20–40 % of stroke patients suffer from varying degrees of aphasia [20–23], this may represent a substantial source of bias in the research findings. It is also difficult to delineate the direction of causation between depression and disability. For instance, while patients with PSD suffer a greater burden of illness than those without PSD, more severe stroke deficits also increase the risk of PSD [2]. This suggests a complex interaction between depression and disability in patients with PSD, in which the disorder may compound the physical, psychological, and social impacts of stroke, and vice versa.

In conclusion, diagnosis of depression among stroke patients has been limited by the overlap of symptoms of depression and stroke and by the lack of specific diagnostic criteria. Nevertheless, several studies have demonstrated that the DSM-IV-TR criteria for major depression have adequate sensitivity and validity for use in stroke [2]. Future studies should develop specific diagnostic interviews and depression rating scales for use in patients with stroke, including those with different severities of aphasia.

Consequences of Post-stroke Depression

Physical and Functional Consequences

There is a wealth of evidence to support an association between the severity of physical impairment after stroke and the severity of concomitant depressive symptoms [24–26]. Fifteen of 18 studies reviewed by Robinson [2] also found a significant association between PSD and deficits in activities of daily living (ADLs). In a study that examined the predictive relationship between PSD and deficits on ADLs during the first 2 years after stroke, Robinson [2] found that in-hospital PSD had a significant correlation with ADLs at 3 and 6 months post-stroke. On the other hand, in-hospital ADL scores correlated significantly with depression scores at 3 and 6 months post-stroke, suggesting a bimodal association between depression and deficits on ADL after stroke. One of the first studies to explore this association in more detail showed that patients with PSD demonstrated significantly less recovery on ADLs than nondepressed patients comparable on demographic variables and baseline impairments [27]. More recently, the Auckland Stroke Outcomes Study demonstrated a 3.5-fold increased risk of dependency, a threefold increased risk of moderate to severe neurological impairment, and significantly reduced ADL activity in patients with PSD at 5 years post-stroke [3]. Several studies have examined the impact of improvement from depression on functional deficits. Chemerinski et al. [28] showed that patients with PSD who remitted from depression 6 months later improved significantly more from their functional deficits than patients without remission.

Effect of Antidepressant Medication

Several studies have demonstrated that antidepressant medication and timing of treatment may significantly influence functional recovery. Reding et al. [29] conducted a randomized clinical trial that compared trazodone (a serotonin antagonist and reuptake inhibitor) at a dose of 50–200 mg/day with placebo treatment. After 6 weeks, patients on active treatment had a significant functional improvement as compared to those on placebo. Robinson et al. [30] compared nortriptyline (a tricyclic antidepressant [TCA]), fluoxetine (a selective serotonin reuptake inhibitor [SSRI]), and placebo treatment in another randomized clinical trial and found that patients on fluoxetine had less recovery on the Functioning Independence Measure than patients on nortriptyline or placebo treatment. Another randomized clinical trial using fluoxetine as active treatment also failed to demonstrate improvements for the group on the active drug [31]. Narushima and Robinson [32] compared 34 patients who received antidepressant medication during the first month after stroke with 28 patients who received antidepressants during the fifth month after stroke and found that the early treatment group had a gradual recovery over 2 years, whereas the late treatment group showed a significant functional decline during the same period. Mikami et al. [33] have also shown that stroke patients who were on antidepressants for 3 months after the acute event had a significantly better recovery from stroke deficits than stroke patients on placebos, regardless of the presence of PSD. These important findings suggest that antidepressants may improve stroke recovery even in the absence of depression and that treatment should be ideally initiated early after acute events for better results.

Cognitive Consequences

Although some studies have suggested that patients with PSD are no more cognitively impaired than stroke patients without depression [24, 34], the weight of evidence supports the observation that patients with PSD have a much greater degree of cognitive impairment than stroke patients without depression [6, 35–42].

PSD is significantly associated with cognitive deficits. Robinson et al. [43] found significantly lower Mini-Mental State Exam (MMSE) scores among patients who developed depression after a left hemisphere stroke compared to nondepressed patients with left hemisphere strokes. To control for lesion variables, Starkstein et al. [44] compared 13 pairs of patients (one with and the other without depression) matched for lesion size and location. Patients with depression still demonstrated significantly lower MMSE scores than nondepressed lesion-matched patients. In a meta-analysis of all 357 patients included in his acute PSD studies, Robinson [2] also demonstrated that cognitive impairment, as defined by an MMSE score of <23, was significantly more frequent in patients with major depression (59 %) as compared to those with minor (43 %) or no depression (36 %).

Effect of Antidepressant Medication

Kimura et al. [45] carried out a merged analysis of patients from several treatment trials for PSD and found that patients with a positive response to antidepressants or placebo treatment showed a significant post-stroke cognitive improvement than nonresponders. In a study that compared 43 stroke patients without depression treated with escitalopram (an SSRI), 45 patients treated with placebos, and 41 patients treated with problem-solving therapy, Jorge et al. [46] found that at 1-year post-stroke, patients treated with escitalopram had greater global cognitive improvement and performed better on memory-specific tasks than either of the other two groups.

Social Consequences

The interaction between PSD and social function is complex. Even prior to stroke, poor social function has been reported as a significant independent risk factor for poor rehabilitation outcomes [47]. Morris et al. [48] examined social support in a series of 76 acute stroke patients and found a significant association between more severe depression and perceived lower social support. Moreover, inadequate social support was a significant predictor of a longer duration of PSD. Patients with poor social support post-stroke are more likely to develop depression, and PSD in turn is clearly associated with impairment in social functioning [3, 49]. Kotila et al. [50] compared the efficacy of active programs including rehabilitation, support, and social activities with treatment as usual. The main finding was that, at both 3 and 12 months after stroke, the frequency of depression was significantly lower among patients with active treatment vs. a more conservative treatment.

PSD is also associated with increased length of hospital stay, increased number of outpatient visits during follow-up, and increased rates of future hospitalization [51]. Moreover, caregivers of patients with PSD experience greater burden [52] and depression [53].

Mortality

Several studies have demonstrated an increase in both short-term (12–24 months) and long-term (5–10 years) mortality in patients with PSD [3, 44, 54, 55]. Reports of death due to suicide are rare, but about 10–15 % of patients who have suffered a stroke will subsequently develop suicidal ideation and it is an important risk to consider [56, 57].

Effect of Antidepressant Treatment

Robinson [2] reanalyzed data from his treatment studies and found that 59 % of patients assigned to receive antidepressants were still alive 9 years after the acute stroke as compared with 36 % of those assigned to receive placebo ($p<0.05$). A logistic regression analysis including the main variables associated with post-stroke mortality demonstrated that both antidepressant use and the presence of diabetes mellitus were independent predictors of survival.

In conclusion, there is a consensus that depression in stroke is significantly associated with more severe functional, cognitive, and social deficits, as well as higher mortality. Longitudinal studies suggest that PSD is a significant predictor of poor functional and social recovery, and successful treatment of PSD may have a positive impact on the long-term recovery of patients.

Treatment of Post-stroke Depression

Pharmacological Treatment

Two meta-analyses examined the efficacy of antidepressant medication to treat PSD. Chen et al. [58] included 1,320 patients with PSD identified from 16 randomized controlled trials. Six of the studies were drawn from Chinese language publications. Twelve studies evaluated SSRIs, two evaluated TCAs, and three evaluated other antidepressants. Pooled response rates demonstrated a significantly greater improvement in depressive symptoms with active treatment as compared to placebo treatment, with improvements from baseline to endpoint in 65 % of the active group and in 44 % of controls. Benefits became significant after 3–4 weeks of treatment, and longer duration of treatment was positively correlated with the degree of improvement in depressive symptoms. On the other hand, there were no significant treatment effects for the severity of neurological impairment with the exception of a modest treatment effect on ADLs. Based on these findings, the authors recommend that antidepressant treatment should be implemented for all patients with PSD, after considering contraindications and potential side effects.

Hackett et al. performed a meta-analysis that included 1,655 patients with PSD identified from 16 randomized controlled trials [10]. Twelve trials examined the efficacy of 13 pharmacological interventions: seven trials evaluated SSRIs, two evaluated TCAs, and four evaluated other antidepressant medications. Of the 1,121 participants who received antidepressant medication, there was evidence that pharmacotherapy was effective both in improving depressive symptoms (i.e., significant response) and treating depression (i.e., significant remission). The authors noted, however, that there was significant heterogeneity in outcomes between different studies and that confidence intervals were wide for both response and remission.

There was no evidence of improvement in cognitive function, ADLs, or disability. Increased rates of adverse events (affecting the neurological and gastrointestinal systems in particular) were also noted in the treatment groups.

In summary, there is significant evidence that the treatment of PSD with antidepressant medications leads to an improvement in depressive symptoms. The evidence for treatment to remission is less compelling and must therefore be weighed more carefully against the potential for adverse medication effects. Whether treatment contributes to improved outcomes in terms of degree of neurological impairment and ADLs is a question that requires more research. Further studies are also required to evaluate the effectiveness of antidepressant treatment for PSD.

Psychostimulants

Psychostimulants include methylphenidate, norepinephrine-dopamine reuptake inhibitors (such as bupropion and atomoxetine), and others. Only three studies have assessed the efficacy of psychostimulants for PSD, two retrospective chart reviews [59, 60] and one small randomized clinical trial [61]. The two chart reviews found psychostimulants to be safe and well tolerated with symptom improvements observed within a few days after stroke. This compares favorably with the longer period of onset for traditional antidepressants and suggests a potential role for psychostimulants in treating PSD during the early post-stroke period. Despite this, no definitive conclusions can be drawn as the studies were retrospective, lacked randomization, and lacked adequate blinding controls. The randomized clinical trial was slightly more robust but failed to demonstrate significant treatment efficacy.

Non-pharmacological Treatments

Behavioral Interventions

While behavioral treatment of PSD is theoretically appealing, few controlled trials have been carried out. Williams et al. conducted a randomized trial of 188 ischemic stroke survivors comparing care management intervention (Activate-Initiate-Monitor) to usual care [62]. Active treatment consisted of acceptance of stroke treatment, initiation of antidepressant treatment, and monitoring treatment effectiveness. Stroke patients randomized to usual care received an identical number of sessions as a control treatment. The main finding was that active management of PSD produced a significantly greater rate of both response (51 % vs. 30 %) and remission (39 % vs. 23 %) of depression than usual care alone. Joubert et al. [63] have also presented an integrated care model of depression treatment consisting of the following: a collaboration between a specialist stroke service and primary care physicians using telephone tracking, a bidirectional information feedback loop, management of vascular risk factors, and regular screening for depressive symptoms. At 12 months post-stroke, they found that patients receiving active management had significantly fewer depressive symptoms than controls [63]. Finally, the impact of a physical exercise program upon depressive symptoms in stroke survivors was examined in a single study [64]. Treatment consisted of a progressive exercise program targeting strength, balance, endurance, and upper extremity function. The study took place in the patients' homes 3 times a week for 36 sessions and was supervised by a physical or occupational therapist. The main finding was that physical exercise during the subacute recovery phase of stroke had a beneficial effect on depressive symptoms.

Psychotherapy

There is only a single meta-analysis that examined the efficacy of psychotherapy as a treatment for PSD [10]. Outcome data were analyzed from four trials that included 902 participants with interventions consisting of supportive psychotherapy, cognitive behavior therapy, problem-solving therapy, and motivational interviewing. Treatment duration varied from 4 weeks to 1 year. All of the studies used "standard care" as a comparator. Unfortunately, no treatment effect on PSD was demonstrated for any of the endpoints measured.

Repetitive Transcranial Magnetic Stimulation

Jorge et al. examined the efficacy of repetitive transcranial magnetic stimulation (rTMS) on a series of 20 patients with PSD who did not respond to antidepressant medication [65]. When compared to sham stimulation, the investigators found that ten sessions of active rTMS of the left dorsolateral prefrontal cortex was associated with a significant reduction of depressive symptoms in 30 % of patients and remission in 10 %. The authors concluded rTMS might be an effective treatment that is safe and well tolerated in stroke patients with refractory depression. Additional studies are needed to replicate these findings and evaluate functional and long-term outcomes.

Electroconvulsive Therapy

Two retrospective chart review studies have examined electroconvulsive therapy (ECT) for PSD [66, 67]. The first study included 14 patients with PSD, of whom 12

(86 %) demonstrated a significant improvement after an average of 7.7 treatments. None of the patients experienced an exacerbation of their neurological symptoms or suffered a recurrence of stroke. A more recent study included 20 patients with PSD, of whom 19 demonstrated a significant improvement after an average of ten treatments. Five patients in this study developed complications requiring medical intervention, including hypertension, prolonged post-ictal confusion, pulmonary edema, and cardiac arrhythmia. In conclusion, preliminary evidence suggests that ECT may be an effective treatment for PSD, but adverse effects may limit its usefulness.

Prevention

Given the high prevalence of PSD and its negative impact on patient outcome, it is imperative to find effective preventive treatments. In recent years there have been an increasing number of studies examining the efficacy of pharmacological and psychological interventions in the prevention of PSD. The main results are discussed as follows.

Palomaki et al. [68] carried out a randomized clinical trial using mianserin (a tetracyclic antidepressant) at 60 mg/day as the active treatment. One hundred consecutive acute stroke patients were recruited and treated during 1 year. The main result was a lack of difference in the frequency of major depression between patients on mianserin and placebo-treated patients at any time point. More recently, Narushima et al. [69] compared the effect of nortriptyline vs. fluoxetine to prevent PSD in a series of 48 nondepressed patients with acute stroke. Patients were treated for 3 months and had an additional 21-month follow-up. The authors' main finding was that both nortriptyline and fluoxetine were effective in preventing depression after stroke among patients who completed the treatment trial. However, after stopping the antidepressants, there was an increase in both the frequency and severity of depressive symptoms, particularly among nortriptyline-treated patients [69]. Rasmussen et al. [70] have also examined the effect of sertraline (an SSRI) in the prevention of PSD among 137 nondepressed patients with an ischemic stroke who were randomly assigned to 12 months of double-blind treatment with either an active drug ($n=70$) or a placebo ($n=67$). The main finding was that sertraline had superior prophylactic efficacy compared to the placebo, with approximately 10 % of the sertraline-treated group developing depression as compared to 30 % of the placebo-treated group. Treatment was well tolerated and patients experienced few adverse events [70]. A contrasting finding was reported by Almeida et al. who randomly assigned 111 acute stroke patients to treatment with either a placebo ($n=56$) or sertraline ($n=55$) at 50 mg/day. There was no significant difference in the frequency of depressive symptoms during the 24 weeks of treatment (17 % sertraline vs. 22 % placebo) [71]. A major limitation of this study, however, was the relatively short follow-up period.

In a multisite randomized clinical trial, Robinson et al. [72] recruited 176 nondepressed patients within 3 months following acute stroke who were followed-up over 12 months. Patients were randomized to receive escitalopram ($n = 59$), placebo ($n = 58$), or a non-blinded problem-solving therapy group ($n = 59$). Patients on placebo were significantly more likely to develop depression than individuals on either escitalopram (adjusted hazard ratio [HR], 4.5; $P < 0.001$) or problem-solving therapy (adjusted HR, 2.2; $P < 0.001$). These results remained significant after adjusting for history of mood disorders, age, gender, treatment site, and severity of impairment [72].

A Cochrane review examined the evidence for the efficacy of pharmacological and psychological interventions to prevent PSD and to improve physical and psychological post-stroke outcomes in participants [73]. Twelve trials involving 1,245 subjects were included. Nine trials compared antidepressants with placebos and showed no clear effect of pharmacological therapy on the prevention of depression. One limitation of this study is the comparison of studies that used different antidepressants, different follow-up periods, and different outcome measures. Hackett et al. performed a systematic review that examined the efficacy of psychotherapy in preventing depression [7]. As previously mentioned, outcome data were analyzed from four trials including 902 participants with interventions consisting of supportive psychotherapy, cognitive behavior therapy, problem-solving therapy, and motivational interviewing. Treatment duration varied from 4 weeks to 1 year. All of the studies used "standard care" as a comparator, and one study also included an "attention-control group" in which subjects spent time in focused conversation with a trained therapist but did not receive structured psychotherapy. Pooled results demonstrated a small effect for the prevention of depression as well as an improvement in psychological distress scores using the Scaled General Health Questionnaire (GHQ-28). There was no evidence for improved functional outcome although fewer subjects reported adverse events in the intervention groups compared to controls. One major limitation with these reviews and meta-analyses is that they compared studies with important methodological differences, such as treatment agent, time since stroke, duration of treatment, assessment methods, and outcome measures.

Summary

PSD affects approximately one-third of stroke survivors and is strongly associated with increased morbidity and mortality from as early as 1-month post-event. The diagnosis of PSD is made based on the presence of depressed mood or loss of interest or pleasure and four other depressive symptoms of at least 2 weeks duration. The consequences of PSD include increased short- and long-term mortality, greater degree of physical and cognitive impairment, increased risk of dependency and social dysfunction, as well as increased consumption of medical services. Research has demonstrated significant efficacy for tricyclics and SSRIs in treating PSD pharmacologically, although their effectiveness in routine clinical practice has rarely been addressed. Psychotherapy may be useful for patients with PSD in whom antidepressants are contraindicated or who develop significant side effects with their use. While several prevention studies have demonstrated the efficacy of SSRIs, negative studies have also been reported and methodologically sound studies are still required.

References

1. Hackett ML, Yapa C, Parag V, Anderson CS. Frequency of depression after stroke: a systematic review of observational studies. Stroke. 2005;36(6):1330–40.
2. Robinson R. The clinical neuropsychiatry of stroke. 2nd ed. Cambridge: Cambridge University Press; 2006.
3. Feigin VL, Barker-Collo S, Parag V, Senior H, Lawes CM, Ratnasabapathy Y, et al. Auckland Stroke Outcomes Study. Part 1: Gender, stroke types, ethnicity, and functional outcomes 5 years poststroke. Neurology. 2010;75(18):1597–607.
4. House A, Knapp P, Bamford J, Vail A. Mortality at 12 and 24 months after stroke may be associated with depressive symptoms at 1 month. Stroke. 2001;32(3):696–701.
5. Ferri CP, Schoenborn C, Kalra L, Acosta D, Guerra M, Huang Y, et al. Prevalence of stroke and related burden among older people living in Latin America, India and China. J Neurol Neurosurg Psychiatry. 2011;82(10):1074–82.
6. Morris PL, Robinson RG, Andrzejewski P, Samuels J, Price TR. Association of depression with 10-year poststroke mortality. Am J Psychiatry. 1993;150(1):124–9.
7. Hackett ML, Anderson CS, House A, Halteh C. Interventions for preventing depression after stroke. Cochrane Database Syst Rev. 2008;(3):CD003689.
8. Hackett ML, Anderson CS, House AO, Xia J. Interventions for treating depression after stroke. Stroke. 2009;40:487–8.
9. Price A, Rayner L, Okon-Rocha E, Evans A, Valsraj K, Higginson IJ, et al. Antidepressants for the treatment of depression in neurological disorders: a systematic review and meta-analysis of randomised controlled trials. J Neurol Neurosurg Psychiatry. 2011;82(8):914–23.
10. Hackett ML, Anderson CS, House A, Xia J. Interventions for treating depression after stroke. Cochrane Database Syst Rev. 2008;(4):CD003437.
11. American Psychiatric Association. Diagnostic and statistical manual of mental disorders: DSM IV-TR. 4th ed. Washington, DC: American Psychiatric Association; 2000. p. 943.
12. Folstein MF, Maiberger R, McHugh PR. Mood disorder as a specific complication of stroke. J Neurol Neurosurg Psychiatry. 1977;40(10):1018–20.
13. Robinson RG, Szetela B. Mood change following left hemispheric brain injury. Ann Neurol. 1981;9(5):447–53.
14. Whyte EM, Mulsant BH. Post stroke depression: epidemiology, pathophysiology, and biological treatment. Biol Psychiatry. 2002;52(3):253–64.
15. Spalletta G, Ripa A, Caltagirone C. Symptom profile of DSM-IV major and minor depressive disorders in first-ever stroke patients. Am J Geriatr Psychiatry. 2005;13(2):108–15.
16. Robinson RG, Spalletta G. Poststroke depression: a review. Can J Psychiatry. 2010;55(6):341–9.
17. Beblo T, Driessen M. No melancholia in poststroke depression? A phenomenologic comparison of primary and poststroke depression. J Geriatr Psychiatry Neurol. 2002;15(1):44–9.
18. House A. Mood disorders after stroke: a review of the evidence. Int J Geriatr Psychiatry. 1987;2(4):211–21.
19. Williams LS, Brizendine EJ, Plue L, Bakas T, Tu W, Hendrie H, et al. Performance of the PHQ-9 as a screening tool for depression after stroke. Stroke. 2005;36(3):635–8.
20. Pedersen PM, Jorgensen HS, Nakayama H, Raaschou HO, Olsen TS. Aphasia in acute stroke: incidence, determinants, and recovery. Ann Neurol. 1995;38(4):659–66.

21. Wade DT, Hewer RL, David RM, Enderby PM. Aphasia after stroke: natural history and associated deficits. J Neurol Neurosurg Psychiatry. 1986;49(1):11–6.
22. Berthier ML. Poststroke aphasia: epidemiology, pathophysiology and treatment. Drugs Aging. 2005;22(2):163–82.
23. Inatomi Y, Yonehara T, Omiya S, Hashimoto Y, Hirano T, Uchino M. Aphasia during the acute phase in ischemic stroke. Cerebrovasc Dis. 2008;25(4):316–23.
24. Eastwood MR, Rifat SL, Nobbs H, Ruderman J. Mood disorder following cerebrovascular accident. Br J Psychiatry. 1989;154:195–200.
25. Astrom M, Adolfsson R, Asplund K. Major depression in stroke patients. A 3-year longitudinal study. Stroke. 1993;24(7):976–82.
26. Robinson RG, Starr LB, Kubos KL, Price TR. A two-year longitudinal study of post-stroke mood disorders: findings during the initial evaluation. Stroke. 1983;14(5):736–41.
27. Parikh RM, Robinson RG, Lipsey JR, Starkstein SE, Fedoroff JP, Price TR. The impact of poststroke depression on recovery in activities of daily living over a 2-year follow-up. Arch Neurol. 1990;47(7):785–9.
28. Chemerinski E, Petracca G, Sabe L, Kremer J, Starkstein SE. The specificity of depressive symptoms in patients with Alzheimer's disease. Am J Psychiatry. 2001;158(1):68–72.
29. Reding MJ, Orto LA, Winter SW, Fortuna IM, Di Ponte P, McDowell FH. Antidepressant therapy after stroke. A double-blind trial. Arch Neurol. 1986;43(8):763–5.
30. Robinson RG, Schultz SK, Castillo C, Kopel T, Kosier JT, Newman RM, et al. Nortriptyline versus fluoxetine in the treatment of depression and in short-term recovery after stroke: a placebo-controlled, double-blind study. [see comment]. Am J Psychiatry. 2000;157(3):351–9.
31. Fruehwald S, Gatterbauer E, Rehak P, Baumhackl U. Early fluoxetine treatment of post-stroke depression–a three-month double-blind placebo-controlled study with an open-label long-term follow up. J Neurol. 2003;250(3):347–51.
32. Narushima K, Robinson RG. The effect of early versus late antidepressant treatment on physical impairment associated with poststroke depression: is there a time-related therapeutic window? J Nerv Ment Dis. 2003;191(10):645–52.
33. Mikami K, Jorge RE, Adams Jr HP, Davis PH, Leira EC, Jang M, et al. Effect of antidepressants on the course of disability following stroke. Am J Geriatr Psychiatry. 2011;19(12):1007–15.
34. Morris PL, Robinson RG, Raphael B. Prevalence and course of depressive disorders in hospitalized stroke patients. Int J Psychiatry Med. 1990;20(4):349–64.
35. Parikh RM, Robinson RG, Lipsey JR, Starkstein SE, Fedoroff JP, Price TR. The impact of poststroke depression on recovery in activities of daily living over a 2-year follow-up. Arch Neurol. 1990;47(7):785–9.
36. Jorge RE, Robinson RG, Arndt S, Starkstein S. Mortality and poststroke depression: a placebo-controlled trial of antidepressants. Am J Psychiatry. 2003;160(10):1823–9.
37. Everson SA, Roberts RE, Goldberg DE, Kaplan GA. Depressive symptoms and increased risk of stroke mortality over a 29-year period. Arch Intern Med. 1998;158(10):1133–8.
38. Bolla-Wilson K, Robinson RG, Starkstein SE, Boston J, Price TR. Lateralization of dementia of depression in stroke patients. Am J Psychiatry. 1989;146(5):627–34.
39. Starkstein SE, Parikh RM, Robinson RG. Post-stroke depression and recovery after stroke. Lancet. 1987;1(8535):743.
40. House A, Dennis M, Mogridge L, Warlow C, Hawton K, Jones L. Mood disorders in the year after first stroke. Br J Psychiatry. 1991;158:83–92.
41. Spalletta G, Caltagirone C. Sertraline treatment of post-stroke major depression: an open study in patients with moderate to severe symptoms. Funct Neurol. 2003;18(4):227–32.
42. Kauhanen M, Korpelainen JT, Hiltunen P, Brusin E, Mononen H, Maatta R, et al. Poststroke depression correlates with cognitive impairment and neurological deficits. Stroke. 1999;30(9):1875–80.
43. Robinson RG, Bolla-Wilson K, Kaplan E, Lipsey JR, Price TR. Depression influences intellectual impairment in stroke patients. Br J Psychiatry. 1986;148:541–7.

44. Starkstein SE, Robinson RG, Price TR. Comparison of patients with and without poststroke major depression matched for size and location of lesion. Arch Gen Psychiatry. 1988;45(3):247–52.
45. Kimura M, Robinson RG, Kosier JT. Treatment of cognitive impairment after poststroke depression: a double-blind treatment trial. Stroke. 2000;31:1482–6.
46. Jorge RE, Starkstein SE, Robinson RG. Apathy following stroke. Can J Psychiatry. 2010;55(6):350–4.
47. Shimoda K, Robinson RG. The relationship between social impairment and recovery from stroke. Psychiatry. 1998;61(2):101–11.
48. Morris PL, Robinson RG, Raphael B, Bishop D. The relationship between the perception of social support and post-stroke depression in hospitalized patients. Psychiatry. 1991;54(3):306–16.
49. Robinson RG, Starr LB, Lipsey JR, Rao K, Price TR. A two-year longitudinal study of post-stroke mood disorders. In-hospital prognostic factors associated with six-month outcome. J Nerv Ment Dis. 1985;173(4):221–6.
50. Kotila M, Numminen H, Waltimo O, Kaste M. Depression after stroke: results of the FINNSTROKE Study. Stroke. 1998;29(2):368–72.
51. Jia H, Damush TM, Qin H, Ried LD, Wang X, Young LJ, et al. The impact of poststroke depression on healthcare use by veterans with acute stroke. Stroke. 2006;37(11):2796–801.
52. Andersen G, Vestergaard K, Ingemann-Nielsen M, Lauritzen L. Risk factors for post-stroke depression. Acta Psychiatr Scand. 1995;92(3):193–8.
53. Dennis M, O'Rourke S, Lewis S, Sharpe M, Warlow C. A quantitative study of the emotional outcome of people caring for stroke survivors. Stroke. 1998;29(9):1867–72.
54. Wade DT, Legh-Smith J, Hewer RA. Depressed mood after stroke. A community study of its frequency. Br J Psychiatry. 1987;151:200–5.
55. Morris PL, Robinson RG, Andrezejewski P, Samuels J, Price TR. Association of depression with 10-year post-stroke mortality. Stroke. 1993;150:124–9.
56. Kishi Y, Kosier JT, Robinson RG. Suicidal plans in patients with acute stroke. J Nerv Ment Dis. 1996;184(5):274–80.
57. Pohjasvaara T, Vataja R, Leppavuori A, Kaste M, Erkinjuntti T. Cognitive functions and depression as predictors of poor outcome 15 months after stroke. Cerebrovasc Dis. 2002;14(3–4):228–33.
58. Chen Y, Guo JJ, Zhan S, Patel NC. Treatment effects of antidepressants in patients with post-stroke depression: a meta-analysis. Ann Pharmacother. 2006;40(12):2115–22.
59. Masand P, Murray GB, Pickett P. Psychostimulants in post-stroke depression. J Neuropsychiatry Clin Neurosci. 1991;3(1):23–7.
60. Lazarus LW, Moberg PJ, Langsley PR, Lingam VR. Methylphenidate and nortriptyline in the treatment of poststroke depression: a retrospective comparison. Arch Phys Med Rehabil. 1994;75(4):403–6.
61. Grade C, Redford B, Chrostowski J, Toussaint L, Blackwell B. Methylphenidate in early post-stroke recovery: a double-blind, placebo-controlled study. Arch Phys Med Rehabil. 1998;79(9):1047–50.
62. Williams LS, Kroenke K, Bakas T, Plue LD, Brizendine E, Tu W, et al. Care management of poststroke depression: a randomized, controlled trial. Stroke. 2007;38(3):998–1003.
63. Joubert J, Joubert L, Reid C, Barton D, Cumming T, Mitchell P, et al. The positive effect of integrated care on depressive symptoms in stroke survivors. Cerebrovasc Dis. 2008;26(2):199–205.
64. Lai SM, Studenski S, Richards L, Perera S, Reker D, Rigler S, et al. Therapeutic exercise and depressive symptoms after stroke. J Am Geriatr Soc. 2006;54(2):240–7.
65. Jorge RE, Robinson RG, Tateno A, Narushima K, Acion L, Moser D, et al. Repetitive transcranial magnetic stimulation as treatment of poststroke depression: a preliminary study. Biol Psychiatry. 2004;55(4):398–405.
66. Currier MB, Murray GB, Welch CC. Electroconvulsive therapy for post-stroke depressed geriatric patients. J Neuropsychiatry Clin Neurosci. 1992;4(2):140–4.

67. Murray GB, Shea V, Conn DK. Electroconvulsive therapy for poststroke depression. J Clin Psychiatry. 1986;47(5):258–60.
68. Palomaki H, Kaste M, Berg A, Lonnqvist R, Lonnqvist J, Lehtihalmes M, et al. Prevention of poststroke depression: 1 year randomised placebo controlled double blind trial of mianserin with 6 month follow up after therapy. J Neurol Neurosurg Psychiatry. 1999;66(4):490–4.
69. Narushima K, Kosier JT, Robinson RG. Preventing poststroke depression: a 12-week double-blind randomized treatment trial and 21-month follow-up. J Nerv Ment Dis. 2002;190(5):296–303.
70. Rasmussen A, Lunde M, Poulsen DL, Sorensen K, Qvitzau S, Bech P. A double-blind, placebo-controlled study of sertraline in the prevention of depression in stroke patients. Psychosomatics. 2003;44(3):216–21.
71. Almeida OP, Waterreus A, Hankey GJ. Preventing depression after stroke: results from a randomized placebo-controlled trial. J Clin Psychiatry. 2006;67(7):1104–9.
72. Robinson RG, Jorge RE, Moser DJ, Acion L, Solodkin A, Small SL, et al. Escitalopram and problem-solving therapy for prevention of poststroke depression: a randomized controlled trial. JAMA. 2008;299(20):2391–400.
73. Anderson CS, Hackett ML, House AO. Interventions for preventing depression after stroke. Cochrane Database Syst Rev. 2004;(2):CD003689.

Chapter 13
Better Dead than Alive? Quality of Life After Stroke

Thomas Schenk and Adam J. Noble

Introduction

Stroke can be a life-threatening disease, but it is more often a disease resulting in chronic disability. In fact, it is the most likely cause of chronic disability in the elderly [1]. Stroke can affect many aspects of our behavior and personality. It can impair our ability to move, see, speak, memorize, or think and can thus affect whether and how we can enjoy living. The focus of stroke research and stroke rehabilitation has long been to increase survival and reduce disability—disability understood as the degree to which the affected patient depends on external help for accomplishing tasks of daily living. This is still a valid perspective, but it is not the only one. From a patient's perspective disability matters but sometimes not as much as other consequences of a stroke. Pain is a case in point. Patients with upper limb plegia frequently develop contractures. These contractures in turn can lead to pain. Regular physiotherapy can often prevent those contractures and reduce the concomitant pain. This will not reduce the required help or reduce the disability but can significantly affect a patient's quality of life. There are other examples that illustrate that disability and life quality can dissociate. Patients with hemiplegia will often require more help than patients with isolated aphasia and are thus more disabled. But from the patient's perspective the inability to communicate their thoughts, worries, and wishes can be much more soul-destroying than the inability to move. It seems obvious that for a system, which aims to improve and maintain the well-being of all patients in its care, the perspective of those patients should have the highest priority and at the very least should be taken

T. Schenk, B.Sc., M.Sc., Ph.D. (✉)
Department of Neurology, University Clinic, Friedrich-Alexander University Erlangen-Nuremberg, Erlangen, Bavaria, Germany
e-mail: Thomas.Schenk@uk-erlangen.de

A.J. Noble, B.Sc., M.Sc., Ph.D.
Department of Clinical Neuroscience, Unit of Neurology and General Practice, Institute of Psychiatry, King's College London, London, UK

T.A. Schweizer and R.L. Macdonald (eds.), *The Behavioral Consequences of Stroke*, 241
DOI 10.1007/978-1-4614-7672-6_13, © Springer Science+Business Media New York 2014

into consideration. Current trends suggest that in the future those perspectives will not only have to be taken into consideration, but might become a major force in shaping our health system. Government organizations with the task to decide which treatments will receive funding are already taking the quality of the life saved or extended into the equation [2].

Quality of life is not only about the patient's perspective. Disabled patients need help. Most of this help is provided by close family members, typically partners and children. Stroke therefore dramatically affects the life of the patients but also, to a significant extent, the life of the patients' family. The quality of life of the caregivers has to be taken into account for their own sake and for the sake of the patient. If unsupported, those caregivers might become patients themselves, and without the full support of their caregivers, patients might have to be moved from home care into a care home.

Thus, quality of life in stroke is concerned with both the patient's perspectives of their disease and the caregiver's perspective on how the disease of their loved ones has changed their life. In this chapter we will deal with both aspects. But first we need to describe how quality of life is measured.

Quantifying Quality

Estimates of a patient's quality of life are increasingly used to determine which treatments are worth funding and which are not [2]. This is particularly true for new drugs and new treatment options. In an ideal world, all drugs and treatment options should be funded and made available to appropriate patients, but, in reality, health budgets are stretched and health providers worldwide face difficult choices when deciding which drugs and treatment options merit the costs that they incur. What criteria can be used to make this decision? Clearly a drug that can cure a common disease will always be a strong contender for funding. However, in many cases, particularly in the case of stroke, a cure is not available. Available treatments for stroke may at best reduce the likelihood of further strokes, slow down or stop the further development of neuropathology, reduce neurological symptoms, or allow patients to compensate more effectively for acquired neurological disabilities. A crude way of assessing the impact of the various treatment options on the course of a chronic disorder, such as stroke, is to compare the number of life years saved by administering a given treatment. However, this is too crude. Most people would agree that a treatment that promises to provide 5 more years of normal, symptom- and pain-free life is clearly superior to a treatment that can extend life by the same amount but only in a pain-ridden, disabled status. This example shows that quantity alone is not enough. The quality of the life saved or extended needs to be taken into account as well. This insight has led to the development of the quality-adjusted life-year (QALY) measure [3]. To calculate QALYs we take the number of years saved or extended and multiply those with an estimate for the quality of that life. These estimates for the quality of life score can vary between 0 (equals death) and

1 (equals perfect health). In the following we will give a hypothetical example of how QALYs can be used to compare the value of two treatments.

Let's say there are two drugs—drug A and drug B. Drug A increases life expectancy by 4 years but the life saved will be associated with pain and disability, accordingly its quality of life is low and has a score of 0.4. Drug B increases life expectancy by only 3 years but the life saved can be lived in normal health; accordingly the quality of life score is high (i.e., 1). We can now calculate the QALYs for both drugs:

Drug A: $4 \times 0.4 = 1.6$
Drug B: $3 \times 1 = 3$

By taking quality of life into account, drug B appears to be the more valuable drug despite the fact that it increases life expectancy by 1 year less than drug A. In this case the health provider will possibly decide to fund drug B but not drug A. Clearly with so much at stake it is important that we use reliable methods to estimate the quality of life associated with a given medical condition.

Measuring Quality of Life

Before measuring quality of life, it has to be defined, which is problematic. There is some agreement that health-related quality of life (HRQoL) is related to the concepts of disability and handicap as outlined in the International Classification of Impairments, Disabilities and Handicaps [4] and it is argued that HRQoL lies beyond both concepts [5]. There is further agreement that HRQoL is a multidimensional concept presumably including physical, mental, and social aspects of life [6], but there is little agreement about anything else. For this reason it is attractive to find ways of measuring HRQoL that do not require its explicit definition but rely on the implicit understanding of patients and other respondents.

There are several ways to produce a number for a patient's HRQoL without having to resort to a formal definition of quality of life [2]. The resulting measures are called utility measures and are primarily used to calculate QALYs and, thus, inform health-economic decisions. The standard-gamble method will ask the respondent to state the odds of a gamble between death and perfect health that they regard as equivalent to accepting a given health state. For example, a respondent might indicate that a gamble where there is an 80 % chance of perfect health and a 20 % chance of death is equivalent to accepting a minor stroke. In this case we assign a utility index of 80 % or 0.8 to minor stroke. The time-tradeoff method uses a similar approach. The respondent is asked to trade lifetime against health. For example, if a respondent states that living for only 90 % of their remaining life span is equivalent to living the total of their remaining life but with a minor stroke, then this is taken as an indication that they value life with a minor stroke at 90 % (or 0.9) of the value of life without stroke. Finally respondents can be simply asked to assign to a given health state (e.g., stroke) a value between 0 (death) and 1 (perfect health). The same assignment can be made by an expert or a panel of experts.

Utility measures have the advantage of presenting a single number that characterizes the value of a person's life. This may be sufficient to calculate QALYs and thereby inform health-economic decisions. However, this approach ignores the fact that quality of life is a subjective property and patients suffering from the same disease vary significantly in their subjective experience of disease. It is therefore important to understand how and why patients with the same health status differ in their assessment of their life's quality. Health profiles, which are based on standardized questionnaires, provide much more detailed information about how a given health status affects a patient's life. Respondents are typically asked to indicate how their health affects their ability to fulfill their work or household duties, how it affects their mood or their ability to enjoy specific pastimes, and how it affects their capacity for communication and participation in social events. Responses to these scales will provide a detailed picture of how a disease impacts on somebody's life but they can also be used to calculate a single score, which can then be plugged into the formula for calculating QALYs.

Evaluation of HRQoL Measures

A comprehensive review of methods to estimate utility measures was published in 2001 [2]. Tengs and colleagues reviewed 67 articles and obtained 161 HRQoL estimates related to stroke. On the basis of those estimates, they explored the consistency of those estimates, their relation to the underlying disorder (e.g., minor stroke versus major stroke), and the popularity of specific methods as compared to other available methods. First, they found that the measures, even for the same type of stroke, varied considerably. For example, in the case of major stroke, the reported utility indices ranged from –0.02 (in this case the respondents felt that major stroke is worse than death) to 0.71. They also found that there was considerable overlap between the range of indices for major, moderate, and minor strokes. Interestingly, the median utility indices for minor and general stroke were almost identical. In some cases, the utility index for major strokes could be higher [7] than those for moderate or minor strokes [8]. This demonstrates that the reliability of many of those estimates is highly questionable. This is not surprising since the most commonly used method to assess HRQoL was expert judgment (i.e., an expert simply rates the HRQoL that is associated with a given health status by assigning a number between 0 and 1). Typically the experts offering this estimate were the authors of the papers themselves. The sample size was thus reduced to the number of authors (in some cases a single person). In other studies this problem was further compounded by the fact that authors did not create their own estimates but relied on the judgment of authors from previous studies. In quite a few cases, the severity of the stroke for which the original estimate was derived differed from the severity of the stroke for which it was applied. Given that these estimates will ultimately determine which treatment will be funded, the unreliability of these estimates should give rise to concern.

A second review [9] focused on health profiles. In the meantime a wide range of such scales are available. The authors reviewed the six scales that have been used most frequently in stroke research, namely the sickness impact profile (SIP) [10], the Nottingham Health Profile (NHP) [11], the Short-Form Health Survey (MOS SF36) [12], the EuroQol [13], the Health Utility Index (HUI) [14], and the London Handicap Scale (LHS) [15]. All of these scales are self-report measures. They do, however, vary quite dramatically with respect to the time required to complete the scale. The SIP requires on average 30 min to complete and the EuroQol can be completed in 3 min. Apart from the HUI, all scales have good reliability and validity, but only the SIP and NHP are sufficiently sensitive to changes and can be recommended for intervention studies.

It should be noted that establishing validity is not an uncontroversial process given that no consensus on a definition for quality of life has been reached. The authors of this review paper assumed validity when patients were involved in the development of the scale and its questions and/or when the dimensions of the scale matched those factors that patients and their relatives identified as particularly critical for their perceived life quality. In this context, it is relevant that none of the six scales were developed for stroke patients and therefore all tend to neglect some aspects of changes in quality of life that are specific to stroke patients, such as the impact of language or visual impairments. However, there are now several stroke-specific HRQoL scales [16–22]. The Stroke Impact Scale (SIS 3.0) [18] and the Stroke-specific Quality of Life Scale (SS-QoL2.0) [22] that are widely used and that were found to have adequate psychometric properties [23, 24].

Challenges remain, however, due to the difficulty in gaining a personal perspective from stroke patients with severe cognitive impairments. Language problems, profound memory disorders, or signs of general cognitive decline can make it impossible to get reliable and useful answers from patients themselves. In this case, researchers have to rely on proxies (i.e., relatives, nurses, or other persons who know the patient). Unfortunately, proxy ratings are to some extent biased and produce unreliable results in some domains [25–27]. As compared to self-rating by patients, proxies tend to exaggerate the disability and the negative impact on a patient's quality of life, and this tendency to paint an unrealistically bleak picture increases with increasing stroke severity. Moreover, for more subjective domains of HRQoL, such as a patient's mood or fatigue, there is little agreement between different proxy raters.

In summary, researchers can now choose from a wide pool of available HRQoL measures. Utility measures will be used when it is important to inform decisions that affect a whole group of patients. As we have seen, some methods used to produce such measures are questionable. Given the significant implications of decisions, which might be informed by those measures, it would be wise to stick to measures that are free of bias (standard-gamble or time-tradeoff method) and are based on a large sample of respondents. However, whenever we are concerned about an individual patient and wish to gain information that might allow us to improve support for this person, utility measures are of little use and health profiles will have to be employed. Health profiles provide a detailed picture of how an illness affects the life of affected persons and will inform us about areas of needs and opportunities for support. Several reliable, sensitive, and specific measures are now available for stroke patients and research. Using those measures has provided us with new insights into the factors determining the well-being of patients and their caregivers. Those findings will be reviewed in the next two sections.

What Determines a Stroke patient's Quality of Life?

For moderate to severe stroke the HRQoL scores range between 0.45 and –0.02. Even if we take the best score, quality of life after stroke is still less than half of what it was prior to the insult. Is this dramatic drop in life quality an inevitable consequence of this neurological disease or are there additional factors at work? Stroke severity certainly goes a long way in explaining the reduced HRQoL. Most studies on HRQoL after stroke picked out stroke severity (typically operationalized as the modified Rankin Scale) as a significant predictor [2]. The role of stroke severity or neurological disability can be seen by comparing utility estimates for patient groups with different patterns of neurological syndromes; whereas for patients with non-severe hemiparesis a utility index of 0.50 is reported, this measure drops to 0.45 in the case of severe hemiparesis, to 0.28 for hemiplegia, and to 0.06 in the case of global aphasia [2, 28, 29]. Thus, stroke severity is clearly an important predictor of HRQoL, but even patients who are considered to be fully recovered frequently show severely depressed HRQoL scores [30]. Other factors must be at work. Work-status, family problems, depression, anxiety, posttraumatic stress disorder (PTSD), and coping strategies are among some of the factors that can also have detrimental effects on post-stroke HRQoL, even in patients who otherwise suffer little neurological disability [30–32].

Another cause of reduced HRQoL is pain. About a third of all stroke patients develop shoulder pain [33] and roughly 20 % still have moderate to severe pain 1 year after stroke [34]. Sexual dysfunction and dissatisfaction with sexual life are found in roughly 70 % of all pairs where one partner has suffered a stroke. Post-stroke impotence is found in 50 % of all male stroke patients. Problems in sexual functioning are linked to the onset of post-stroke depression. Other social and psychosocial factors also exert a strong influence on sexual satisfaction after stroke [35, 36]. Unsurprisingly, sexual function has a significant impact on HRQoL after stroke. A surprising but consistent finding is that female patients have a significantly worse HRQoL after stroke than male patients [37, 38]. A multitude of factors are probably responsible for this finding. On average female stroke patients are more disabled [23]. Female patients are also less likely to be discharged home, as male partners are less likely to care for their spouses [23]. Moreover, differences in coping strategies and the prevalence of depression in male and female patients might also contribute to the lower HRQoL in female patients [37]. Coping strategies and depression play an important role in the long-term HRQoL, not just of female patients. In the initial stage, a feeling of uncontrollability prevails, frequently triggering the onset of depression. Later, positive coping strategies allow patients to regain some control of the situation leading to improved HRQoL [39]. Strategies such as tenacious goal pursuit and flexible goal adjustment are linked to a positive development [40]. This link between HRQoL, coping strategies, and depression might suggest that depression is another important predictor of HRQoL after stroke (see Chap. 12). This expectation is certainly confirmed by empirical data. Depression is a common consequence of stroke [41] and has a significant impact on HRQoL [37, 42]. The relationship between depression and HRQoL is quite complex. Depression affects functional recovery, cognitive function, and healthcare use, and leads to a worse functional outcome [43]. We can therefore assume that the reduced HRQoL found in stroke patients with depression reflects both a worse objective situation and a negatively biased subjective appraisal of the patient's current situation. Given that appraisal and coping strategies change over time with more positive coping strategies observed 6 months after stroke [44], one might also expect to see parallel changes in the occurrence of depression. However, a recent large-scale (more than 8,000 patients), 5-year longitudinal study on post-stroke depression found stable rates of depression across all 5 years [41].

In younger stroke patients, who are still at work and raising a family, the effect of a stroke can have a particularly detrimental impact on their life. In a recent systematic review on the social consequences of stroke for working-aged adults, it was found that only 44 % of patients return to work after stroke [45]. Financial problems will therefore be frequent in households where one of the members has suffered a stroke (24–33 %), increasing tensions within the family and ultimately causing severe family problems (such as separation or divorce). The frequency with which these problems are found varies considerably between studies, with one study reporting a deterioration of spousal relationship in only 5 % of their sample [46] and another study reporting a prevalence for partnership conflicts of 38 % [47]. Despite this variability there is no doubt that such psychosocial factors have a significant impact on post-stroke HRQoL.

Psychosocial factors are of increased relevance in the case of patients with less severe forms of stroke and in particular when comparatively young age and reduced severity come together, as is the case for subarachnoid hemorrhage (SAH) (see Chap. 10). SAH is a relatively rare type of stroke with a worldwide incidence of 9/100,000 per year [48]. Its economic and social impact is out of proportion to its incidence, as it predominately affects younger people who are still working and may have small children [49]. The average age at which people suffer SAH is 52–55 years as compared to an average of 73–75 years for stroke patients [50–52]. Half of those patients who experience a SAH will stop working, work shorter hours, or accept a position with less responsibility [53]. Consequently, HRQoL is significantly reduced post-SAH. Up to 55 % of SAH patients experience substantial reductions in their HRQoL [54, 55]. However, as we found in a recent meta-analysis, traditional predictors for post-stroke HRQoL—such as severity of hemorrhage, gender, physical and cognitive disabilities, and age—do not predict HRQoL very well

[56]. The only factor that proved significant was physical disability. Physical disability could account for 40 % of physical QoL and 10 % mental QoL. This means that in the case of mental QoL 90 % of the variability is unaccounted for. Two candidates deserve our attention: hormonal dysfunction [55, 57] and PTSD [56, 58, 59]. Up to 47 % of SAH patients can experience hypopituitarism [57]—a dysfunction that impacts both mental and physical HRQoL [55]. PTSD should also be taken into consideration. SAH is a traumatic experience. Half of all patients die; those who survive will often undergo surgery [60]. In fact, PTSD—which is characterized by intrusive thoughts and memories, emotional numbing, and persistent overarousal—can be found in about a third of all patients (as compared to a lifetime prevalence of 1–8 % in the general population) [56, 58, 59, 61]. We recently found in a prospective sample of 105 patients, that PTSD is by far the best predictor of the patients' mental HRQoL, explaining up to 48 % of its variance [61]. Importantly, PTSD is not restricted to SAH but can also occur after other types of stroke [31].

Identifying predictors of post-stroke HRQoL is important because these factors might be modulated by prevention or intervention programs, thereby reducing their detrimental effect on the patient's HRQoL. Post-stroke depression is an obvious target for treatment and in fact a few randomized placebo-controlled studies have demonstrated the benefit of antidepressant medication for patients with post-stroke depression [62, 63]. PTSD in SAH patients is another candidate. We found that it is typically those events and experiences, which occur in the hours and days after the onset of the SAH (such as the waking up in a neurological ward, facing up to a life with disabilities, or learning about the diagnosis), that are experienced as most traumatic by SAH patients [49]. This suggests that supervision and counseling of patients during that period might reduce the likelihood of later developing PTSD. Furthermore stress-coping skills have been identified as the prime determinant of the later development of PTSD in SAH patients [61]. Again training in coping skills (already available for other groups of patients with PTSD) might also be beneficial for SAH patients. Currently these suggestions are speculative but they offer specific opportunities for interventions that should be explored in future studies.

HRQoL in Primary Caregivers of Stroke Victims

On discharge from the hospital, many stroke victims require continued care. This care is overwhelmingly provided by informal caregivers such as family members [64]. The majority of caregivers are female (typically the spouses of male stroke victims) and they tend to be younger than the patients in their care (on average between 50 and 60 years old) [65]. Since patients can require support for many years [64, 66], the care provided by family members constitutes a considerable commitment in personal and economic terms. It has been estimated that in 23 % of cases, informal caregivers lose up to 30 weeks from work, and in 77 % of cases, they lose up to 16 weeks [67]. Less quantifiable, but more problematic, are the emotional costs of caring for a stroke patient. Anxiety and depression are common in caregivers of stroke patients, with an estimated prevalence of 20 % [68]. As a consequence of their role as a caregiver they often feel emotionally drained, socially isolated, and stressed [35]. It is not surprising, therefore, that the caregiver's HRQoL is significantly reduced [69–73]. Furthermore, poor psychological health in caregivers translates into a worse functional outcome for the patients in their care [74]. Unsupported caregivers might in the long term have to withdraw from their role as caregivers and stop providing a service that saves society substantial economic costs [75].

The psychological health and well-being of family members and friends who care for stroke patients should be an important concern for the health system and society at large. However, to provide appropriate support it is important to understand what causes psychological distress in informal caregivers of stroke patients. In this section we will review what is known about the factors influencing the caregivers' well-being and psychological health. It should be noted that the terms "HRQoL" and "well-being" are used quite loosely in this section. Studies on psychological health of caregivers of stroke patients have used a variety of different outcome measures, including classical HRQoL measures, but also other scales that primarily address emotional problems or the burden of caring [65]. Studies using these different outcome measures point to similar conclusions and the term "well-being" will therefore be used to encompass findings obtained with any of those measures.

In a recent systematic review, Greenwood et al. [65] reviewed 39 studies incorporating 4,181 caregivers. They distinguished between caregiver and stroke survivor factors. Self-efficacy, coping strategies, mastery, and self-esteem were the caregiver factors that had the most consistent impact on caregivers' well-being. An interesting relationship was also found for the caregiver's age. Increasing age increased the perceived strain and burden, but at the same time reduced the negative impact of caring on the caregiver's HRQoL. In the case of patient/survivor factors, only one aspect was consistently found to impact on caregiver's well-being: the patient's disability and dependency. This seems to suggest that the burden of caring for a patient with stroke, which is directly related to the patient's disability and degree of dependency, is the main cause of caregiver's reduced well-being.

While this explanation might be appropriate for many forms of strokes, it is unconvincing when applied to caregivers of SAH patient [76]. Caregivers of SAH patients report that the patient's physical impairment causes them less distress than alterations in the patient's personality, interpersonal behavior, and cognition [77]. Furthermore the correlation between the patient's disability and the caregiver's well-being is not as strong as expected [64, 66, 78] and patient disability accounts in some studies for less than 10 % of variability in caregiver's well-being. It is also interesting that other friends and relatives of patients who are not directly involved in the patient's care can be equally affected [66, 79, 80]. This suggests that burden of care on its own is insufficient to account for the increased emotional distress found in caregivers of SAH patients. Additional factors need to be taken into consideration to account for the fact that the psychopathological outcome of caregivers of SAH patients is significantly worse than that of other patients with similar needs and disabilities [81]. In this context, it is interesting to observe that significant others (i.e., close relatives and friends) are frequently more affected by the patient's illness than the patient themselves [79, 81, 82]. Significant others report that they live in perpetual fear that another SAH might occur despite expert reassurance that this is unlikely [79]. In a study by Pritchard et al. [82], a quarter of all significant others who were interviewed required medication to deal with their increased anxiety levels. This suggests that observing or learning about a loved one's SAH can be experienced by significant others as a traumatic event in itself and may cause persistent fears and other psychological problems. As we have already mentioned this phenomenon is recognized as an independent psychiatric illness called PTSD. In a recent study on 86 caregivers of SAH patients, it was found that almost a third met diagnostic criteria for PTSD [83]. Furthermore, PTSD accounts for a substantial proportion of the variability in the caregiver's HRQoL [83]. Interestingly, their stress-coping skills are the main factor predicting whether caregiver's will develop PTSD or not [83]. This last finding could provide the starting point for the development of specific interventions, which can target the prevention or reduction of PTSD in stroke caregivers. Brief training procedures to help modify a person's stress-coping skills are available [84, 85] and could provide an effective tool to improve the HRQoL of stroke caregivers. That improved support for stroke caregivers can improve their HRQoL has already been demonstrated in two studies [71, 86]. It is hoped that more specific interventions, which target some of the main determinants of caregiver's well-being, could be even more effective.

Summary and Open Questions

The concept of QALYs, which are used in health-economic decisions, and the increased recognition of the need for patient-centered medicine has led to a dramatic increase of research measuring the impact of disease on the quality of life of stroke patients and their caregivers. There are numerous methods to estimate HRQoL, although a widely agreed upon definition of QoL does not exist. As we have seen some of the methods are more reliable than others, but the researcher can now choose from a number of generic or stroke-specific HRQoL scales with satisfactory psychometric properties.

In general, studies show that HRQoL is reduced in patients with stroke. It also is important that those who care for patients with stroke are as much crippled by this experience as the patients themselves. This is particularly true in the case of SAH where the sudden and dramatic onset can trigger PTSD in relatives or friends who witnessed the SAH and its sequelae. Also, a somewhat unexpected finding is that a patient's disability typically accounts for less than 50 % of the variability in patients' and caregiver's HRQoL. In fact, for some aspects (e.g., mental HRQoL) and in some forms of stroke (e.g., SAH) it accounts for less than 10 % of the variability. This shows clearly that other factors beyond disability and burden of care have a substantial impact on the well-being of the patient and the caregiver. This partial dissociation between disability and HRQoL provides some retrospective justification of the drive to develop and use measures that go beyond a patient's disability and take the patient's own perspective into account. More importantly, the insight that emotional, psychological, and social factors are important determinants of patients' and caregivers' well-being creates new opportunities for intervention. This is particularly relevant for stroke where many disabilities tend to be chronic. It is therefore encouraging to see that there are other aspects determining the well-being of affected patients and caregivers that might respond to well-targeted interventions. Additional targets for interventions directed at improving HRQoL include post-stroke depression and PTSD [41].

As discussed, PTSD in SAH patients and their caregivers is particularly common. Evidence suggests that this stems from a lack of positive coping skills. Training stress-coping skills could provide an effective procedure to improve the well-being of SAH patients and their caregivers. However, despite recent advances, research on HRQoL in stroke is still a young research field and many questions remain. For example, we have seen that different studies produce quite different estimates for the HRQoL that is associated with stroke. The source of these inconsistencies has not yet been identified. But in our view there are two strong contenders. First, a variety of different measures have been used and some of those measures are of questionable reliability (for a review, see [2]). Second, it needs to be recognized that stroke is a crude diagnostic label for a very heterogeneous set of outcomes and etiologies. It is now clear that the patient's HRQoL is both determined by the specific symptoms and the specific etiology. Stroke leading to global aphasia and stroke leading to hemiplegia will both be classed as major stroke but the HRQoL for patients with global aphasia is four times less than that for patients with hemiplegia [23]. We have also stressed the importance of distinguishing between different etiologies. SAH with its sudden onset, life-threatening quality, and low age of onset has a traumatic potential, which is not always found in other forms of stroke. It thus creates a different set of problems and has a different impact on a patient's well being [32]. In HRQoL research stroke is still often treated as one entity; however, as the previous examples show, this might lead us to neglect important differences.

Probably the most important shortcoming of current HRQoL research concerns the dearth of studies that evaluate specific interventions aimed at improving the well-being of patients and their caregivers. Without such research, and the subsequent clinical implementation of feasible and efficient intervention strategies, HRQoL research remains a sterile academic exercise. The last decade of HRQoL research has suggested new avenues to improve the well-being of stroke patients and their caregivers. It is now time to put this knowledge into practice and develop effective support and intervention strategies.

References

1. Bonita R, Solomon N, Broad JB. Prevalence of stroke and stroke-related disability—estimates from the Auckland stroke studies. Stroke. 1997;28(10):1898–902.
2. Tengs TO, Yu M, Luistro E. Health-related quality of life after stroke—a comprehensive review. Stroke. 2001;32(4):964–71.

3. Gold MR, Franks P, McCoy KI, Fryback DG. Toward consistency in cost-utility analyses—using national measures to create condition-specific values. Med Care. 1998;36(6):778–92.
4. World Health Organization. The international classification of impairments, disabilities and handicaps. Geneva: World Health Organization; 1980.
5. Tennant A, Geddes JM, Fear J, Hillman M, Chamberlain MA. Outcome following stroke. Disabil Rehabil. 1997;19:278–84.
6. Indredavik B, Bakke F, Slørdahl SA, Rokseth R, Håheim LL. Stroke unit treatment improves long-term quality of life—a randomized controlled trial. Stroke. 1998;29(5):895–9.
7. Shin AY, Porter PJ, Wallace MC, Naglie G. Quality of life of stroke in younger individuals—utility assessment in patients with arteriovenous malformations. Stroke. 1997;28(12):2395–9.
8. Gage BF, Cardinalli AB, Owens DK. The effect of stroke and stroke prophylaxis with aspirin or warfarin on quality of life. Arch Intern Med. 1996;156(16):1829–36.
9. Buck D, Jacoby A, Massey A, Ford G. Evaluation of measures used to assess quality of life after stroke. Stroke. 2000;31(8):2004–10.
10. Bergner M, Bobbitt RA, Carter WB, Gilson BS. The sickness impact profile—development and final revision of a health-status measure. Med Care. 1981;19(8):787–805.
11. Hunt SM, McKenna SP, McEwen J, Backett EM, Williams J, Papp E. A quantitative approach to perceived health-status—a validation study. J Epidemiol Community Health. 1980;34(4):281–6.
12. Ware JE, Sherbourne CD. The MOS 36-item short-form health survey (SF-36). 1. Conceptual-framework and item selection. Med Care. 1992;30(6):473–83.
13. The EuroQol Group. EuroQol: a new facility for the measurement of health related quality of life. Health Policy. 1990;16:199–208.
14. Feeny D, Torrance GW, Furlong WJ. Health utility index. In: Spilker B, editor. Quality of life and pharmacoeconomics in clinical trials. Philadelphia, PA: Lippincott-Raven; 1996.
15. Harwood RH, Gompertz P, Ebrahim S. Handicap one-year after a stroke—validity of a new scale. J Neurol Neurosurg Psychiatry. 1994;57(7):825–9.
16. Buck D, Jacoby A, Massey A, Steen N, Sharma A, Ford GA. Development and validation of NEWSQOL (R), the Newcastle Stroke-Specific Quality of Life Measure. Cerebrovasc Dis. 2004;17(2–3):143–52.
17. Doyle PJ, McNeil MR, Bost JE, Ross KB, Wambaugh JL, Hula WD, et al. The Burden of Stroke Scale (BOSS) provided valid, reliable, and responsive score estimates of functioning and well-being during the first year of recovery from stroke. Qual Life Res. 2007;16(8):1389–98.
18. Duncan PW, Wallace D, Lai SM, Johnson D, Embretson S, Laster LJ. The stroke impact scale version 2.0—evaluation of reliability, validity, and sensitivity to change. Stroke. 1999;30(10):2131–40.
19. Hamedani AG, Wells CK, Brass LM, Kernan WN, Viscoli CM, Maraire JN, et al. A quality-of-life instrument for young hemorrhagic stroke patients. Stroke. 2001;32(3):687–95.
20. Hilari K, Byng S, Lamping DL, Smith SC. Stroke and aphasia quality of life scale-39 (SAQOL-39)—evaluation of acceptability, reliability, and validity. Stroke. 2003;34(8):1944–50.
21. vanStraten A, deHaan RJ, Limburg M, Schuling J, Bossuyt PM, van den Bos GA. A stroke-adapted 30-item version of the sickness impact profile to assess quality of life (SA-SIP30). Stroke. 1997;28(11):2155–61.
22. Williams LS, Weinberger M, Harris LE, Clark DO, Biller J. Development of a stroke-specific quality of life scale. Stroke. 1999;30(7):1362–9.
23. Carod-Artal FJ, Egido JA. Quality of life after stroke: the importance of a good recovery. Cerebrovasc Dis. 2009;27:204–14.
24. Carod-Artal FJ, Coral LF, Trizotto DS, Moreira CM. The stroke impact scale 3.0—evaluation of acceptability, reliability, and validity of the Brazilian version. Stroke. 2008;39(9):2477–84.
25. Duncan PW, Lai SM, Tyler D, Perera S, Reker DM, Studenski S. Evaluation of proxy responses to the Stroke Impact Scale. Stroke. 2002;33(11):2593–9.

26. Hilari K, Owen S, Farrelly SJ. Proxy and self-report agreement on the Stroke and Aphasia Quality of Life Scale-39. J Neurol Neurosurg Psychiatry. 2007;78(10):1072–5.
27. Williams LS, Bakas T, Brizendine E, Plue L, Tu W, Hendrie H, et al. How valid are family proxy assessments of stroke patients' health-related quality of life? Stroke. 2006;37(8):2081–5.
28. Haacke C, Althaus A, Spottke A, Siebert U, Back T, Dodel R. Long-term outcome after stroke—evaluating health-related quality of life using utility measurements. Stroke. 2006;37(1):193–8.
29. Post PN, Stiggelbout AM, Wakker PP. The utility of health states after stroke—a systematic review of the literature. Stroke. 2001;32(6):1425–9.
30. Lai SM, Studenski S, Duncan PW, Perera S. Persisting consequences of stroke measured by the Stroke Impact Scale. Stroke. 2002;33(7):1840–4.
31. Bruggimann L, Annoni JM, Staub F, von Steinbüchel N, Van der Linden M, Bogousslavsky J. Chronic posttraumatic stress symptoms after nonsevere stroke. Neurology. 2006;66(4):513–6.
32. Noble AJ, Schenk T. Psychosocial outcome following subarachnoid haemorrhage: an under-researched problem. Br J Neurosci Nurs. 2009;5(8):372–80.
33. Lindgren I, Jönsson AC, Norrving B, Lindgren A. Shoulder pain after stroke—a prospective population-based study. Stroke. 2007;38(2):343–8.
34. Jönsson AC, Lindgren I, Hallström B, Norrving B, Lindgren A. Prevalence and intensity of pain after stroke: a population based study focusing on patients' perspectives. J Neurol Neurosurg Psychiatry. 2006;77(5):590–5.
35. Carod J, Egido J, González JL, Varela De Seijas E. Poststroke sexual dysfunction and quality of life. Stroke. 1999;30(10):2238–9.
36. Korpelainen JT, Nieminen P, Myllyla VV. Sexual functioning among stroke patients and their spouses. Stroke. 1999;30(4):715–9.
37. Carod-Artal FJ, González-Gutiérrez JL, Herrero JA, Horan T, De Seijas EV. Functional recovery and instrumental activities of daily living: follow-up 1-year after treatment in a stroke unit. Brain Inj. 2002;16(3):207–16.
38. Gargano JW, Reeves MJ; Paul Coverdell National Acute Stroke Registry Michigan Prototype Investigators. Sex differences in stroke recovery and stroke-specific quality of life—results from a statewide stroke registry. Stroke. 2007;38(9):2541–48.
39. Forsberg-Warleby G, Moller A, Blomstrand C. Psychological well-being of spouses of stroke patients during the first year after stroke. Clin Rehabil. 2004;18(4):430–7.
40. Darlington AS, Dippel DW, Ribbers GM, van Balen R, Passchier J, Busschbach JJ. Coping strategies as determinants of quality of life in stroke patients: a longitudinal study. Cerebrovasc Dis. 2007;23(5–6):401–7.
41. Ayerbe L, Ayis S, Rudd AG, Heuschmann PU, Wolfe CD. Natural history, predictors, and associations of depression 5 years after stroke the South London Stroke Register. Stroke. 2011;42(7):1907–11.
42. Kwok T, Lo RS, Wong E, Wai-Kwong T, Mok V, Kai-Sing W. Quality of life of stroke survivors: a 1-year follow-up study. Arch Phys Med Rehabil. 2006;87(9):1177–82.
43. Carod-Artal FJ. Post-stroke depression (I). Epidemiology diagnostic criteria and risk factors. Rev Neurol. 2006;42(3):169–75.
44. Rochette A, Bravo G, Desrosiers J, St-Cyr Tribble D, Bourget A. Adaptation process, participation and depression over six months in first-stroke individuals and spouses. Clin Rehabil. 2007;1(6):554–62.
45. Daniel K, Wolfe CD, Busch MA, McKevitt C. What are the social consequences of stroke for working-aged adults? A systematic review. Stroke. 2009;40(6):E431–40.
46. Mackay A, Nias BC. Strokes in the young and middle-aged—consequences to the family and to society. J R Coll Physicians Lond. 1979;13(2):106–12.
47. Teasell RW, McRae MP, Finestone HM. Social issues in the rehabilitation of younger stroke patients. [Article]. Arch Phys Med Rehabil. 2000;81(2):205–9.

48. de Rooij NK, Linn FH, van der Plas JA, Algra A, Rinkel GJ. Incidence of subarachnoid haemorrhage: a systematic review with emphasis on region, age, gender and time trends. J Neurol Neurosurg Psychiatry. 2007;78(12):1365–72.
49. Baisch SB, Schenk T, Noble AJ. What is the cause of post-traumatic stress disorder following subarachnoid haemorrhage? Post-ictal events are key. Acta Neurochir. 2011;153(4):913–22.
50. Anderson C, Anderson N, Bonita R, et al. Epidemiology of aneurysmal subarachnoid hemorrhage in Australia and New Zealand—incidence and case fatality from the Australasian Cooperative Research on Subarachnoid Hemorrhage Study (ACROSS). Stroke. 2000;31(8):1843–50.
51. Bamford J, Sandercock P, Dennis M, Burn J, Warlow C. A prospective-study of acute cerebrovascular-disease in the community—the Oxfordshire Community Stroke Project 1981–86.2. Incidence, case fatality rates and overall outcome at one year of cerebral infarction, primary intracerebral and subarachnoid hemorrhage. J Neurol Neurosurg Psychiatry. 1990;53(1):16–22.
52. Kolominsky-Rabas PL, Weber M, Gefeller O, Neundoerfer B, Heuschmann PU. Epidemiology of ischemic stroke subtypes according to TOAST criteria: incidence, recurrence, and long-term survival in ischemic stroke subtypes—a population-based study. Stroke. 2001;32(12):2735–40.
53. Wermer MJ, Kool H, Albrecht KW, Rinkel GJ; Aneurysm Screening after Treatment for Ruptured Aneurysms Study Group. Subarachnoid hemorrhage treated with clipping: long-term effects on employment, relationships, personality, and mood. Neurosurgery. 2007;60(1):91–7.
54. Hütter BO, Gilsbach JM, Kreitschmann I. Quality-of-life and cognitive deficits after subarachnoid hemorrhage. Br J Neurosurg. 1995;9(4):465–75.
55. Kreitschmann-Andermahr I, Poll E, Hutter BO, Reineke A, Kristes S, Gilsbach JM, et al. Quality of life and psychiatric sequelae following aneurysmal subarachnoid haemorrhage: does neuroendocrine dysfunction play a role? Clin Endocrinol. 2007;66(6):833–7.
56. Noble AJ, Schenk T. Which variables help explain the poor health-related quality of life after subarachnoid hemorrhage? A meta-analysis. Neurosurgery. 2010;66(4):772–83.
57. Schneider HJ, Kreitschmann-Andermahr I, Ghigo E, Stalla GK, Agha A. Hypothalamopituitary dysfunction following traumatic brain injury and aneurysmal subarachnoid hemorrhage—a systematic review. JAMA. 2007;298(12):1429–38.
58. Berry E. Post-traumatic stress disorder after subarachnoid haemorrhage. Br J Clin Psychol. 1998;37:365–7.
59. Hütter BO. Neuropsychological sequelae of subarachnoid haemorrhage and its treatment. Vienna: Springer; 2000.
60. Nieuwkamp DJ, Setz LE, Algra A, Linn FH, de Rooij NK, Rinkel GJ. Changes in case fatality of aneurysmal subarachnoid haemorrhage overtime, according to age, sex, and region: a meta-analysis. Lancet Neurol. 2009;8(7):635–42.
61. Noble AJ, Baisch S, Mendelow AD, Allen L, Kane P, Schenk T. Posttraumatic stress disorder explains reduced quality of life in subarachnoid hemorrhage patients in both the short- and long-term. Neurosurgery. 2008;63:1095–104.
62. Lipsey JR, Robinson RG, Pearlson GD, Rao K, Price TR. Nortriptyline treatment of post-stroke depression—a double-blind-study. Lancet. 1984;1(8372):297–300.
63. Wiart L, Petit H, Joseph PA, Mazaux JM, Barat M. Fluoxetine in early poststroke depression—a double-blind placebo-controlled study. Stroke. 2000;31(8):1829–32.
64. Mezue W, Mathew B, Draper P, Watson R. The impact of care on carers of patients treated for aneurysmal subarachnoid haemorrhage. Br J Neurosurg. 2004;18(2):135–7.
65. Greenwood N, Mackenzie A, Cloud GC, Wilson N. Informal carers of stroke survivors: factors influencing carers—a systematic review of quantitative studies. Disabil Rehabil. 2008;30(18):1329–49.
66. Hellawell DJ, Taylor R, Pentland B. Persisting symptoms and carers' views of outcome after subarachnoid haemorrhage. Clin Rehabil. 1999;13(4):333–40.

67. Pritchard C, Foulkes L, Lang DA, Neil-Dwyer G. Two-year prospective study of psychosocial outcomes and a cost-analysis of 'treatment-as-usual' versus an 'enhanced' (specialist liaison nurse) service for aneurysmal sub arachnoid haemorrhage (ASAH) patients and families. Br J Neurosurg. 2004;18(4):347–56.
68. Béthoux F, Calmels P, Gautheron V, Minaire P. Quality of life of the spouses of stroke patients: a preliminary study. Int J Rehabil Res. 1996;19:291–9.
69. Forsberg-Wärleby G, Möller A, Blomstrand C. Spouses of first-ever stroke patients—psychological well-being in the first phase after stroke. Stroke. 2001;32(7):1646–51.
70. Jönsson AC, Lindgren I, Hallström B, Norrving B, Lindgren A. Determinants of quality of life in stroke survivors and their informal caregivers. Stroke. 2005;36(4):803–8.
71. McCullagh E, Brigstocke G, Donaldson N, Kalra L. Determinants of caregiving burden and quality of life in caregivers of stroke patients. Stroke. 2005;36(10):2181–6.
72. Rochette A, Desrosiers J, Bravo G, Tribble DS, Bourget A. Changes in participation level after spouse's first stroke and relationship to burden and depressive symptoms. Cerebrovasc Dis. 2007;24(2–3):255–60.
73. Rombough RE, Howse EL, Bagg SD, Bartfay WJ. A comparison of studies on the quality of life of primary caregivers of stroke survivors: a systematic review of the literature. Top Stroke Rehabil. 2007;14(3):69–79.
74. Kwakkel G, Wagenaar RC, Kollen BJ, Lankhorst GJ. Predicting disability in stroke—a critical review of the literature. Age Ageing. 1996;25(6):479–89.
75. Dewey HM, Thrift AG, Mihalopoulos C, Carter R, Macdonell RA, McNeil JJ, et al. Informal care for stroke survivors—results from the North East Melbourne Stroke Incidence Study (NEMESIS). Stroke. 2002;33(4):1028–33.
76. Noble AJ, Schenk T. Looking beyond the spontaneous subarachnoid haemorrhage patient: a review of the impact upon family and friends. Br J Neurosci Nurs. 2008;4(6):278–85.
77. Hellawell DJ, Pentland B. Relatives' reports of long term problems following traumatic brain injury or subarachnoid haemorrhage. Disabil Rehabil. 2001;23(7):300–5.
78. Buchanan KM, Elias LJ, Goplen GB. Differing perspectives on outcome after subarachnoid hemorrhage: the patient, the relative, the neurosurgeon. Neurosurgery. 2000;46(4):831–8.
79. Hop JW, Rinkel GJ, Algra A, van Gijn J. Quality of life in patients and partners after aneurysmal subarachnoid hemorrhage. Stroke. 1998;29(4):798–804.
80. Ogden JA, Utley T, Mee EW. Neurological and psychosocial outcome 4 to 7 years after subarachnoid hemorrhage. Neurosurgery. 1997;41(1):25–34.
81. Pritchard C, Clapham L, Foulkes L, Lang DA, Neil-Dwyer G. Comparison of cohorts of elective and emergency neurosurgical patients: psychosocial outcomes of acoustic neuroma and aneurysmal sub arachnoid hemorrhage patients and carers. Surg Neurol. 2004;62(1):7–16.
82. Pritchard C, Foulkes L, Lang DA, Neil-Dwyer G. Psychosocial outcomes for patients and carers after aneurysmal subarachnoid haemorrhage. Br J Neurosurg. 2001;15(6):456–63.
83. Noble AJ, Schenk T. Post-traumatic stress disorder in family and friends of spontaneous subarachnoid hemorrhage patients. J Neurosurg. 2008;109:1027–33.
84. Keefe FJ, Blumenthal J, Baucom D, Affleck G, Waugh R, Caldwell DS, et al. Effects of spouse-assisted coping skills training and exercise training in patients with osteoarthritic knee pain: a randomized controlled study. Pain. 2004;110(3):539–49.
85. Kennedy P, Duff J, Evans M, Beedie A. Coping effectiveness training reduces depression and anxiety following traumatic spinal cord injuries. Br J Clin Psychol. 2003;42:41–52.
86. Mant J, Carter J, Wade DT, Winner S. Family support for stroke: a randomised controlled trial. Lancet. 2000;356(9232):808–13.

Chapter 14
MRI Methods Applied to Stroke

Bradley J. MacIntosh and Simon J. Graham

Introduction

Diagnostic imaging is an invaluable aspect of clinical stroke medicine, providing the location, volume, and the nature of the stroke lesion. Anatomical stroke imaging typically has been done with computed tomography (CT), but ever increasingly is supplanted by the superb soft tissue contrast provided by magnetic resonance imaging (MRI). It is more cumbersome to obtain MRI in acute stroke patients; so at many centers, CT remains the initial screening test, with or without contrast administration (CT angiography, CT perfusion). This rapidly differentiates ischemic from hemorrhagic stroke and can then guide acute management, for example, with thrombolytics. The versatility of MRI methods, however, also enables much more detailed biophysical information to be obtained about stroke physiology, above and beyond lesion structure. For example, just after stroke onset, diffusion-weighted imaging (DWI) and perfusion-weighted imaging (PWI) provide information about diffusion of water molecules and microvascular blood flow within brain tissue, respectively. The DWI and PWI methods help to evaluate the ischemic zone surrounding infarcted tissue that is potentially salvageable by recanalization therapies. Magnetic resonance angiography (MRA) approaches are also available to characterize larger-scale vasculature. In the post-acute and chronic phases, functional MRI (fMRI) offers the ability to detect alterations in brain activation patterns post-stroke,

B.J. MacIntosh, Ph.D. (✉) • S.J. Graham, Ph.D., P.Eng.
Department of Medical Biophysics, University of Toronto, Toronto, ON, Canada

Sunnybrook Research Institute, Sunnybrook Health Sciences Centre,
2075 Bayview Avenue, Room A453, Toronto, ON, Canada

Heart and Stroke Foundation, Centre of Stroke Recovery, Toronto, ON, Canada
e-mail: bmac@sri.utoronto.ca

T.A. Schweizer and R.L. Macdonald (eds.), *The Behavioral Consequences of Stroke*,
DOI 10.1007/978-1-4614-7672-6_14, © Springer Science+Business Media New York 2014

either associated with a particular behavioral task or during the resting state. Lastly, it is also possible now to image blood flow in the brain noninvasively using arterial spin labeling (ASL) MRI. In this chapter, we provide a biophysical understanding of these and other basic MRI methods and discuss their application to stroke recovery.

Basic Physical Principles

Use of MRI in medicine has evolved from the discovery of nuclear magnetic resonance (NMR) in the 1950s [1, 2]. The term NMR is apt—*nuclear* refers to specific atomic nuclei that, when placed in a large applied static *magnetic* field, can undergo *resonance* (substantial energy absorption and subsequent energy emission) if a pulsed magnetic wave is applied at a certain characteristic frequency. For the NMR effect to occur, the nuclei of interest must have the quantum mechanical property of "spin." Such nuclei simplistically can be thought of as very small objects with spinning charge and therefore exhibiting nuclear magnetic fields. When placed in a large applied magnetic field, the nuclear magnetic fields tend to align with the applied field, much as a compass needle aligns with the magnetic field of the earth. However, in the case of NMR-capable nuclei, a small number of preferred alignments are possible. The simplest scenario involves the hydrogen nucleus, or proton, which can take on two different energy states: a low energy "spin-up" state in which the nuclear magnetic field partially aligns to the external field and a higher-energy "spin-down" state in which the nuclear field aligns partially antiparallel. The word "partial" refers to the fact that in each state, the alignment is not perfect. Rather, the nuclear field "precesses" around the direction of external field, like a spinning top wobbles about its axis.

Within a material containing NMR-capable nuclei, the nuclear magnetic fields for the given energy states can be thought of as "adding up," resulting in a quantity known as net nuclear magnetization. Stated another way, such a material becomes very slightly magnetized when placed in a large applied magnetic field. The magnetization is typically small because at room temperature there is sufficient thermal energy to promote changes between spin energy states so that the states become almost equally populated. However, if the external magnetic field is made strong enough, the population of spin energy states becomes more unbalanced and nuclear magnetization increases to the point that it can be easily observed above the noise background of a detector circuit. For this signal-to-noise ratio (SNR) consideration, current MRI systems operate using superconductor technology to generate very large static magnetic fields of 1.5–3.0 Tesla (T), with ultrahigh field systems at 7.0 T and beyond under development (1 T is approximately 2×10^4 times the earth's magnetic field).

The conventional manner for observing nuclear magnetization can be described in three steps. First, the NMR-capable sample is placed in a large applied magnetic field, creating appreciable magnetization. Second, a magnetic radio frequency (RF) pulse is applied, causing energy absorption that perturbs the energy states and, consequently, the magnetization. The RF pulse has the macroscopic effect of "tipping" the magnetization out of alignment with the applied field, as a result a vector component oriented transverse (orthogonal) to the direction of the applied magnetic field and oscillates at a characteristic rate, known as the Larmor frequency. Lastly, the oscillating transverse magnetization component is detected as an NMR signal using an electrical circuit known as a "receiver coil."

Early in the development of NMR, it was recognized that hydrogen exhibits the largest nuclear magnetic field and thus the largest NMR signal of all the elements. This is particularly advantageous for MRI, given that hydrogen is by far the most biologically abundant of nuclei, present primarily within the human body as constituents of water molecules.

Main Sources of NMR Signal Contrast

In the early 1970s, it became increasingly apparent that biological tissues—both abnormal and abnormal—often have markedly different NMR signal characteristics [3]. There are four main mechanisms whereby NMR signals provide tissue contrast (i.e., the ability to detect different tissue types based on the associated NMR signal differences). First, the number of protons per unit volume, or proton density, is a factor that weights all NMR signals. Typically, the proton density of tissues is rather uniform and a relatively poor source of signal contrast, as most tissues have a water content ranging from approximately 70 to 80 %. Much improved signal contrast is obtained based on several NMR parameters that characterize how magnetization tends to recover toward its equilibrium condition after the RF pulse is turned off. The time that magnetization takes to recover in the longitudinal direction is parameterized by the time constant T1, whereas the time that magnetization takes to decay in the transverse plane is parameterized by either the time constant T2* or T2, depending on the specific details of the NMR measurement method. The process of T2* decay is governed by spatial variations in magnetic field primarily on a microscopic scale within and between tissues. Another source of tissue contrast is dynamic proton–proton magnetic field interactions from within water molecules and between neighboring water molecules as they tumble and translate with respect to one another, a process that is influenced by factors such as temperature, viscosity, and the presence of ions, macromolecules, and tissue microstructures such as cell membranes. This latter source of contrast plays a substantial role in the T1 and T2 characteristics of tissues.

Development of MRI

The initial period of investigating NMR signals from tissues focused on the ability to distinguish cancerous tissue from normal tissue, and to distinguish malignant from benign tissue, in bulk excised specimens. This work provided some of the impetus for developing methods to encode NMR signals spatially in the form of images. Key innovators in these subsequent developments were Lauterbur and Mansfield, who received the Nobel Prize in physiology and medicine in 2003. Today, the term "nuclear" has been dropped to avoid any potential misconception that NMR methods involve use of ionizing radiation—hence the acronym MRI. The first whole-body MRI systems were developed for commercial use in the late 1980s, and there has been a steady increase in technical capabilities and MR image quality since this time. Today, multifaceted MRI "protocols" are achieved with approximately 1 mm spatial resolution and exquisite image contrast.

The additional technological concept that led to the development of MRI from NMR is the "imaging gradient" for spatial encoding. The imaging gradient is an electrical circuit that produces a linear change in magnetic field as a function of position. All commercial MRI systems provide magnetic field gradients in the three orthogonal directions x, y, and z. The application of these gradients, either alone or in combination, enables magnetization to be spatially encoded through ingenious manipulations of the Larmor frequency as a function of position. Both two-dimensional (2D) "multi-slice" MRI (a stack of image slices, each with a specific slice thickness) and three-dimensional (3D) MRI (signals encoded as a matrix of specific x, y, and z values) are commonly available. Both 2D and 3D MRI require multiple repetitions of RF pulse excitation, data acquisition, and recovery periods to obtain all the necessary spatial encoding information required for image reconstruction. Both types of imaging require scan times in the range of approximately 1–10 min. Typically, 3D MRI provides improved SNR for a given signal contrast, but requires longer acquisition time.

Beyond typical 2D and 3D MRI methods, there is also a class of "fast" MRI techniques capable of providing a fully encoded image slice in approximately 100 ms, or multi-slice datasets that encompass the brain volume in approximately 1–2 s. These techniques, the most common of which is known as echo planar imaging (EPI) [4], can provide full spatial encoding within a single RF excitation episode that is followed by data acquisition during rapidly varying gradient waveforms. EPI can be used to image time-dependent physiological processes within the body that occur on the timescale of seconds, such as the wash-in and wash-out of injectable contrast agents. However, there are inevitable tradeoffs: reduced SNR, reduced spatial resolution, and regions of signal loss and substantial geometric distortion arising from sensitivity to magnetic field uniformity, that result in an overall decrement in image quality compared to what can be achieved by conventional 2D and 3D MRI. Despite these challenges, however, the continued advancement of MRI system technology is now making EPI a feasible and useful part of clinical imaging protocols.

MRI of Acute Stroke

Multifaceted MRI protocols (i.e., a series of MRI acquisitions with different signal contrast weightings, tailored to detect pathology) are becoming increasingly important in the acute stroke setting to confirm diagnosis and to direct the use of acute interventions [5, 6]. Using MRI, it is possible to identify regions of active ischemia, hemorrhage, and occluded vessels, and to gain insight regarding the size and location of the core infarct and the extent of surrounding tissue that potentially can be salvaged (i.e., the ischemic penumbra). The typical MRI protocol for acute stroke has five main components: diffusion-weighted imaging (DWI), fluid-attenuated inversion recovery (FLAIR), MRA, perfusion-weighted imaging (PWI), and T2*-weighted imaging. Some representative examples of these techniques are shown in Fig. 14.1, and the role of each imaging component is subsequently described.

Standard MRI sequences that provide T1-weighting or T2-weighting, and their derivatives, are well known to be insensitive to the immediate effects of cerebral ischemia. However, DWI is highly sensitive and the preferred noninvasive imaging method for detecting acute ischemic stroke [7]. The technique is based on the use of a pair of counter-balanced, pulsed imaging gradients applied in the same direction. The first gradient pulse is used to encode protons as a function of position, whereas the second is used to remove this encoding. If protons are completely static, the counter-balanced gradient pulses have no effect and the resultant MRI signal is proportional to the net magnetization within a volume element (voxel). However, if the protons move between the applications of the gradient pulses, as is the case for water molecules diffusing in biological tissue, then the spatial encoding is not fully removed and is measured as a signal attenuation providing diffusion-weighted contrast. If the degree of diffusion weighting is manipulated in a series of DWI experiments, simple mathematical modeling enables calculation of an apparent diffusion coefficient (ADC) image, which has information distinct from that provided by proton density or relaxation time parameters.

Within minutes of vessel occlusion, failure of the sodium–potassium pump and associated cytotoxic edema leads to an influx of water from the extracellular space to the intracellular space. This impedes the diffusion of water molecules, as cell membranes act as more of a barrier. The restricted diffusion of water molecules results in increased signal intensity on DWI. The use of DWI is approximately 4–5 times more sensitive for detecting acute stroke than is non-contrast computed tomography [8].

The combined use of FLAIR and DWI helps to distinguish acute from subacute and chronic stroke lesions. Given that cerebrospinal fluid (CSF) tends to pool within the infarct zone as time progresses, use of FLAIR imaging (which provides T1-weighted tissue contrast while suppressing signal from CSF) combined with DWI can improve the identification of new lesions near sites of prior ischemic injury, potentially providing insight into stroke physiology and subtype. As FLAIR images typically can detect the presence of an ischemic region by approximately 3 h after stroke [9], normal FLAIR appearance in the presence of DWI hyperintensity provides potential for assessing the time of injury. This is of critical importance given that the only clinically approved therapy for acute stroke at present, recombinant tissue plasminogen activator (rt-PA), requires administration within several hours of stroke onset [10]. Maximizing the use of rt-PA places an increasing need on the availability and capability of MRI for acute stroke.

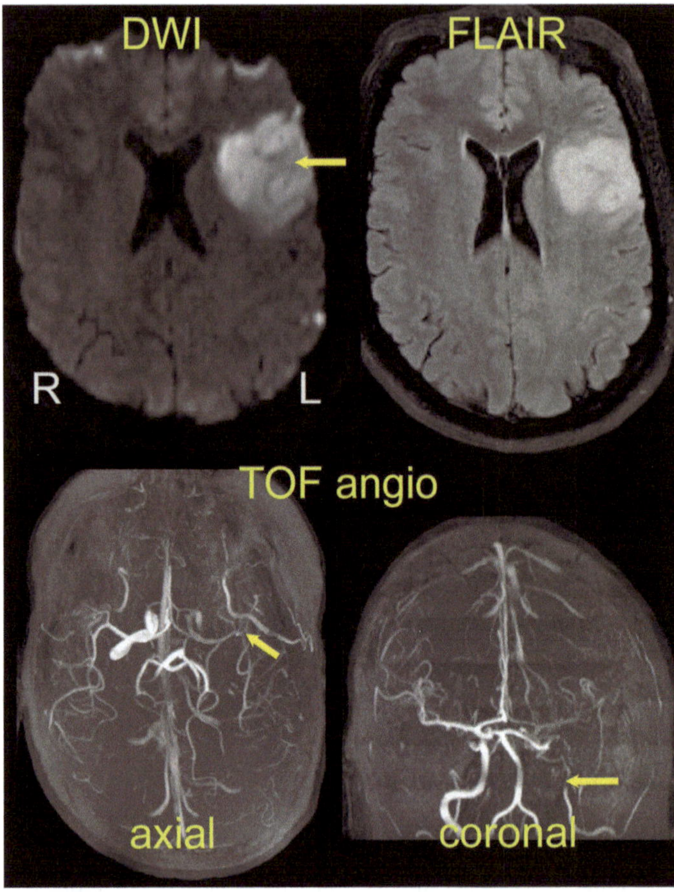

Fig. 14.1 MR images from a 48-year-old woman with a near occlusion in the left internal carotid artery and an acute stroke. *Top left*: Diffusion-weighted image (DWI) shows an area of restricted diffusion and consequent hyperintensity in the left frontal lobe. *Top right*: The fluid-attenuated inversion recovery (FLAIR) image shows hyperintensity in the corresponding location when compared to the DWI scan. *Bottom*: Time of flight angiography (TOF angio) images in the axial and coronal plane show reduced flow through the arteries in the left hemisphere. *Arrows* indicate location of near occlusion

MRA provides a supportive, noninvasive method to screen for vessel occlusions, stenosis, or malformations. There are at least two MRI techniques available for this purpose, with each performing slightly less sensitively than the gold standard X-ray imaging method, digital subtraction angiography. Three-dimensional time of flight angiography (3DTOF) [11] provides a noninvasive method of assessing the intracranial circulation. The technique is based on the fact that if RF pulses are applied rapidly in succession, there is little time for T1 recovery of longitudinal magnetization. Over time, a steady-state of longitudinal magnetization is obtained that is much lower than the equilibrium magnetization. However, if blood flows into the imaging plane it still retains its equilibrium value and thus exhibits positive image contrast with respect to the background static tissues. This effect enables imaging of the vascular lumen, but only for through-plane flow. Three-dimensional MRI is used to maximize the effect by keeping slice thickness to a minimum. The alternative technique is known as contrast-enhanced magnetic resonance angiography (CE-MRA) [12], involving the intravenous injection of a paramagnetic contrast agent known as gadolinium diethyl-triamine-penta-acetic acid (Gd-DTPA). This agent reduces the T1 value of blood and the enhanced T1 recovery results in increased signal intensity for blood on T1-weighted images, with respect to the signal from static tissues. This technique provides more contrast than 3DTOF and is much less sensitive to flow dynamics within the vasculature. Consequently, CE-MRA can be used more effectively to image extracranial vessels in addition to intracranial vessels. If scan time permits, the two angiographic methods are complementary and use of both 3DTOF and CE-MRA can improve diagnostic ability [13]. However, CE-MRA requires contrast agent administration, which is contraindicated in patients with poor renal function [14].

PWI is similar to CE-MRA in that a bolus of Gd-DTPA is typically administered, although in this case fast imaging (i.e., EPI or a variant) is employed to generate a time series of images to track reductions in T2*-weighted signal intensity of tissues as the contrast agent travels through the microvasculature. Semiquantitative perfusion maps are obtainable from this examination that estimate the cerebral blood volume (CBV)—the blood volume per unit of brain; the mean transit time (MTT)—the average time required by the bolus of contrast agent to cross the capillary network; the cerebral blood flow (CBF)—the volume of blood flowing per brain mass and per unit of time (i.e., mL/100 g tissue/min); and several other parameters that quantify the minimal signal intensity and latency of the bolus within the capillary network [15]. When combined with DWI, both imaging datasets provide information about the ischemic core and about the ischemic penumbra. It was originally thought that the difference between the spatial extent of perfusion deficit and DWI hyperintensity reflected the ischemic penumbra. It is now known that DWI hyperintensity can resolve in many stroke patients, and that the extent of perfusion deficits does not necessarily reflect the true extent of the penumbra. Irrespective of these issues, however, the imaging sequences have been demonstrated to be of benefit for identifying patients that are good candidates for rt-PA therapy [16] and for clinical trials aimed at developing new acute stroke therapies, such as mechanical embolectomy [17].

Lastly, T2*-weighted imaging has multiple applications in acute stroke. These applications stem from the fact that the T2* of blood varies with oxygenation content, based on the magnetic susceptibility characteristics of hemoglobin. Oxygenated hemoglobin is diamagnetic, causing a slight decrease in the applied magnetic field that is supported within blood, whereas deoxygenated hemoglobin is paramagnetic, causing increased magnetic field within red blood cells. The abnormal accumulation of deoxygenated blood thus provides hypointensity on T2*-weighted images as an indicator of vascular pathology. Studies have shown that T2*-weighted images are capable of detecting acute hemorrhage with equivalent accuracy to CT [8]. Microbleeds, indicative of multiple types of micro-angiopathy, are also detected on T2*-weighted images [18] that are not visualized on CT due to insufficient signal contrast and spatial resolution. T2*-weighted images are also capable of depicting hemorrhagic transformations of ischemic infarcts, as well as depicting thrombosed veins or sinuses. Lastly, T2*-weighted applications have been enhanced in recent years through use of a technique known as susceptibility-weighted imaging (SWI), which provides improved characterization of susceptibility changes in the brain microvasculature for angiography, venography, and detection of atherosclerosis and thrombosis [19].

MRI in Stroke Recovery

MRI may have an expanded role related to stroke in the future. Although MRI is often viewed as not widely accessible and expensive, its versatility and non-invasiveness make it well-suited for neuroscience studies involving stroke patients to understand more about stroke pathology and the process of recovery (in the presence or absence of therapeutic interventions). Ultimately, there is the possibility that if useful new interventions arise from such studies, then MRI could be used as a biomarker in the clinical setting. The neuroscience studies involving MRI of stroke patients are numerous. In the remainder of this chapter, several representative examples are provided that illustrate the breadth of the opportunity, and also the challenges that lie ahead.

MRI for Neuroanatomical Correlates of Stroke Symptoms

Given the spectrum of deficits exhibited by stroke patients, the variability in location and extent of lesions caused by stroke, and the current imperfect understanding of normal brain function, it is not surprising that there are a substantial number of studies that investigate how stroke lesions in specific brain locations correlate with behavioral deficits. The use of stroke patients as a "knock-out" model of brain function can help to clarify the role of a given brain region; however, many strokes do not affect only one or two discrete brain functional areas, so definitive conclusions and findings must be corroborated through replication, using MRI and other methods. For stroke patients, improved knowledge of how damaged neuroanatomical structures relate to behavioral deficits can potentially be important in the refinement of cognitive and physical rehabilitation strategies.

An interesting example of structural analysis comes from a study that used high resolution T1-weighted and FLAIR images to overlay lesion locations from a cohort of 44 ischemic stroke survivors [20]. Using a method known as voxel-based lesion symptom mapping (VLSM), Molenberghs and colleagues identified brain lesions that were correlated with cognitive deficits observed when participants performed

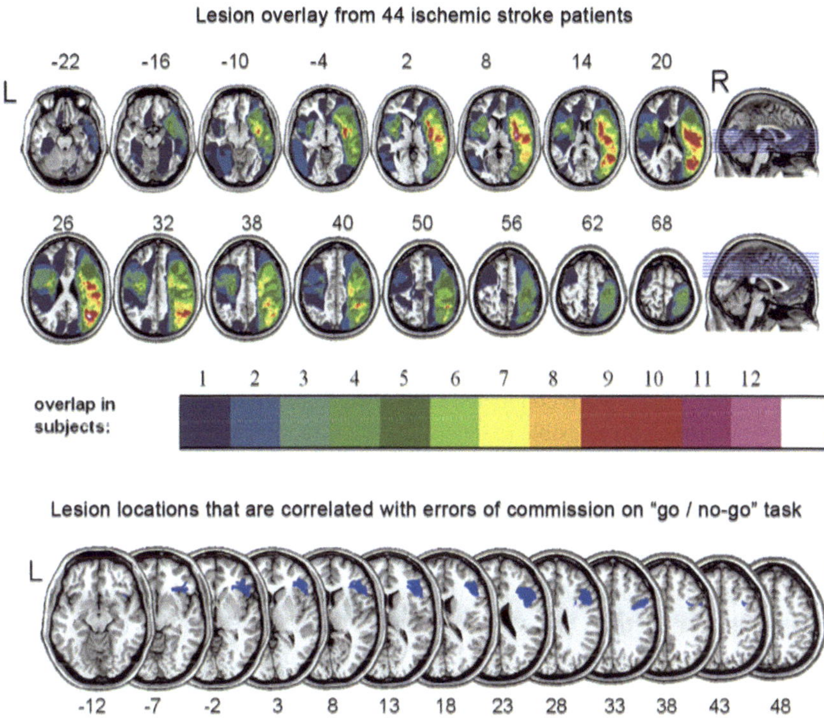

Fig. 14.2 *Top*: Color overlay of lesions from a group of 44 stroke survivors. *Bottom*: Results of voxel-based lesion symptom mapping (VLSM) identifying the inferior frontal gyrus as positively correlated with errors of commission when performing the sustained-attention to response task (SART). In other words, individuals that had a lesion in this location tended to have a higher number of errors compared to others in the cohort (based on published work) [20]

the sustained-attention to response task (SART). For example, lesions in the right inferior frontal gyrus were associated with an increase in the number of errors of commission, reflecting impairments of inhibition and sustained attention (Fig. 14.2).

Other examples of work in this area involve use of MRI-derived voxel-based morphometry (VBM) or cortical thickness measures estimated by sophisticated image-processing techniques. Based on differences in signal intensity in high resolution T1-weighted images, it is possible to "segment" or classify cortical grey matter from CSF and cortical grey matter from subcortical white matter. The tissue boundaries can then be represented as convoluted surfaces, with point-to-point thickness estimates obtained by projecting a surface normal (a line oriented 90° from a given surface) such that the other surface is intersected. A recent study used VBM to help classify stroke patients that responded positively to epidural motor cortex stimulation compared to nonresponders [21]. Another recent study measured changes in cortical thickness associated with surgical revascularization in patients with severe cerebrovascular steno-occlusive disease [22]. At present, the latter methodology can be limited by the available signal contrast and SNR at 3.0 T, but in future developments it may be possible to resolve multiple cortical layers using ultrahigh field MRI [23].

Diffusion Tensor Imaging

It has been known since the 1990s that water diffusion in certain tissues is anisotropic [24]. The effect is pronounced in white matter tracts with diffusion preferentially occurring along the length of axons and more restricted in the orthogonal directions. The myelination of axons, the axonal membrane, and microtubules within axons are all thought to contribute to water diffusing in a preferred direction. If a series of DWI scans are conducted with diffusion weighting in a variety of different orientations, it is possible to determine the direction and magnitude of diffusion and the correlations between diffusion in different directions, which can be modeled mathematically using an anisotropic diffusion tensor—hence the term "diffusion tensor imaging" (DTI) [25]. A minimum of seven DWI experiments with different diffusion orientations (including one experiment with no diffusion weighting) are required to calculate basic diffusion tensor parameters, although encoding more diffusion directions improves data quality and enables more sophisticated modeling. Thus, DTI experiments are often time-consuming (approximately 10 min or longer), even when EPI techniques are used for spatial encoding. The technique can be sensitive to a variety of different image artifacts, including head motion, as the signal attenuation measured in DWI experiments does not differentiate well between movement of water molecules on a microscopic or a macroscopic scale.

Data provided by DTI enable calculation of "rotationally invariant" signal contrast parameters that are theoretically independent of which gradient orientations are chosen for diffusion weighting (as long as they are well-spaced in orientation). The mean diffusivity provides contrast similar to that of the ADC, by providing a weighted average of diffusion in three orthogonal directions. A parameter known as the fractional anisotropy (FA) quantifies the extent that there is a preferred diffusion direction within a particular voxel. The FA value for grey matter is thus lower than that for white matter, potentially providing a rich source of information relating to brain connectivity.

It is well established that the location of lesions due to stroke will strongly influence the potential for recovery. From lesion analysis work, it is known that a stroke in the internal capsule is particularly problematic for motor recovery because the output of motor signals converges on this subcortical brain structure [26]. One of the related, major appeals of DTI for stroke recovery research is the ability to visualize the integrity of the corticospinal tract. Studies have shown that DTI can be used to probe the integrity of the corticospinal tract within the first weeks after stroke and that the early DTI measures are highly correlated with residual motor function in the acute and chronic stages [27]. Thus, DTI measures can help to determine whether a patient has the potential to recover a particular behavior or not, depending on the integrity of the descending white matter fibers.

DTI data can also be used for tractography, which is a noninvasive means for mapping the paths of white matter fibers in the brain [25]. Although such maps can be generated in MRI acquisition times of approximately 10 min, even longer acquisitions continue to be beneficial at present. This is, in part, because the diffusion tensor model for white matter is rather simplistic and does not account well for regions of the brain that have crossing fiber tracts. More sophisticated models require more gradient orientations, and high spatial resolution is paramount. Thus, DTI is a promising methodology for use in stroke patients, but the most practical additional information at present is obtained from use of FA information. For example, Schweizer and colleagues [28] reported a patient with subarachnoid hemorrhage who exhibited memory and executive functioning impairment 1 year post-hemorrhage. DTI revealed a significant reduction in white matter integrity (as indicated by reduced FA values) in the right hippocampus and the left dorsolateral prefrontal region, which are the precise areas expected to be involved in a patient with these cognitive deficits. Lastly, Fig. 14.3 shows DTI tractography results implicating the involvement of specific fronto-parietal and fronto-occipital pathways in stroke patients with left hemi-spatial neglect [29].

Functional MRI

Blood oxygenation level-dependent (BOLD) fMRI [30–32] has revolutionized neuroscience through its capability to measure signal changes associated with neuronal activity generated by sensory stimuli, or behavioral tasks involving memory, cognition, action, or emotion. Neurons communicate with glial cells and the nearby microvasculature to signal for the delivery of additional blood through vasodilation, when neuronal activity levels increase. The neurovascular unit that governs such processes represents the ultimate spatial resolution achievable with hemodynamic measures of brain activity, in the range of hundreds of microns [33], although typically fMRI studies are conducted with voxel dimensions of several millimeters. Researchers have determined that the local field potentials arising from healthy populations of neurons are strongly correlated with BOLD fMRI signals [34].

In normal brain, the BOLD fMRI response to neuronal activity is characterized by three distinct phases: (1) a fast response lasting 1–2 s when there is a small decrease in the BOLD signal; (2) a relatively large amplitude hyperemia that is caused by the inflow of more highly oxygenated blood, peaking approximately 4–5 s after the stimulus; and (3) a refractory period lasting 5–10 s where the BOLD signal undershoots and then reaches the baseline [35]. The period of hyperemia provides the most robust means to detect increases in brain activity [36]. These signal characteristics are a complex function of multiple physiological parameters such as CBF, CBV, cerebral rate of metabolic oxygen consumption, and hematocrit. However, the key properties used for generating fMRI, through which these other parameter changes are viewed, are the amounts of oxy- and deoxyhemoglobin within the neurovascular unit and consequently the local magnetic field

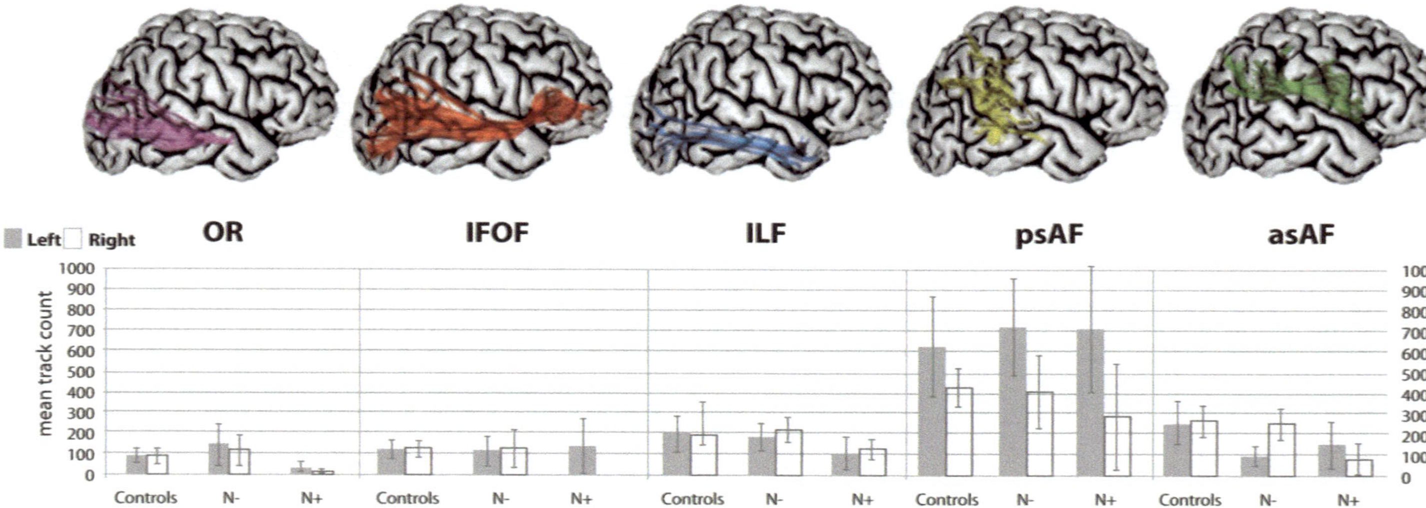

Fig. 14.3 *Top*: Diffusion tensor imaging (DTI) tractography of six fiber tracts shown for 3D-reconstructed healthy brain: the optic radiation (OR); the inferior fronto-occipital fasciculus (IFOF); the inferior longitudinal fasciculus (ILF); the posterior segment of the arcuate fasciculus (psAF); and the anterior segment of the arcuate fasciculus (asAF). *Bottom*: Mean track count (number of estimated fiber tracts) with 95 % confidence intervals for the left (*grey*) and right (*white*) hemisphere for all tracts for a group of 12 healthy controls and groups of stroke patients without (N–) and with (N+) left hemi-spatial neglect (six and six patients, respectively). The stroke patients show substantial left–right asymmetries in the psAF and asAF compared to controls, with a substantially reduced track count in the right asAF for patients with neglect (based on work published elsewhere) [29]

at the microscopic level. Thus, the predominant method for measuring BOLD fMRI signal changes at 1.5 and 3.0 T involves using T2*-weighted EPI technique to generate time-series data during periods of neuronal activity. The resulting signal changes are typically quite small, only a few percent from signal baseline.

Task-Based BOLD fMRI

The majority of the fMRI literature includes task-based experiments, whereby participants alternate between different behavioral or stimulus conditions to induce a measurable BOLD signal change. For example, the simplest approach is to alternate between one specific stimulus/task and a resting condition (often visual fixation on an image with a centrally located cursor). When the stimulus/task in question is of very brief duration (approximately 100 ms to several seconds), the experimental design is described as "event-related," whereas longer duration task and control condition periods (typically 15–30 s) are used in "block designs." In stroke recovery research the majority of fMRI studies have used a block design primarily because the block design affords greater detection sensitivity in terms of the BOLD time course data [37]. Irrespective of the experimental design, the task of interest is undertaken in multiple repetitions to improve the ability to detect brain activity under limited SNR conditions.

The first task-based fMRI experiment involving stroke patients dates back to 1997, when Cramer and colleagues [38] visualized the brain regions involved when stroke survivors moved their affected limb. Since this pioneering work, there have been numerous other fMRI studies that have helped to characterize brain activation patterns that indicate positive or negative outcomes after stroke. Despite the heterogeneous nature of lesions and behavioral deficits in stroke patients, some "rules of thumb" have become evident. One sign of poor recovery is significant activation in the unaffected hemisphere (i.e., contralateral to the side of the lesion) when performing a task that would normally lateralize to the affected hemisphere in normal individuals. For example, a recent study found that activation of the left inferior frontal gyrus was associated with improvement in picture naming and sentence comprehension in stroke patients recovering from aphasia, whereas activation of the right inferior frontal gyrus was associated with poorer behavioral performance, possibly due to up-regulation of nonlinguistic cognitive processing [39]. Calautti and Baron discuss fMRI in stroke recovery in more detail, by looking at cross-sectional and longitudinal fMRI studies [37]. Their review also addresses the laterality index metric, which is a quantitative measure of activation that is detected in contralateral and ipsilateral hemispheres.

One of the challenges of task-based fMRI in stroke recovery research is the issue of task performance. It is important to characterize properly what the patient "does" during fMRI, to rule out the possibility that their activation patterns are different from normal individuals simply because they did the task differently (e.g., speed and extent of a motor task, and force and level of effort associated with a motor task). One strategy to circumvent this important confound is to measure relevant biophysical signals concurrently during fMRI, such as electromyography (EMG) [40] or electrodermal activity (EDA) [41, 42]. EMG provides important information regarding the timing and intensity of muscle contraction, and potentially the muscle groups that are activated if multichannel EMG is conducted. EMG is also important for determining if mirror movements confound the interpretation of brain activation patterns, when unilateral movement is prescribed. Alternatively, EDA recordings provide an indirect probe of the autonomic nervous system and associated aspects of arousal, emotion, and sense of effort. EDA data have been used to characterize sense of effort (and brain activity in the contralesional hemisphere) during the performance of a motor task with the affected limb post-stroke, as represented in Fig. 14.4.

Despite the wealth of information that can be gained in such studies, use of task-based BOLD fMRI to study brain activation in individuals with cerebrovascular disease is not without controversy. Recalling that BOLD signal changes are influenced by factors such as CBV, CBF, and the cerebral rate of metabolic oxygen consumption, pathological changes in these parameters that affect fMRI signals may be mistaken for changes in neuronal activity. For example, a slight reduction in baseline CBF, due to ischemia conditions or chronic hypoperfusion, could manifest as a larger BOLD signal if neurons remain viable and active in the region and still generate a robust hyperemic response. Beyond hemodynamic parameters, medications and various other stimulants are known to influence the BOLD fMRI signal [43]. It is also possible that the temporal characteristics of the BOLD response can be affected after stroke. Patients with significant extracranial atherosclerotic stenoses or occlusions have shown a delayed, abnormal BOLD signal response relative to controls [44].

Another challenging issue associated with a task-based BOLD fMRI is the situation in which the hemodynamic BOLD signal response is completely absent, despite evidence of an underlying neuronal activation recorded by an alternative functional neuroimaging modality. Rossini and colleagues performed a neuroimaging study in chronic stroke using BOLD fMRI, magnetoencephalography (MEG), and transcranial Doppler (TCD) ultrasound [45]. They found that while MEG-evoked responses to somatosensory stimuli were detected among chronic stroke survivors, an equivalent BOLD signal was not seen in a subset of patients. In these patients, the absence of task-related BOLD signal was related to the ability of blood vessels to react to a vasodilatory carbon dioxide stimulus at baseline. Indeed, others have found that cerebrovascular reactivity (i.e., the capacity of the vessels to dilate) was correlated with motor-related BOLD response associated with moving of the affected hand [46]. These studies argue for the importance of understanding the underlying vascular health when conducting task-related BOLD fMRI, and the need to confirm findings through use of multiple functional neuroimaging modalities.

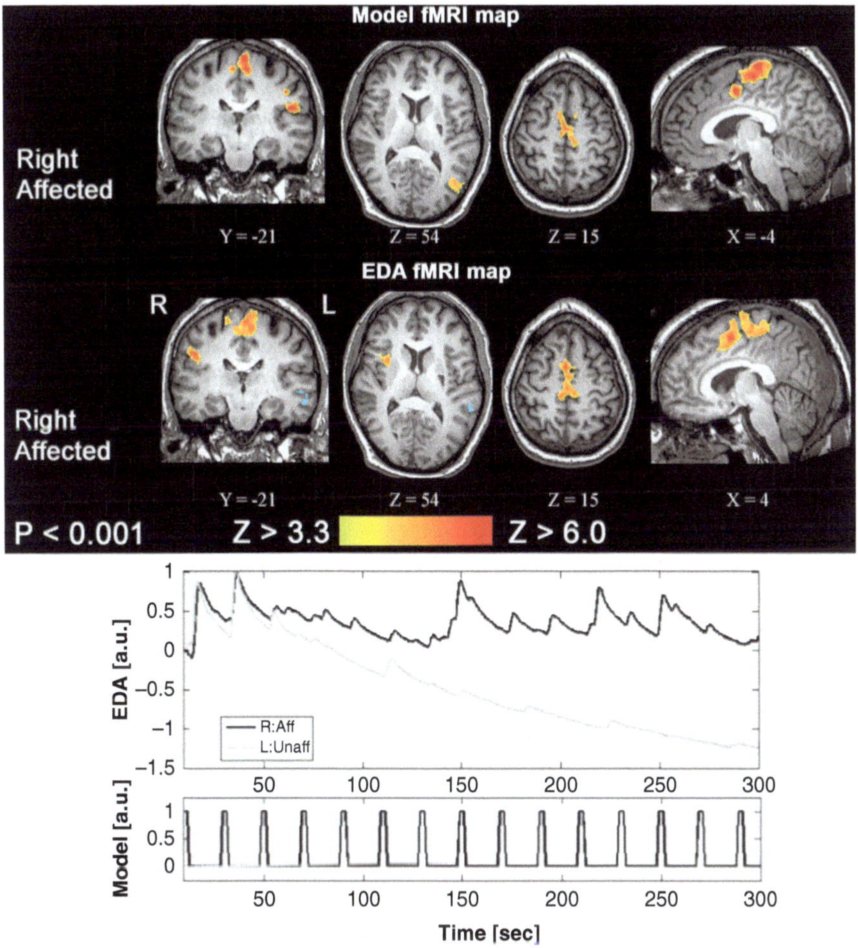

Fig. 14.4 *Top*: Task-based fMRI activation associated with movement of the right affected ankle. The two activation patterns are generated by using a standard fMRI model and as an alternative approach the electrodermal activity (EDA) directly. *Bottom*: The two EDA traces show different features when the stroke patient moved their affected versus their unaffected limb (based on work published elsewhere) [40]

Resting-State BOLD fMRI

Resting-state fMRI (rs-fMRI) is a new branch of human brain mapping that has emerged in recent years [47, 48], although the seminal work occurred in 1995 [49]. A major advantage of rs-fMRI is that it does not require the patient to perform a task or any specified mental processing during the scan. Thus, concerns regarding abnormal behavioral performance of patients are removed in comparison with normal healthy individuals. Brain activity in the resting state is estimated not by task performance, but by temporal correlation in the baseline fMRI signal between voxels, usually in the frequency range between 0.008 and 0.08 Hz. Spatial maps can be obtained by "seed-voxel" approaches, in which the strength of correlations is reported between the signal from one voxel of interest (e.g., in the hippocampus) and the signal from all other voxels; or by multivariate approaches such as independent component analysis, which generate maps of regional brain activity for a series of statistically independent "latent variable" time waveforms [50]. In the latter case, each activation map reflects the voxel-by-voxel strength that a specific latent variable is expressed. In either case, activation maps are commonly identified as "resting-state networks" (RSNs). A given RSN is not necessarily a reflection of anatomical connectivity, but "functional connectivity" determined from BOLD fMRI signals.

Task-based fMRI has been used for stroke recovery research, while rs-fMRI has historically not been used. This trend, however, appears to be changing [51, 52], particularly as rs-fMRI can be conducted more easily in patients with profound impairments. Important ongoing issues in the development of rs-fMRI include aspects of signal processing, as the rs-fMRI signal is even smaller than that of task-based fMRI; as well as exploring the interrelationship between DTI, task-based fMRI, and rs-fMRI parameters.

Arterial Spin Labeling MRI

Another MRI technique that is capable of functional neuroimaging is ASL. Although discovered and developed contemporaneously, ASL is a complex imaging experiment that historically has not been used as widely as BOLD fMRI [53]. Recent technical advances have made ASL more popular. Two images are required; in the first, RF pulses are used to invert magnetization in arterial blood proximal to the imaging region of interest. The "tagged" or "labeled" blood water subsequently is allowed to flow into the imaging region for a sufficient time to reach the microvasculature, causing a small decrease in the measured MRI signal intensity in proportion to the NMR signal properties of the tagged blood in the microvasculature, and the microvasculature volume fraction within a given imaging voxel. The second image (the "control") has no effective labeling but otherwise near identical NMR signal contrast weighting. The difference between the control and tag images provides an image that is CBF-weighted [54] (Fig. 14.5).

As for BOLD fMRI, the ASL perfusion signal is limited by the small volume fraction of microvascular blood in tissue, in this case representing approximately 1–2 % of the available signal in grey matter. Low SNR levels are a major limitation of the technique and good temporal image stability is required in particular, as many images are typically acquired in a time series and then averaged together to yield perfusion maps. Refinements in MRI hardware and acquisition methods have improved ASL image quality in recent years due to: (1) improved gradient

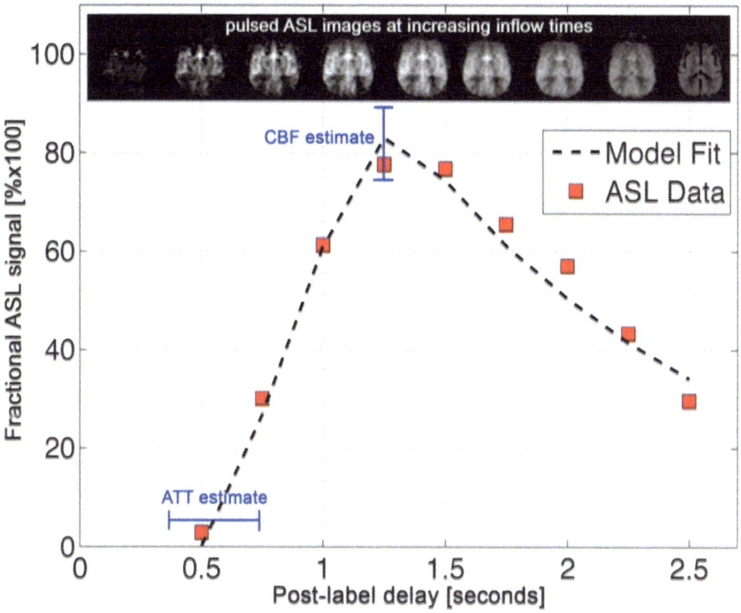

Fig. 14.5 The fractional arterial spin labeling (ASL) signal plotted as a function of the post-label delay time for a pulsed ASL experiment. Images along the *top* show the cerebral blood flow (CBF)-weighted difference images that correspond to the post-label delay. An ASL model can be used to fit for arterial transit time (ATT) and CBF on a voxel-by-voxel basis. Estimates and standard deviation for ATT and CBF are shown. The CBF estimate can be converted to an absolute measure of perfusion in units of mL/100 g/min by measuring or assuming additional parameters; i.e., T1 of grey matter and arterial blood, initial magnetization of arterial blood, etc. (based on work published elsewhere) [53]

technology for fast imaging; (2) availability of higher main magnetic field MR systems (e.g., ≥3 T) [55] that lengthen T1 relaxation times and enhance labeling efficiency; (3) multi-slice capabilities [56] that improve the volume of coverage; and lastly, (4) improved theoretical models from which quantitative estimates of CBF can be derived [57, 58].

The ASL experiment is versatile and can be used either to map brain activity analogous to task-based or rs-fMRI using BOLD signal contrast [59], or to produce baseline CBF images [60] as an alternative to PWI, avoiding the use of Gd-DTPA contrast agent. Two possible advantages of ASL over BOLD fMRI are that the ASL signal is closely associated with CBF whereas the BOLD signal is of much more complex signal origins, and that the hemodynamic response of the ASL signal to neuronal activity has improved spatial resolution compared to that of the BOLD signal. However, the reduced SNR of ASL images in relation to BOLD fMR images remains challenging. Alternatively, the use of ASL to focus on the CBF characteristics of the normal and abnormal cerebral vasculature is promising.

The first example of stroke imaging with ASL dates back a decade [61], in which Chalela and colleagues found that ASL-based CBF in the affected hemisphere correlated with stroke scale on admission and Rankin score 30 days post-stroke.

The role that ASL can play in imaging acute and chronic stroke patients remains open for exploration. One important consideration is the potential use of ASL instead of PWI, particularly for patients in which Gd-DTPA is contraindicated. Thorough comparisons of the two imaging approaches are now being conducted in the literature [62–67] and in diagnostic imaging departments worldwide, as MRI system vendors have recently made ASL sequences commercially available. In general, PWI and ASL have been shown to have good agreement when considering healthy brain tissue, but some divergence between the techniques may occur in the context of stroke and cerebrovascular disease. One option that appears to be feasible within a clinical MRI examination is to include a 6-min quantitative ASL acquisition to improve the quantitative accuracy of the PWI-derived CBF maps [66]. Interestingly, Zaharchuk and colleagues found that among individuals with suspected cerebrovascular disease, roughly half of the patients had an abnormal ASL CBF image while the dynamic susceptibility contrast (DSC) image was viewed as normal [64].

The ASL method also appears to have an experimental role to play in assessing minor stroke or chronic stroke recovery. In the presence of cerebrovascular disease, it may be important to consider the time that it takes for the ASL-tagged bolus to travel from the labeling plane to the imaging plane. This ASL metric has an analogous PWI counterpart and has been described using numerous terms, for example, the bolus arrival time, the arterial arrival time [68], and, perhaps most commonly, the arterial transit time (ATT or t_A) [69]. In general, vascular pathology will delay the arrival of the ASL signal. If unaccounted for, the delay will manifest as underestimated CBF. Some clinical ASL studies have incorporated an ATT estimate, which involves acquiring ASL data at multiple inflow times (i.e., the time interval between labeling and image acquisition). The downside of this approach is the SNR penalty/acquisition time tradeoff, with either the need to compromise in terms of the SNR in individual images or the need to average at each of the multiple inflow times. Nevertheless, there is now data suggesting that the ATT information may be useful in stroke diagnosis (Fig. 14.6) [70, 71]. Future studies should evaluate how ATT can be used to improve understanding of stroke outcomes.

An interesting recent study used ASL to investigate physiological correlates of dementia at 6 years post-stroke, compared to adults with Alzheimer disease (AD) and healthy controls [72]. The authors found that the best MRI-based predictor of AD was hippocampal volume, whereas the ASL-derived CBF expressed as a ratio of grey matter to white matter was the best predictor of dementia in the stroke group. The grey to white matter CBF ratio approach may be a suitable alternative to absolute CBF images in clinical scenarios in which vastly different resting CBF levels are observed across patients. Others have shown that use of the normalized CBF image had higher clinical sensitivity in a study of AD patients [73].

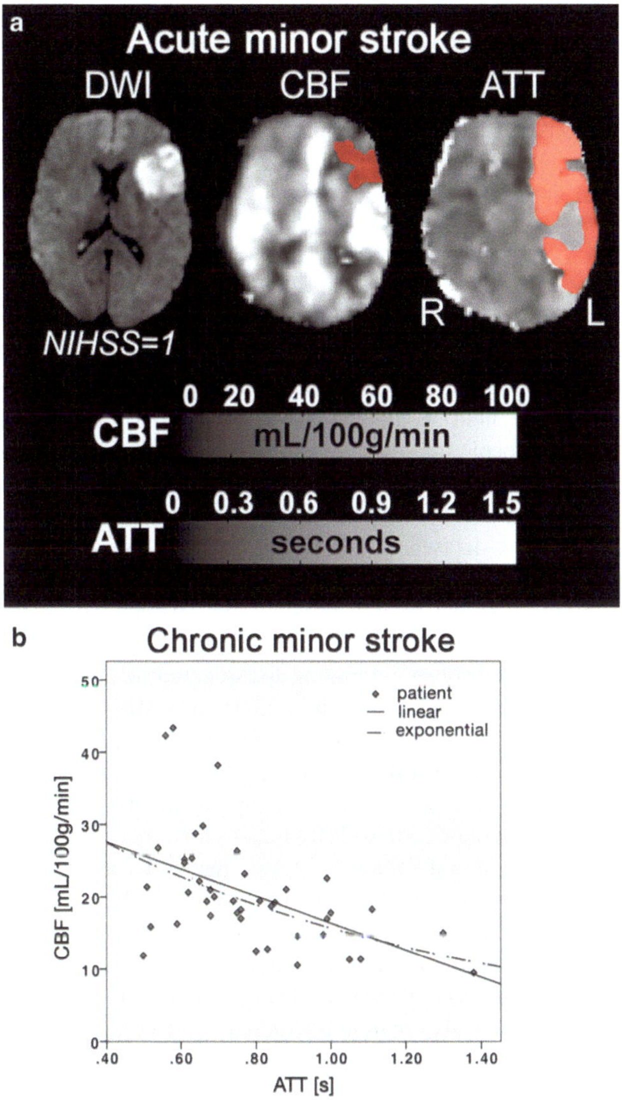

Fig. 14.6 (**a**) An acute stroke patient with mild impairment as measured by the National Institutes of Health (NIH) Stroke Scale (NIHSS = 1). The DWI shows restricted diffusion in the left frontal lobe. Corresponding image slices for the ASL are shown. The CBF image shows a small region of low perfusion (shaded in *dark red*) while ATT image shows a larger region of prolonged ASL bolus transit (shaded in *light red*) (based on work published elsewhere) [69]. (**b**) Among a group of chronic minor stroke and TIA patients there appears to be a significant relationship between CBF and ATT ($P < 0.001$). Reduced CBF was associated with a prolonged ATT, that is, a delayed transit of the ASL bolus (based on published work) [70]

Future Directions for MRI Research in Stroke Recovery

Given the breadth and versatility of MRI methods, there are numerous interesting research directions that can be undertaken to advance the scientific understanding of stroke recovery, ultimately toward improving health care management of chronic stroke patients.

Multi-parametric MRI Protocols

First, similar to the multifaceted MRI protocol that has arisen for imaging acute stroke, the scientific field would benefit from improved integration of multi-parametric MRI protocols for subacute and chronic strokes. For example, to date there are few examples integrating DTI white matter tract data, rs-fMRI functional connectivity data, and task-based fMRI data in stroke patients, yet there should be linkages between these datasets. It will be important to assess the extent that each experiment provides unique, clinically relevant information and whether all experiments are required in clinical research protocols, or whether only the most sensitive indicators are practical for expediency. An additional area requiring expanded multi-parametric MRI involves improving the quantitative interpretation of task-based BOLD fMRI data. With the potential to develop MRI experiments sensitive to parameters such as microvascular CBV and hematocrit, it may be possible to combine such measures with ASL-derived CBF to model the relationship between BOLD fMRI signal changes and cerebral metabolic rate of oxygen ($CMRO_2$) [74]. This offers the prospect of providing $CMRO_2$ maps for stroke patients, which should provide a marker of neuronal pathology. Again, the ability to perform such multi-parametric assessments in a timely fashion remains a challenge. However, this is quite important given the heterogeneity of MRI findings across stroke patients, and because there is a wealth of BOLD fMRI data already collected in stroke patients that would benefit from more detailed evaluation and insight. Lastly, BOLD fMRI studies in stroke patients also require additional validation through supplemental measurements using functional neuroimaging modalities that do not rely on the neurovascular-coupling phenomenon. For assessing cortical activity, electroencephalography has potential for this purpose and is relatively low-cost and widely available. As in the case of fMRI, however, electroencephalography datasets are complex and require careful interpretation.

Second, MRI promises to play a role in the interplay between genetics and stroke recovery. Recently, a number of fMRI studies have been undertaken to assess the effects of various genetic polymorphisms on brain activation patterns. For example, mutations of the apolipoprotein E and brain-derived neurotrophic factor genes may influence outcome after stroke [75]. As the role of various genetic factors becomes better understood, it is possible that part of the work-up for stroke patients will be a genetic profile that provides some insight into the potential for recovery after stroke, and that may help to select the appropriate rehabilitation therapy for specific patients.

MRI to Guide Intervention and Rehabilitation

A last area that deserves brief mention is the use of MRI to guide intervention and rehabilitation. Technology is developing toward expanding the list of options for therapeutic intervention for stroke patients in the acute phase. One example of emerging technology involves use of focused ultrasound. Through use of ultrasound transducer arrays with high channel count, it is becoming possible to distribute heat load across the skull while adjusting the transmit phase of individual elements such that they constructively interfere to generate focused ultrasound pressure fields within the brain [76]. Such systems require detailed measurements of skull thickness (as provided by CT, for example) to adjust transducer phase appropriately, and MRI guidance to assess the effects of focused ultrasound treatment. MRI guidance is critical not only to detect the effects of ultrasound on anatomical structures but also because various NMR parameters are temperature-sensitive [77]. Thus, specialized MRI techniques provide a method of noninvasive thermometry that can be used to control focused ultrasound heat treatments. Completely noninvasive thermal ablation treatments are possible, as demonstrated in patients with brain tumors [78], although future applications to stroke are possible. One such approach involves use of MRI-guided focused ultrasound not for thermal therapy, but for transient disruption of the blood brain barrier that enables targeted delivery of stem cells for neural repair [79].

Summary

MRI is a highly versatile imaging modality that can be used in numerous different applications to probe the anatomy and physiology of the healthy and injured brain. Substantial capacity remains for new technical innovation to expand the wealth of biophysical information that MRI provides. In the coming decades, it is very likely that MRI techniques will play a critical role in the development and subsequent clinical implementation of new therapeutic strategies to improve outcome after stroke.

References

1. Bloch F. Nuclear induction. Phys Rev. 1946;70:460–73.
2. Purcell EM, Torrey HC, Pound RV. Resonance absorption by nuclear magnetic moments in a solid. Phys Rev. 1946;69:37–8.
3. Damadian R. Tumor detection by nuclear magnetic resonance. Science. 1971;171:1151–3.
4. Stehling MK, Turner R, Mansfield P. Echo-planar imaging: magnetic resonance imaging in a fraction of a second. Science. 1991;254(5028):43–50.
5. Burgess RE, Kidwell CS. Use of MRI in the assessment of patients with stroke. Curr Neurol Neurosci Rep. 2011;11(1):28–34.

6. Leiva-Salinas C, Wintermark M, Kidwell CS. Neuroimaging of cerebral ischemia and infarction. Neurotherapeutics. 2011;8(1):19–27.
7. Mascalchi M, Filippi M, Floris R, Fonda C, Gasparotti R, Villari N. Diffusion-weighted MR of the brain: methodology and clinical application. Radiol Med. 2005;109(3):155–97.
8. Chalela JA, Kidwell CS, Nentwich LM, Luby M, Butman JA, Demchuk AM, et al. Magnetic resonance imaging and computed tomography in emergency assessment of patients with suspected acute stroke: a prospective comparison. Lancet. 2007;369(9558):293–8.
9. Thomalla G, Rossbach P, Rosenkranz M, Siemonsen S, Krützelmann A, Fiehler J, et al. Negative fluid-attenuated inversion recovery imaging identifies acute ischemic stroke at 3 hours or less. Ann Neurol. 2009;65(6):724–32.
10. Taussky P, Tawk RG, Daugherty WP, Hanel RA. Medical therapy for ischemic stroke: review of intravenous and intra-arterial treatment options. World Neurosurg. 2011;76(6 Suppl):S9–15.
11. Davis WL, Warnock SH, Harnsberger HR, Parker DL, Chen CX. Intracranial MRA: single volume vs. multiple thin slab 3D time-of-flight acquisition. J Comput Assist Tomogr. 1993;17(1):15–21.
12. Runge VM, Kirsch JE, Lee C. Contrast-enhanced MR angiography. J Magn Reson Imaging. 1993;3(1):233–9.
13. Sohn CH, Sevick RJ, Frayne R. Contrast-enhanced MR angiography of the intracranial circulation. Magn Reson Imaging Clin N Am. 2003;11(4):599–614.
14. Morita S, Masukawa A, Suzuki K, Hirata M, Kojima S, Ueno E. Unenhanced MR angiography: techniques and clinical applications in patients with chronic kidney disease. Radiographics. 2011;31(2):E13–33.
15. Luypaert R, Boujraf S, Sourbron S, Osteaux M. Diffusion and perfusion MRI: basic physics. Eur J Radiol. 2001;38(1):19–27.
16. Donnan GA, Baron JC, Ma H, Davis SM. Penumbral selection of patients for trials of acute stroke therapy. Lancet Neurol. 2009;8(3):261–9.
17. Tenser MS, Amar AP, Mack WJ. Mechanical thrombectomy for acute ischemic stroke using the MERCI retriever and penumbra aspiration systems. World Neurosurg. 2011;76(6 Suppl):S16–23.
18. Fazekas F, Kleinert R, Roob G, Kleinert G, Kapeller P, Schmidt R, et al. Histopathologic analysis of foci of signal loss on gradient-echo T2*-weighted MR images in patients with spontaneous intracerebral hemorrhage: evidence of microangiopathy-related microbleeds. AJNR Am J Neuroradiol. 1999;20(4):637–42.
19. Barnes SR, Haacke EM. Susceptibility-weighted imaging: clinical angiographic applications. Magn Reson Imaging Clin N Am. 2009;17(1):47–61.
20. Molenberghs P, Gillebert CR, Schoofs H, Dupont P, Peeters R, Vandenberghe R. Lesion neuroanatomy of the sustained attention to response task. Neuropsychologia. 2009;47(13):2866–75.
21. Nouri S, Cramer SC. Anatomy and physiology predict response to motor cortex stimulation after stroke. Neurology. 2011;77(11):1076–83.
22. Fierstra J, Maclean DB, Fisher JA, Han JS, Mandell DM, Conklin J, et al. Surgical revascularization reverses cerebral cortical thinning in patients with severe cerebrovascular steno-occlusive disease. Stroke. 2011;42(6):1631–7.
23. Zwanenburg JJ, Hendrikse J, Luijten PR. Generalized multiple-layer appearance of the cerebral cortex with 3D FLAIR 7.0-T MR imaging. Radiology. 2012;262(3):995–1001.
24. Henkelman RM, Stanisz GJ, Kim JK, Bronskill MJ. Anisotropy of NMR properties of tissues. Magn Reson Med. 1994;31:592–601.
25. Nucifora PG, Verma R, Lee SK, Melhem ER. Diffusion-tensor MR imaging and tractography: exploring brain microstructure and connectivity. Radiology. 2007;245(2):367–84.
26. Shelton FN, Reding MJ. Effect of lesion location on upper limb motor recovery after stroke. Stroke. 2001;32(1):107–12.
27. Radlinska B, Ghinani S, Leppert IR, Minuk J, Pike GB, Thiel A. Diffusion tensor imaging, permanent pyramidal tract damage, and outcome in subcortical stroke. Neurology. 2010;75(12):1048–54.

28. Schweizer TA, Al-Khindi T, Macdonald RL. Diffusion tensor imaging as a surrogate marker for outcome after perimesencephalic subarachnoid hemorrhage. Clin Neurol Neurosurg. 2012;114:798–800.
29. Urbanski M, Thiebaut de Schotten M, Rodrigo S, Oppenheim C, Touzé E, Méder JF, et al. DTI-MR tractography of white matter damage in stroke patients with neglect. Exp Brain Res. 2011;208(4):491–505.
30. Ogawa S, Lee TM, Kay AR, Tank DW. Brain magnetic resonance imaging with contrast dependent on blood oxygenation. Proc Natl Acad Sci USA. 1990;87:9868–72.
31. Ogawa S, Tank DW, Menon R, Ellermann JM, Kim SG, Merkle H, et al. Intrinsic signal changes accompanying sensory stimulation: functional brain mapping with magnetic resonance imaging. Proc Natl Acad Sci USA. 1992;89(13):5951–5.
32. Bandettini PA, Wong EC, Hinks RS, Tikofsky RS, Hyde JS. Time course EPI of human brain function during task activation. Magn Reson Med. 1992;25(2):390–7.
33. Menon RS, Goodyear BG. Submillimetre functional localization in human strate cortex using BOLD contrast at 4 Tesla: implications for the vascular point-spread function. Magn Reson Med. 1999;41:230 5.
34. Logothetis NK, Pauls J, Augath M, Trinath T, Oeltermann A. Neurophysiological investigation of the basis of the fMRI signal. Nature. 2001;412(6843):150–7.
35. Norris DG. Principles of magnetic resonance assessment of brain function. J Magn Reson Imaging. 2006;23(6):794–807.
36. Cohen MS. Parametric analysis of fMRI data using linear systems methods. Neuroimage. 1997;6(2):93–103.
37. Calautti C, Baron JC. Functional neuroimaging studies of motor recovery after stroke in adults: a review. Stroke. 2003;34(6):1553–66.
38. Cramer SC, Nelles G, Benson RR, Kaplan JD, Parker RA, Kwong KK, et al. A functional MRI study of subjects recovered from hemiparetic stroke. Stroke. 1997;28(12):2518–27.
39. van Oers CA, Vink M, van Zandvoort MJ, van der Worp HB, de Haan EH, Kappelle LJ, et al. Contribution of the left and right inferior frontal gyrus in recovery from aphasia. A functional MRI study in stroke patients with preserved hemodynamic responsiveness. Neuroimage. 2010;49(1):885–93. Epub 2009.
40. MacIntosh BJ, Baker SN, Mraz R, Ives JR, Martel AL, McIlroy WE, et al. Improving functional magnetic resonance imaging motor studies through simultaneous electromyography recordings. Hum Brain Mapp. 2007;28(9):835–45.
41. MacIntosh BJ, McIlroy WE, Mraz R, Staines WR, Black SE, Graham SJ. Electrodermal recording and fMRI to inform sensorimotor recovery in stroke patients. Neurorehabil Neural Repair. 2008;22(6):728–36.
42. MacIntosh BJ, Mraz R, McIlroy WE, Graham SJ. Brain activity during a motor learning task: an fMRI and skin conductance study. Hum Brain Mapp. 2007;28(12):1359–67.
43. Pattinson KT, Governo RJ, MacIntosh BJ, Russell EC, Corfield DR, Tracey I, et al. Opioids depress cortical centers responsible for the volitional control of respiration. J Neurosci. 2009;29(25):8177–86.
44. Roc AC, Wang J, Ances BM, Liebeskind DS, Kasner SE, Detre JA. Altered hemodynamics and regional cerebral blood flow in patients with hemodynamically significant stenoses. Stroke. 2006;37(2):382–7.
45. Rossini PM, Altamura C, Ferretti A, Vernieri F, Zappasodi F, Caulo M, et al. Does cerebrovascular disease affect the coupling between neuronal activity and local haemodynamics? Brain. 2004;127(Pt 1):99–110.
46. Krainik A, Hund-Georgiadis M, Zysset S, von Cramon DY. Regional impairment of cerebrovascular reactivity and BOLD signal in adults after stroke. Stroke. 2005;36(6):1146–52.
47. Raichle ME, MacLeod AM, Snyder AZ, Powers WJ, Gusnard DA, Shulman GL. A default mode of brain function. Proc Natl Acad Sci USA. 2001;98(2):676–82.
48. Damoiseaux JS, Rombouts SA, Barkhof F, Scheltens P, Stam CJ, Smith SM, et al. Consistent resting-state networks across healthy subjects. Proc Natl Acad Sci USA. 2006;103(37):13848–53.

49. Biswal B, Yetkin FZ, Haughton VM, Hyde JS. Functional connectivity in the motor cortex of resting human brain using echo-planar MRI. Magn Reson Med. 1995;34(4):537–41.
50. Beckmann CF, DeLuca M, Devlin JT, Smith SM. Investigations into resting-state connectivity using independent component analysis. Philos Trans R Soc Lond B Biol Sci. 2005;360(1457):1001–13.
51. Wang L, Yu C, Chen H, Qin W, He Y, Fan F, et al. Dynamic functional reorganization of the motor execution network after stroke. Brain. 2010;133(Pt 4):1224–38.
52. Park CH, Chang WH, Ohn SH, Kim ST, Bang OY, Pascual-Leone A, et al. Longitudinal changes of resting-state functional connectivity during motor recovery after stroke. Stroke. 2011;42(5):1357–62.
53. Williams DS, Detre JA, Leigh JS, Koretsky AP. Magnetic resonance imaging of perfusion using spin inversion of arterial water. Proc Natl Acad Sci USA. 1992;89(1):212–6.
54. MacIntosh BJ, Pattinson KT, Gallichan D, Ahmad I, Miller KL, Feinberg DA, et al. Measuring the effects of remifentanil on cerebral blood flow and arterial arrival time using 3D GRASE MRI with pulsed arterial spin labelling. J Cereb Blood Flow Metab. 2008;28(8):1514–22.
55. Petersen ET, Zimine I, Ho YC, Golay X. Non-invasive measurement of perfusion: a critical review of arterial spin labelling techniques. Br J Radiol. 2006;79(944):688–701.
56. Gunther M, Oshio K, Feinberg DA. Single-shot 3D imaging techniques improve arterial spin labeling perfusion measurements. Magn Reson Med. 2005;54(2):491–8.
57. Buxton RB, Frank LR, Wong EC, Siewert B, Warach S, Edelman RR. A general kinetic model for quantitative perfusion imaging with arterial spin labeling. Magn Reson Med. 1998;40(3):383–96.
58. Chappell MA, MacIntosh BJ, Donahue MJ, Günther M, Jezzard P, Woolrich MW. Separation of macrovascular signal in multi-inversion time arterial spin labelling MRI. Magn Reson Med. 2010;63(5):1357–65.
59. Aguirre GK, Detre JA, Zarahn E, Alsop DC. Experimental design and the relative sensitivity of BOLD and perfusion fMRI. Neuroimage. 2002;15(3):488–500.
60. Alsop DC, Detre JA. Reduced transit-time sensitivity in noninvasive magnetic resonance imaging of human cerebral blood flow. J Cereb Blood Flow Metab. 1996;16(6):1236–49.
61. Chalela JA, Alsop DC, Gonzalez-Atavales JB, Maldjian JA, Kasner SE, Detre JA. Magnetic resonance perfusion imaging in acute ischemic stroke using continuous arterial spin labeling. Stroke. 2000;31(3):680–7.
62. Siewert B, Schlaug G, Edelman RR, Warach S. Comparison of EPISTAR and T2*-weighted gadolinium-enhanced perfusion imaging in patients with acute cerebral ischemia. Neurology. 1997;48(3):673–9.
63. Weber MA, Günther M, Lichy MP, Delorme S, Bongers A, Thilmann C, et al. Comparison of arterial spin-labeling techniques and dynamic susceptibility-weighted contrast-enhanced MRI in perfusion imaging of normal brain tissue. Invest Radiol. 2003;38(11):712–8.
64. Zaharchuk G, Bammer R, Straka M, Shankaranarayan A, Alsop DC, Fischbein NJ, et al. Arterial spin-label imaging in patients with normal bolus perfusion-weighted MR imaging findings: pilot identification of the borderzone sign. Radiology. 2009;252(3):797–807.
65. Kim HS, Kim SY, Kim JM. Underestimation of cerebral perfusion on flow-sensitive alternating inversion recovery image: semiquantitative evaluation with time-to-peak values. AJNR Am J Neuroradiol. 2007;28(10):2008–13.
66. Zaharchuk G, Straka M, Marks MP, Albers GW, Moseley ME, Bammer R. Combined arterial spin label and dynamic susceptibility contrast measurement of cerebral blood flow. Magn Reson Med. 2010;63(6):1548–56.
67. Hunsche S, Sauner D, Schreiber WG, Oelkers P, Stoeter P. FAIR and dynamic susceptibility contrast-enhanced perfusion imaging in healthy subjects and stroke patients. J Magn Reson Imaging. 2002;16(2):137–46.
68. MacIntosh BJ, Filippini N, Chappell MA, Woolrich MW, Mackay CE, Jezzard P. Assessment of arterial arrival times derived from multiple inversion time pulsed arterial spin labeling MRI. Magn Reson Med. 2010;63(3):641–7.

69. Wang J, Alsop DC, Song HK, Maldjian JA, Tang K, Salvucci AE, et al. Arterial transit time imaging with flow encoding arterial spin tagging (FEAST). Magn Reson Med. 2003;50(3):599–607.
70. MacIntosh BJ, Lindsay AC, Kylintireas I, Kuker W, Günther M, Robson MD, et al. Multiple inflow pulsed arterial spin-labeling reveals delays in the arterial arrival time in minor stroke and transient ischemic attack. AJNR Am J Neuroradiol. 2010;31:1892–4.
71. MacIntosh BJ, Marquardt L, Schulz UG, Jezzard P, Rothwell PM. Haemodynamic alterations in vertebrobasilar large artery disease assessed by arterial spin labeling magnetic resonance imaging. AJNR Am J Neuroradiol. 2012;33(10):1939–44. Epub 2012 May 3.
72. Firbank MJ, He J, Blamire AM, Singh B, Danson P, Kalaria RN, et al. Cerebral blood flow by arterial spin labeling in poststroke dementia. Neurology. 2011;76(17):1478–84.
73. Yoshiura T, Hiwatashi A, Noguchi T, Yamashita K, Ohyagi Y, Monji A, et al. Arterial spin labelling at 3-T MR imaging for detection of individuals with Alzheimer's disease. Eur Radiol. 2009;19(12):2819–25.
74. Hoge RD, Atkinson J, Gill B, Crelier GR, Marrett S, Pike GB. Linear coupling between cerebral blood flow and oxygen consumption in activated human cortex. Proc Natl Acad Sci USA. 1999;96(16):9403–8.
75. Cramer SC, Procaccio V. Correlation between genetic polymorphisms and stroke recovery: analysis of the GAIN Americas and GAIN International Studies. Eur J Neurol. 2012;19(5):718–24.
76. Clement GT, Hynynen K. A non-invasive method for focusing ultrasound through the human skull. Phys Med Biol. 2002;47(8):1219–36.
77. Peters RD, Hinks RS, Henkelman RM. Ex vivo tissue-type independence in proton-resonance frequency shift MR thermometry. Magn Reson Med. 1998;40(3):454–9.
78. McDannold N, Clement GT, Black P, Jolesz F, Hynynen K. Transcranial magnetic resonance imaging-guided focused ultrasound surgery of brain tumors: initial findings in 3 patients. Neurosurgery. 2010;66(2):323–32; discussion 332.
79. Burgess A, Ayala-Grosso CA, Ganguly M, Jordão JF, Aubert I, Hynynen K. Targeted delivery of neural stem cells to the brain using MRI-guided focused ultrasound to disrupt the blood-brain barrier. PLoS One. 2011;6(11):e27877.

Chapter 15
Clinical Outcomes, Stroke Trials, and Cognitive Outcome

Benjamin W.Y. Lo, Filip Stojic, Julian Spears, Tom A. Schweizer, and R. Loch Macdonald

Introduction

Stroke as a cause of death in the United States decreased 19 % from 1998 to 2008, and death from stroke declined 35 % over the same time period [1]. This puts stroke as the fourth leading cause of death; down from third, behind heart disease, cancer, and chronic lung disease. The decrease in mortality in the United States differs from the increasing incidence in low- to middle-income countries [2].

Over the past 2 decades, management of ischemic and hemorrhagic stroke has evolved with large multicenter trials providing evidence for treatments such as

B.W.Y. Lo, M.D., Ph.D.(cand.), F.R.C.S.C. (✉) • R.L. Macdonald, M.D., Ph.D.
J. Spears, M.D., S.M. F.R.C.S.C., F.A.C.S
Division of Neurosurgery, Department of Surgery, University of Toronto,
St. Michael's Hospital, Toronto, ON, Canada
e-mail: benjamin.lo@alumni.utoronto.ca

F. Stojic
Keenan Research Centre in the Li Ka Shing Knowledge Institute,
St. Michael's Hospital, Toronto, ON, Canada

T.A. Schweizer, Ph.D.
Division of Neurosurgery, Department of Surgery, University of Toronto, St. Michael's Hospital, Toronto, ON, Canada

Keenan Research Centre in the Li Ka Shing Knowledge Institute, St. Michael's Hospital, Toronto, ON, Canada

Institute of Biomedical Engineering, University of Toronto, Toronto, ON, Canada

Heart and Stroke Foundation, Centre for Stroke Recovery, Toronto, ON, Canada

T.A. Schweizer and R.L. Macdonald (eds.), *The Behavioral Consequences of Stroke*,
DOI 10.1007/978-1-4614-7672-6_15,

acetylsalicylic acid (ASA) for transient ischemic attacks, carotid endarterectomy for symptomatic carotid stenosis, tissue plasminogen activator for acute ischemic stroke, and blood pressure control for prevention of cerebrovascular disease.

Recovery after stroke occurs through a variety of mechanisms that are largely theoretical at this point and based on experimental studies. The cerebral cortex undergoes functional and structural reorganization, a process that may be influenced by rehabilitation strategies. Neuroplasticity, with functional changes within the areas surrounding damaged regions, occurs with (1) resolution of post-stroke edema (as edema disrupts neuronal functioning), (2) reperfusion of ischemic penumbra (salvageable penumbra surrounding core infarcted areas), and (3) resolution of diaschisis (reversible states of depressed neuronal function in surrounding and remote areas connected to sites of tissue damage) [3].

Prognostic Factors

The greatest risk of mortality for stroke patients occurs within the first 30 days post-ictus, with the immediate cause of death in more than 60 % of cases related to the stroke itself. During the first week after ischemic stroke, the most important neurological conditions associated with risk of death include impaired consciousness on admission, posterior circulation infarcts, and transtentorial herniation. After the first month, comorbid complications arising from stroke contribute to mortality in up to 80 % of patients—including cardiac causes, pneumonia, pulmonary embolism, and sepsis. Early recognition of mortality predictors after stroke onset is crucial in preventing its occurrence, with the aim of maximizing duration and quality of patients' survival. However, longitudinal studies of ischemic stroke patients have shown their risk of death to be between 40 and 60 % at 5 years [4].

Prognostic factors for death after ischemic stroke can be classified into demographic, clinical, and biochemical variables. These include age, initial stroke severity (National Institute of Health Stroke Scale [NIHSS] score of 16 or greater), decreased level of consciousness, infarct size, large hemispheric or basilar syndrome, ischemic stroke subtype, fever, hypertension, atrial fibrillation, congestive heart failure, previous strokes, persistent hyperglycemia, and elevation of C-reactive protein [5].

Prognostic factors for poor outcome after aneurysmal subarachnoid hemorrhage (SAH) were summarized by Rosengart and colleagues and include increasing age, worsening admission neurological grade, and increasing amount of SAH on admission computed tomography (CT) [6]. Other factors that have often been found to contribute to poor outcome are increased aneurysm size, posterior circulation aneurysm, history of hypertension, intraventricular hemorrhage, intracerebral hemorrhage, and development of cerebral infarction due to angiographic vasospasm. Some other factors have been found in smaller studies; however, these findings have not been replicated consistently. The prognosis of patients with intracerebral hemorrhage worsens with increasing age, lower Glasgow coma score, increasing hematoma volume, and presence of intraventricular hemorrhage. Other less consistently identified factors include midline shift, fever, and hyperglycemia.

Clinical Outcomes and Comorbid Burden

Stroke morbidity and mortality increase with age. This may be because the severity of stroke increases and/or because reactive neuronal synaptogenesis declines, with less robust sprouting responses and lower synaptic replacement rates. Admission body temperature predicts both short- and long-term mortality after ischemic stroke and in some studies of SAH, with a 1 °C increase of admission temperature predicting a 30 % relative increase (95 % confidence interval of 4–57 %) in long-term mortality. Aggressive blood pressure control in patients with ischemic stroke is important, as elevated blood pressures promote early stroke recurrence, hemorrhagic transformation, and increased cerebral edema. Early mortality increases by 4 % for every 10 mmHg above 150 mmHg. After SAH, the blood pressure is lowered before the ruptured aneurysm is repaired, but permissive hypertension is allowed after repair. Blood pressure parameters for patients with intracerebral hemorrhage are not based on high-quality medical evidence and depend on the admission blood pressure and the patient's intracranial pressure [7]. Hyperglycemia, in both patients with and without diabetes, during acute ischemic stroke, predisposes patients to intracerebral hemorrhage and augments brain injury. Increased inflammatory biochemical markers, such as C-reactive protein and lipoprotein-associated phospholipase A2, signal development of atherothrombotic events and long-term mortality after acute ischemic stroke. Cardiac risk factors for death after first cerebral infarction include congestive heart failure, persistent atrial fibrillation, and ischemic heart disease. Cardiac failure predicts increased morbidity and mortality within 1–5 years after stroke. Thus, increased comorbid burden is associated with decreased functional ability over the course of rehabilitation and decreased rehabilitation efficiency. Indeed, stroke patients experiencing cognitive difficulties in the form of executive dysfunction (e.g., abstract thinking, judgment, recall, memory, and comprehension) often show decreased abilities to perform instrumental activities of daily living (IADLs). Functional and physical impairments are further exacerbated by post-stroke mood disorders, such as depression (see Chap. 12) [8].

Early admission to neuro-rehabilitation units with delivery of interdisciplinary specialized stroke care tailored to the individual stroke patient is believed to be important for achieving greater functional gains, although the level of medical evidence to support this is low (see Chap. 16). Rehabilitation might facilitate both functional and structural cerebral cortical reorganization for weeks to months after a patient's initial ischemic injury.

Health Research Methods in Clinical Stroke Trials

Different types of clinical trials exist, including treatment trials, diagnostic trials, screening trials, and prevention trials for those who have never had a stroke (primary prevention) or those who have had a stroke (secondary prevention). A clinical stroke trial of a certain intervention has typically been performed when clinical equipoise exists (i.e., only when there is uncertainty regarding whether the intervention will carry beneficial, harmful, or no significant effects). This may be relatively straightforward when testing new drugs or treatments that have not yet been studied in humans and when the potential efficacy of new drugs and treatments is unknown; however, there are numerous problems with the concept of equipoise in guiding whether or not to conduct a clinical trial [9].

There are at least four phases of clinical trials. Phase 1 trials attempt to establish a safe dosage range, route of administration, clinical pharmacology, including side effects, and determine if a drug is safe to study further. Fewer than 100 patients may be studied. Phase 2 trials determine feasibility, tolerability, safety, and, sometimes, preliminary efficacy based on surrogate or clinical endpoints in a group of usually between 100 and 300 subjects. Phase 3 trials try to confirm the intervention's effectiveness and may include anywhere from a few hundred to tens of thousands of patients. Phase 4 is a post-marketing surveillance to assess long-term effects of the intervention.

Randomization is the process by which participants are assigned to the different groups in a trial, based on chance alone. In order to avoid imbalances in risk factors, stratification of subjects before randomization may be done to assign them into specific subgroups of similar characteristics. However, sample sizes and number of subgroups increase as the number of strata rises, with little gain in statistical precision once the number of subjects per group exceeds 50 [10].

Generalizability of the results of a trial is affected by its inclusion and exclusion criteria—one that includes only limited subtypes of participants will generate results that may not be generalizable to stroke patients in general.

There are many different designs for phase 3 trials in stroke [11]. The most common is a parallel group design where patients are randomized to receive either a treatment or placebo. The trial may be designed for the treatment to show superiority over the placebo group (which will generally be receiving the standard of care) or to show equivalence or non-inferiority to standard of care. These different designs have different statistical considerations and sample sizes. In traditional or frequentist statistical significance testing, two-tailed p values are used, with conclusive studies being those in which the probability of chance finding is less than 5 % ($p<0.05$), or sometimes 1 %, and the null hypothesis does not fall within the 95 % confidence interval range. For frequentist sample size determination, this type I error (alpha level, calling an ineffective intervention effective, false-positive) is usually set at 0.05. Type II error (beta), on the other hand, is the probability of concluding that a difference exists when it does not (false-negative). Power (1 minus beta, chance of missing true difference and failing to identify a treatment that does work) is usually set at 80 % or more for sample size determination purposes. Frequentist stroke trials determine sample sizes a priori, before conducting the trial itself. Adaptive designs allow sample size re-estimation during the course of the trial, especially if there is strong evidence pointing to futility or efficacy. These continual reassessment methods make use of Bayesian methods of incorporating different prior probabilities of a certain event, when incoming data become available. They are beginning to be used in trials in stroke [12–14]. Yet, the clinical trialist must be cautious of early termination of randomized clinical trials (RCTs) for efficacy and futility, as truncated trials may be associated with greater effect sizes in smaller studies, independent of the presence of statistical stopping rules, when aggregated results have not yet reached a steady state.

Stroke Measurement and Outcome Scales

One consideration for these trials is their endpoints. For most phase 3 RCTs, the endpoint is something that matters to the patient and that patients would tend to want to universally avoid like death, the inability to walk, or other disabilities [15]. Studies using these endpoints (e.g., most commonly in stroke currently is the Modified Rankin Scale [mRS]) [16] often need large sample sizes, and thus, are expensive and time-consuming. Numerous statistical methods have been proposed and used to improve the ability of these scales to detect treatment effects. The other approach is to use surrogate endpoints, which are laboratory tests or physical signs that are a substitute for the clinically meaningful outcome but that are expected to reflect or correlate (although correlation alone is an inadequate criteria) with the clinical outcome [17]. Examples in stroke might be infarct size or recanalization of an occluded artery. There are numerous limitations to such surrogates and they have infrequently been the primary outcomes in phase 3 trials [15].

Methodology

When reviewing stroke trial outcome measures, Quinn and colleagues noted that the World Health Organization (WHO) international classification of functioning, disability, and health described that physical recovery includes physical impairment, functional activity (disability), and societal participation (handicap) [18]. There are outcomes in stroke studies that are considered to measure these. For example, the NIHSS measures impairment, the mRS measures activity, and the Stroke Impact Scale (SIS) measures participation. These instruments are designed to measure meaningful changes in function or capture impairment over time. They can be used for purposes of (1) patient recruitment for intervention, (2) patient monitoring and treatment based on changes in measurement, and (3) patient follow-up and long-term prognostication. In general, a discriminative health measurement instrument is critiqued based on its validity, reliability, responsiveness, and convenience of statistical properties [19].

A health measurement scale's validity denotes the degree to which it actually measures what it is intended to, including subdomains of content validity (if it covers specific content domains), predictive validity (if it can predict future criterion score), and concurrent validity (if its outcome is in accordance with another established functional measurement administered concurrently). The instrument's reliability denotes the extent to which repeated measurements give the same outcome, including subdomains of test–retest reliability (if consistent results are found with re-administration by same person), inter-rater reliability (if consistent results are found with re-administration by different people), and internal consistency reliability (if items measuring various constructs deliver consistent results). A responsive scale is one that is sensitive to clinically meaningful changes over time in a patient with and without intervention effects. Finally, a practical stroke scale is feasible, in that it does not require excessive training, is easy to use by the participants, can be used across a variety of settings, and does not take too long to complete [20].

Common Stroke Measurement Scales

A variety of types of stroke measurement scales exist and can be classified into (1) those pertaining to physical impairment and functional activity such as the Glasgow Outcome Scale (GOS), Barthel Index Score, Chedoke-McMaster Stroke Assessment, NIHSS, and mRS and (2) those pertaining to societal participation such as the WHO International Classification of Functioning, Disability and Health; SIS; American Heart Association Stroke Outcome Classification; and the Frenchay Activities Index [21–23].

Quinn and colleagues conducted a systematic review of stroke functional outcome scales used in clinical trials [18]. They identified 126 trials conducted between 2001 and 2006. The trials used 47 different outcome measures, on average 2 per trial. The most commonly used was the mRS (in 64 % of studies) followed by the Barthel Index (41 %) and NIHSS (28 %). Seventy-seven studies used a measure of physical impairment (most commonly the NIHSS), 103 used a measure of activity (most commonly the mRS), and 11 used a measure of participation (most commonly the SIS). Outcome was assessed a median of 90 days after stroke. Cognitive, mood, language, and quality of life scores were not used very often [19]. The National Institutes of Health common data elements group recommended the mRS, NIHSS, Barthel Index, and EuroQol for acute stroke studies, with conditional support of the Glasgow outcome score and Functional Independence Measure.

Physical Impairment and Functional Activity Scales

The NIHSS is a stroke deficit scale that scores from 0 to 15 based on neurological deficits in a level of consciousness, motor function, sensory function, and other functions. The Barthel Index measures activities of daily living on a scale from 0 to 100. The mRS (from 0 being no symptoms to 6 being dead) measures disability or dependency in daily activities [16]. The EuroQol includes the EuroQol-5D and assesses health-related quality of life. All of these scales are easy to administer and are reasonably valid and reliable but have variable sensitivity and responsiveness. Although easy to administer, these scales do not capture the neurocognitive aspects of outcome in patients with stroke. Indeed, stroke survivors with significant neurocognitive dysfunction may still score well on these scales.

Impairment (Disability) Scales

Common impairment (disability) scales with good psychometric properties include the WHO International Classification of Functioning, Disability and Health; Stroke Impact Scale; American Heart Association Stroke Outcome Classification; Chedoke-McMaster Stroke Assessment; and Frenchay Activities Index. Each of these scales takes between 20 and 60 min to administer.

WHO International Classification of Functioning, Disability and Health

The WHO International Classification of Functioning, Disability and Health encompasses both function and disability—including body functions and structures; physiological functions; activities by the individual; participation in life situations, including learning and applying knowledge; general tasks and demands; communication; mobility; self-care; domestic life; interpersonal interactions and relationships; major life areas; and community, social, and civic life. It incorporates health-related quality of life.

Stroke Impact Scale

The SIS is comprehensive, evaluating behavior rather than subjective health. It consists of eight domains, including four physical domains (strength, hand function, mobility, activities of daily living/IADLs), and four psychosocial domains (emotion, communication, memory, social participation). The SIS also includes an individual's global assessment of the percentage of recovery on a visual analog scale.

American Heart Association Stroke Outcome Classification

The American Heart Association Stroke Outcome Classification covers all three domains of body structures and functions, activities, and participation. Its global classification (AHA.SOC) identifies levels of severity, extent of neurologic impairments (motor, sensory, vision, affect, cognition, language), and level of independence according to basic and IADLs (independent to completely dependent).

Chedoke-McMaster Stroke Assessment Scale

The Chedoke-McMaster Stroke Assessment Scale involves a direct performance examination that assesses the domains of physical impairment and activity inventory (disability inventory). In the physical impairment domain, body structure and function measures include shoulder pain and postural control of the arms, hands, legs, and feet. The activity impairment domain assesses changes in mobility function, including motor function and walking (gross motor and walking subscales).

Frenchay Activities Index

The Frenchay Activities Index is a 15-item questionnaire assessing activities inside and outside the home (domestic, work, outdoor activity domains), including one mobility, seven IADLs, and seven indoor or outdoor social activities. This evaluative instrument obtains information on premorbid lifestyle up to 3–6 months after strokes to help determine rehabilitation goals and record progress in activities.

Cognitive, Mood, Language, Quality of Life Scales for Stroke Outcome

About 25 % of survivors of ischemic stroke develop depression or dementia that affects their quality of life, functional outcome, and even mortality [24]. The frequency of cognitive and mood dysfunction after SAH is even higher (around 50 %) [25]. Scales used in stroke studies include the Stroke-Specific Sickness Impact Profile, Short Form-36 (SF-36), Beck Depression Inventory (BDI), SIS, EuroQoL, Mini-Mental State Exam (MMSE), and Montreal Cognitive Assessment (MoCA) (see Table 15.1) [26–171].

Table 15.1 Common outcome scales used in stroke patients

Measure	Abbreviation	Overview	Time to administer	Strengths	Weaknesses
Modified Rankin Scale	mRS	Degree of handicap is assigned on a scale depending on how severely affected activities of daily living are [16]. The scale ranges from no symptoms (a score of 0) to no significant disability despite symptoms (1), slight disability (2), moderate disability (3), moderately severe disability (4), severe disability (5), and death (6)	5–15 min	High test–retest reliability has been reported for the mRS [26, 27]. In addition to this, high correlations between mRA score and stroke size have been found [28, 29]	Some consider the categories in mRS to be broad/poorly defined [30]. There is high inter-rater variability [31], although a structured version (mRS-SI) is available [30] that decreases inter-rater variability [27]. A questionnaire is also available that improves inter-rater reliability [32]. The mRS is not ideal for proxy use due to it's potential for variability [33]
National Institutes of Health Stroke Scale	NIHSS	Measures neurological impairment using a 15-item scale that considers levels of consciousness, pupillary response, gaze palsy, hemianopia, facial palsy, motor ability, limb ataxia, sensory loss, inattention and extinction (formerly neglect), dysarthria, and aphasia [34]	5–8 min	NIHSS can be estimated using medical charts with high correlations to actual NIHSS scores [35]. Moderate to high correlations with lesion volume [34, 36–41]. Can be administered by both neurologists and nurses with similar reliabilities [42]	Some items are considered to be redundant or to not contribute meaningful information although a modified NIHSS (mNIHSS) has been created that is considered to be simpler and have improved reliability and continued validity [43, 44]. The mNIHSS can also be better estimated from medical charts than the original NIHSS [45]. It is not a hierarchical scale. Some ceiling effects have been noted with the NIHSS [46–48]

(continued)

Table 15.1 (continued)

Measure	Abbreviation	Overview	Time to administer	Strengths	Weaknesses
Functional Independence Measure	FIM	Thirteen measures of motor function and five measures of cognitive function including self-care, sphincter control, transfers, locomotion, communication, and social cognition [49]	30–40 min	High test–retest reliability [50–54]. Is able to predict discharge destination [55, 56]. Ideal for proxy use and telephone administration [57]. Good predictor of hospital length of stay [58]	Ceiling effects discovered for cognitive section [59, 60]; however, for the total FIM score, there are no ceiling effects [59–61]
Glasgow Outcome Scale	GOS	The GOS is a measure of global outcome categorized into 5 groups (death, persistent vegetative state, severe disability, moderate disability, and good recovery) [62]. The extended GOS divides severe disability, moderate disability, and good recovery into "upper" and "lower" subcategories	<5 min	Considered to be a fast and simple measure of morbidity/mortality. Both GOS and GOSE scores strongly correlate to self-report measures of outcome and therefore represent the perspective of the patient well [63]. GOS scores have been correlated to psychological test scores [64], and both GOS/GOSE scores have been found to correlate to quality of life [64, 65]. Reliable scores can also be obtained via GOS/GOSE questionnaires sent out by post [66]	Some categories in the GOS are considered broad/open to interpretation and insensitive to subtle changes in function [67–69]. However, the extended version of the GOS (GOSE) has eight distinct outcome categories (instead of 5) that increase sensitivity, validity, and efficacy [70, 71], and, more importantly, structured interviews that increase inter-rater reliability and comprehensiveness [72, 73]

EuroQoL	EQ-5D	Measure of health-related quality of life covering five dimensions: mobility, self-care, main activity, social relationships, pain, and mood [74]. Each dimension has 3 possible responses for the EQ-5D-3L ("no problems," "some problems," and "extreme problems") and 5 responses for the EQ-5D-5L ("no problems," "slight problems," "moderate problems," "severe problems," and "unable to")	<10 min	Reasonable concurrent and discriminant validity [75]. Can be used by proxy respondents with moderate reliability [76]. Although it is a general measure of quality of life (i.e., not stroke-specific), the EQ-5D has been found to correlate well with severity of stroke [77, 78]	Not suitable for serial assessment [79]. Ceiling effects and insensitivity have been noted in the EQ-5D-3L [80–83] so the EQ-5D-5L has been created to correct these issues [84] with some success [85]

(continued)

Table 15.1 (continued)

Measure	Abbreviation	Overview	Time to administer	Strengths	Weaknesses
Barthel Index	BI	Measures disability in ten categories: feeding, bathing, grooming, dressing, bowels, bladder, toilet use, transfers, mobility, and stairs [86]	2–5 min (self-report), 20 min (observation)	High test–retest reliability [87]. The modified and original versions predict length of hospital stay [88–90] and the modified version predicts stroke recovery [91]. Can be used in telephone interviews reliably [92, 93]. The BI has been validated as an adequate measure of long-term outcome in stroke patients [94]	Ceiling/floor effects have been found [61, 95–97]. Cutoff scores are often chosen arbitrarily, although some have been suggested [98, 99]. The original BI (with a three-point scale per category) was considered to be insensitive, so a modified version that expanded the scoring of each category into a five-point scale was developed and was found to increase sensitivity and reliability [100]
WHO International Classification of Functioning, Disability, and Health	ICF	Health-related quality of life measured by three dimensions: body, activities, and participation	>60 min	Provides an exhaustive and wholesome international standard for assessing outcome. Good content validity [101]	Lengthy administration time. Moderate inter-rater reliability for the stroke core set of the WHO-ICF, though standardization has been suggested to increase the inter-rater agreement [102]
Stroke Impact Scale	SIS	Measures eight domains including strength, hand function, activities of daily living and instrumental activities of daily living, mobility, communication, emotion, memory and thinking, and participation [103]	15–20 min	High reliability measured by internal consistency [103–105]. Suitable for proxy use [106, 107]. The latest version (SIS 3.0) was found to detect change better and have higher responsiveness than the SS-QoL [108]	Ceiling effects have been reported in the communication, memory and emotion domains [103, 104], memory, and emotion [104], and a floor effect has been reported in the hand function domain [103]. Although the measure can be mailed, telephone administration has been found to yield less bias and higher test–retest reliability than postal administration [109]

American Heart Association Stroke Outcome Classification	AHA.SOC	Creates an outcome score based off of neurological impairment (motor, sensory, vision, language, cognition, affect)—which is calculated using other outcome scales—severity, and function [110]	Varies	High inter-rater reliability [110]. Comprehensive measure of activities of daily living and instrumental activities of daily living that captures impairments, disabilities, and handicap after stroke well [111]	It lacks specific guidelines as to which tests and scales to use for the neurological impairment domain. Does not capture differences in mental health after stroke [111]
Chedoke-McMaster Stroke Assessment	Chedoke Assessment	Measure of physical impairment and disability that assesses shoulder pain, postural control, arm, hand leg, and foot based on motor recovery [112]	45–60 min	High inter-rater, intra-rater, and test–retest reliabilities [112, 113]	Cannot be used in patients younger than 19 years of age [114]

(continued)

Table 15.1 (continued)

Measure	Abbreviation	Overview	Time to administer	Strengths	Weaknesses
Frenchay Activities Index	FAI	Measures instrumental activities of daily living covering domestic chores, leisure/work, and outdoor activities [115]	5 min	Possible use with proxy respondents—there is high total and subscale agreement between patients and proxies [116]—which may lead to reduction in subjectivity [117], although caution is advised (see weaknesses). No ceiling effects [118, 119]. The FAI has been determined to provide information on activities of daily living not obtained from basic scales like the BI [120]. Can be administered as a postal version with good reliability [121]	Some evidence of gender bias [115, 122, 123] and potential for proxy biases that should be considered [116]. Two items from the original FAI (gainful work and reading books) have been considered to yield little informative and discriminative value; it has been suggested that removing the two would increase reliability [118]. Inter-rater reliability can be improved by specifying and clarifying scoring instructions [124, 125]
Stroke-Adapted Sickness Impact Profile	SA-SIP30	Measures quality of life by covering domains of body care and movement, social interaction, mobility, communication, emotional behavior, household management, alertness behavior, and ambulation [126]	5–10 min	Shorter than the original Sickness Impact Profile (SIP-136) and catered specifically to stroke victims [126]. Less skewed toward good health outcome than the SIP-136 (i.e., can better discriminate between patients in good health) [127]. Can be self- and interview-administered [128]	May not be suitable for patients with severe stroke [126]. The total scores also rely heavily on the physical dimension and not health-related quality of life, which should be taken into account [127]

Medical Outcomes Study Short Form-36	SF-36	Assessment of eight domains: limitations in physical activities due to health problems, limitations in social activities due to physical or emotional problems, limitations in usual role activities due to physical health problems, bodily pain, general mental health, limitations in usual role activities due to emotional problems, vitality, and general health perceptions [129]	10 min	Simple and easy to administer, no training is required. Can discriminate well between stroke and healthy population health-related qualities of life [130]. Adequate reliability and validity [131–133]	Not adequate for serial assessments of individual patients or for proxy respondents [79]. Does not measure social functioning effectively [134]. Ceiling and/or floor effects [130, 132, 135, 136]

(continued)

Table 15.1 (continued)

Measure	Abbreviation	Overview	Time to administer	Strengths	Weaknesses
Beck Depression Inventory –second version	BDI-II	Screening tool used as an indicator of depression (according to DSM-IV), though not intended as a tool for diagnosis (see weaknesses). Evaluation of 21 symptoms of depression covering emotion, behavioral changes, and somatic symptoms [137]	10–15 min	Sensitive to change in depression [138, 139]. Good internal consistency for assessing post-stroke depression [140]. The BDI is one of the most commonly used measures in post-stroke depression and has the advantage of relying more so on cognitive affective than somatic symptoms [141]. Can be used with proxies, although with caution since they tend to rate the patients as more depressive than the patients rate themselves [142]	There is a noted gender effect that should be taken into account; the BDI has greater accuracy in screening depression in males than females [140]. The BDI has been reported to have low specificity and high sensitivity, which leads to more false-positives [142, 143]. However, according to Berg et al. (2009), higher sensitivity is more important than high specificity in screening [142]. High specificity is more important for diagnosis and so the BDI is not suitable for diagnosing depression

Mini-Mental State Examination	MMSE	A brief measure of cognitive impairment that specifically examines arithmetic, memory, language, registration, attention, and orientation [144]	5–15 min	Both the MMSE and the modified MMSE (3MS) are adequate in detecting cognitive impairment in left-sided strokes [145]. The modified version is also considered to be more sensitive and a better predictor of functional outcome in stroke populations than the original [145]. Use of the MMSE in stroke patients has been advocated due to its simplicity to administer and brevity compared to other measures such as the FIM Cognitive Subscale and Loewenstein Occupational Therapy Cognitive Assessment [146, 147]. Lack of long-term change in MMSE score has previously been linked to lacunar infarction [148]. Poor performance on the MMSE is also predictive of long-term cognitive outcome [149]	Scores are affected by education, age, and cultural background [150–152], as well as language (minimal English is associated with low score). The MMSE and 3MS have previously been reported as inadequate for patients with right-sided cerebrovascular accidents [145], although the impact of hemisphere localization has been challenged [153]. The MMSE has been found to be insensitive (particularly in abstract reasoning, executive function, and visual perception/construction) when detecting cognitive impairment in acute stroke patients [154]. Ceiling effects have also been reported for the MMSE [155]

(continued)

Table 15.1 (continued)

Measure	Abbreviation	Overview	Time to administer	Strengths	Weaknesses
Montreal Cognitive Assessment	MoCA	Tool used to assess cognitive impairment in short-term memory, visuospatial abilities, executive function, attention, concentration, working memory, language, and orientation [156]	5–10 min	Higher sensitivity than the MMSE at detecting subtle cognitive impairment [156–160]. Lower ceiling effects and higher internal reliability than the MMSE have been reported for the MoCA in stroke patients [158]. The MoCA has also been able detect cognitive impairment in aneurysmal subarachnoid hemorrhage patients in domains including verbal learning, executive function, working memory, visuospatial function, and motor function wherein the MMSE failed to detect any [160], and similar findings were reported in stroke and transient ischemic attack patients [155]	The MoCA has been found to have lower specificity than the MMSE for stroke patients [161] as well as other patient groups [159], although the opposite has been reported as well for cognitively impaired patients [157]. Scores are affected by age and education [162]. The development of an abbreviated version has been suggested, with the naming domain removed (due to evidence of a ceiling effect) and the word recall removed (due to evidence of a floor effect) [157]

Stroke-Specific Quality of Life	SS-QOL	A measure of health-related quality of life by assessing 49 items in domains covering energy, family roles, language, mobility, mood, personality, self-care, social roles, thinking, upper extremity function, vision, and work/productivity	10–15 min	Suitable for both aneurysmal subarachnoid hemorrhage patients and ischemic stroke patients [163–165]. High test–retest and inter-rater reliability [166]. The abbreviated version with 12 items instead of 49 (SS-QoL-12) has been developed with acceptable validity and reliability [165, 167, 168]	The SS-QoL has been deemed not suitable for use with proxy respondents; proxies tend to report more dysfunction than patients [169], though another study found the opposite in a different cultural group [170]. Ceiling and floor effects have been reported [164, 171]

Key Stroke Trials

Ginsberg categorized stroke trials into two categories [172]. The first category tested drugs or strategies that were trying to achieve neuroprotection, which he defined as a treatment designed to slow or stop the biochemical and molecular events leading to irreversible ischemic brain injury [172]. There were about 160 ongoing or completed as of 2007. The other category of stroke trials includes those that target the blood supply to the brain where the treatment is designed to maintain patency of or prevent or reverse occlusion or reduction in blood supply to the brain. There were about 100 listed by Ginsberg in 2008. They include treatments such as thrombolytics (tissue plasminogen activator), anti-thrombotic, and antiplatelet drugs. Another example is devices to remove clots or dilate stenosis. The Internet stroke center lists 2,344 clinical trials in stroke [173] (www.Strokecenter.org) and www.clinicaltrials.gov returned 3,516 studies when searching "stroke" [174].

Trials with Functional Activity Measures Alone

International Stroke Trial

The International Stroke Trial was a randomized trial of ASA, subcutaneous heparin, both, or neither among 19,435 patients with acute ischemic stroke [175]. The primary endpoint was death from any cause within 14 days and death or dependency (defined as needing help from another person with daily activities) at 6 months. The trial found no difference in mortality between groups, although there were fewer recurrent ischemic strokes and more hemorrhages with heparin or ASA at 6 months. Secondary endpoints were symptomatic intracranial hemorrhage or ischemic strokes within 14 days, major extracranial hemorrhage, death from any cause by 6 months, and dependency or incomplete recovery from stroke at 6 months. The main conclusion was that ASA should be started as soon as possible after ischemic stroke.

British Aneurysm Nimodipine Trial

The British Aneurysm Nimodipine Trial randomized 554 adults with aneurysmal SAH at four medical centers in the United Kingdom [176]. The primary endpoint was the incidence of cerebral infarction and secondary endpoints were poor functional outcome on the GOS 3 months after SAH and incidence of rebleeding. The GOS was dichotomized into good recovery/moderate disability vs. severe disability/vegetative/dead. The study found that patients who received nimodipine had a 34 % reduction in cerebral infarction from 33 % in the placebo to 22 % in the nimodipine group. Poor outcome on the GOS also was reduced by nimodipine by 40 %. There were no important side effects observed. This trial was important in establishing oral nimodipine for use in almost all patients with aneurysmal SAH.

International Subarachnoid Aneurysm Trial

The International Subarachnoid Aneurysm Trial (ISAT) was a RCT that compared neurosurgical clipping to endovascular coiling in 2,143 patients with ruptured intracranial aneurysms from 42 centers [177]. The primary endpoint was death or dependence at 1 year (mRS of 3–6) and secondary endpoints were rate of rebleeding and seizures. Endovascular coiling was associated with a 7 % absolute and 23 % relative risk reduction in poor outcome to 24 %, compared to 31 % in the neurosurgical-clipping group at 1 year. The EuroQol was collected also, as well as neurocognitive outcomes in a subset of patients from the eight United Kingdom centers who had mRS of 0–2 12 months after SAH [178]. Cognitive impairment was less likely in patients treated by coiling compared to clipping (odds ratio 0.58, 95 % confidence interval 0.38–0.87, $p=0.0055$). The outcomes on the EuroQol seemed similar between groups.

Carotid Surgery Trials

The North American Symptomatic Carotid Endarterectomy Trial (NASCET) included 659 patients from 50 center, with included patients experiencing transient ischemic attack or non-disabling stroke within 120 days due to 70–99 % carotid stenosis [179]. Patients were randomized to undergo medical management plus carotid endarterectomy or medical management only. The primary endpoint was fatal or nonfatal ipsilateral stroke, any stroke, or death with follow-up to 12 months. Significantly fewer strokes (any, major or fatal) were noted in the surgical arm.

The Asymptomatic Carotid Surgery Trial randomized 3,120 patients from 126 centers to carotid endarterectomy or medical management with endarterectomy only if it became necessary [180]. Patients were symptomatic or asymptomatic. The results showed that younger patients with asymptomatic carotid stenosis of greater than 70 % benefitted from reduction in stroke risk after carotid endarterectomy, if the perioperative stroke and mortality rate from surgery was low (<3 %). This trial excluded patients with previous ipsilateral carotid endarterectomies, poor surgical risk (including coronary artery disease, potential embolic strokes), and poor life expectancy precluding long-term follow-up. The primary endpoints were any perioperative morbidity and mortality (including stroke and myocardial infarction). The results in asymptomatic patients were similar to the Asymptomatic Carotid Atherosclerosis Study (ACAS) [181].

The European Carotid Surgery Trial was similar to NASCET and randomized patients to medical management plus endarterectomy versus medical management for recently symptomatic carotid stenosis [182]. The trial enrolled 3,204 patients from 96 centers, including those of any age, who recently had a transient ischemic attack or mild stroke in the distribution of the carotid artery (within previous 6 months) and internal carotid stenosis responsible for the patients' symptoms. Those with cardiac conditions likely causing embolic events and more severe disease of distal rather than proximal ICA were excluded. The primary endpoint was major stroke or death. Secondary endpoints were death or major stroke within 30 days of carotid endarterectomy, estimated gain in stroke-free life expectancy. Significantly fewer strokes or deaths were noted in the surgical arm.

The carotid endarterectomy studies have not included cognitive outcomes. Emerging evidence suggests carotid stenosis, endarterectomy, and carotid stenting are associated with cognitive impairment and that endarterectomy may affect this [183].

Trials with Both Functional Activity and Other Measures

International Surgical Trial in Intracerebral Hemorrhage

The International Surgical Trial in Intracerebral Hemorrhage (STICH) entered 1,033 patients at 107 centers with spontaneous supratentorial intracerebral hematomas who were randomized to early surgery (within 24 h of randomization) vs. initial conservative treatment [184]. The primary endpoint was the extended GOS, with secondary endpoints being mortality, Barthel Index, and mRS at 6 months. Surgery did not confer any functional or survival benefit. An individual patient data analysis of 2,186 cases from eight RCTs of surgery for intracerebral hemorrhage found that clinical outcome was better if patients underwent surgery within 8 h of hemorrhage, the hematoma volume was 20–50 mL, the Glasgow coma score was 9 to 12, or the patient was 50 to 69 years old [185].

Acute Stroke Ischemic NXY-059 Treatment

The Acute Stroke Ischemic NXY-059 Treatment (SAINT) Trial investigated the free radical trapping agent, NXY-059, in 1,722 adult patients with acute ischemic stroke in 158 hospitals, enrolling those with acute stroke within 6 hours who were conscious, had limb weakness, and NIHSS score of at least 6 [186]. The agent was administered within 6 hours of ischemic stroke onset. The study had multiple outcome measures, including the mRS, Barthel Index, SIS, and EuroQol EQ-5D, with the primary outcome being the mRS. The mRS was not dichotomized but analyzed by Cochran-Mantel-Haenszel test across the whole range of scores, adjusting for various baseline variables. NXY-059 significantly improved the primary outcome (reduced disability at 90 days on the mRS), but did not significantly improve secondary outcomes (NIHSS to 90 days). The second major clinical trial of NXY-059 for ischemic stroke, however, showed no evidence of efficacy based on the mRS [187].

Decompressive Craniectomy Trials for Ischemic Stroke

Major published decompressive craniectomy trials included the French DECIMAL [188], German DESTINY [189], and Dutch HAMLET trials [190], with a subsequent pooled analysis that was possible due to the similar design of the studies.

When combined, 93 patients were available for analysis. The primary outcome was the mRS, which was dichotomized at 0–3 vs. 4–6, which is different from that used in many ischemic stroke trials that used 0–2 vs. 3–6. The surgical technique used was very similar in all three trials, with all patients having at least a 12-cm diameter bone flap removed from the side of stroke. Medical management was also very similar in all three trials. Patients were intubated in the intensive care unit (ICU) with no initial intracranial pressure monitoring. Management of cerebral edema included restriction of intravenous fluids to 500 mL per day, use of osmotic diuretics, antihypertensives to keep systolic blood pressure <220 and diastolic blood pressure <120, treatment of hyperthermia and hyperglycemia, elevation of the head of the bed, and the option of anticonvulsant use.

The pooled analysis showed that significantly fewer patients who underwent craniectomy had unfavorable outcome (mRS > 3 at 12 months). Other important data to emerge from these studies were that the quality of life as measured by the SIS was the same in patients who did or did not have decompressive craniectomy in DECIMAL. There was no statistically significant difference in depression between groups in HAMLET.

Summary

The most common primary endpoint for stroke trials is the mRS, although different authorities have recommended various other scales. The advantages of the mRS include that it had the lowest amount of ceiling effect compared to other outcome scales [24]. The mRS focuses on physical deficits but does address some cognitive and communication issues. Because of this, it may not detect important cognitive, mood, and quality of life deficits, especially in severely affected stroke patients. There is no outcome scale with good psychometric properties, which captures both functional and cognitive behavioral outcomes and that is validated and meets other criteria as an outcome measure for stroke. While recent reviews of stroke outcome scales emphasize use of the mRS, these trials were mostly ischemic stroke and intracerebral hemorrhage. There is less information on outcome scales for SAH, and since these patients tend not to have the same focal neurological deficits, cognitive behavioral outcomes might be important to assess after SAH.

References

1. Roger VL, Go AS, Lloyd-Jones DM, Adams RJ, Berry JD, Brown TM, et al. American Heart Association Statistics Committee and Stroke Statistics Subcommittee. Heart disease and stroke statistics—2011 update: a report from the American Heart Association. Circulation. 2011;123:e18–209.
2. Feigin VL, Lawes CM, Bennett DA, Barker-Collo SL, Parag V. Worldwide stroke incidence and early case fatality reported in 56 population-based studies: a systematic review. Lancet Neurol. 2009;8:355–69.

3. Mohr JP, editor. Stroke: pathophysiology, diagnosis and management. 5th ed. Philadelphia: Elsevier/Saunders; 2011.
4. Gillen G, editor. Stroke rehabilitation: a functional-based approach. 3rd ed. St. Louis, Mo: Mosby/Elsevier; 2011.
5. Umphred D, editor. Neurological rehabilitation. 5th ed. St. Louis, Mo: Mosby; 2007.
6. Rosengart AJ, Huo JD, Tolentino J, Novakovic RL, Frank JI, Goldenberg FD, et al. Outcome in patients with subarachnoid hemorrhage treated with antiepileptic drugs. J Neurosurg. 2007;107:253–60.
7. Morgenstern LB, Hemphill JC III, Anderson C, Becker K, Broderick JP, Connolly ES Jr, et al. American Heart Association Stroke Council and Council on Cardiovascular Nursing. Guidelines for the management of spontaneous intracerebral hemorrhage: a guideline for healthcare professionals from the American Heart Association/American Stroke Association. Stroke. 2010;41:2108–29.
8. Lew HL, Lee E, Date ES, Zeiner H. Influence of medical comorbidities and complications on FIM change and length of stay during in patient rehabilitation. Am J Phys Med Rehabil. 2002;81(11):830–7.
9. Miller FG, Joffe S. Equipoise and the dilemma of randomized clinical trials. N Engl J Med. 2011;364:476–80.
10. Guyatt G, editor. Users' guides to the medical literature: a manual for evidence-based clinical practice. 2nd ed. New York: McGraw-Hill; 2008.
11. Ahn C, Ahn D. Randomized clinical trials in stroke research. J Investig Med. 2010;58:277–81.
12. Elkind MS, Sacco RL, MacArthur RB, Fink DJ, Peerschke E, Andrews H, et al. The Neuroprotection with Statin Therapy for Acute Recovery Trial (NeuSTART): an adaptive design phase I dose-escalation study of high-dose lovastatin in acute ischemic stroke. Int J Stroke. 2008;3:210–8.
13. Grieve AP, Krams M. ASTIN: a Bayesian adaptive dose-response trial in acute stroke. Clin Trials. 2005;2:340–51.
14. Selim M, Yeatts S, Goldstein JN, Gomes J, Greenberg S, Morgenstern LB, et al.; Deferoxamine Mesylate in Intracerebral Hemorrhage Investigators. Safety and tolerability of deferoxamine mesylate in patients with acute intracerebral hemorrhage. Stroke. 2011;42:3067–74.
15. Fleming TR, DeMets DL. Surrogate end points in clinical trials: are we being misled? Ann Intern Med. 1996;125:605–13.
16. van Swieten JC, Koudstaal PJ, Visser MC, Schouten HJ, van Gijn J. Interobserver agreement for the assessment of handicap in stroke patients. Stroke. 1988;19:604–7.
17. Nelson L, editor. Neuroepidemiology: from principles to practice. Oxford: Oxford University Press; 2004.
18. Quinn TJ, Dawson J, Walters MR, Lees KR. Functional outcome measures in contemporary stroke trials. Int J Stroke. 2009;4:200–5.
19. Lees KR, Bath PM, Schellinger PD, Kerr DM, Fulton R, Hacke W, et al.; European Stroke Organization Outcomes Working Group. Contemporary outcome measures in acute stroke research: choice of primary outcome measure. Stroke. 2012;43:1163–70.
20. Streiner DL, Norman GR. Health measurement scales: a practical guide to their development and use. 4th ed. Oxford: Oxford University Press; 2008.
21. Duncan PW, Jorgensen HS, Wade DT. Outcome measures in acute stroke trials: a systematic review and some recommendations to improve practice. Stroke. 2000;31:1429–38.
22. Kim DH, Haney CL, Van Ginhoven G. Utility of outcome measures after treatment for intracranial aneurysms: a propective trial involving 520 patients. Stroke. 2005;36:792–6.
23. Roberts L, Counsell C. Assessment of clinical outcomes in acute stroke trials. Stroke. 1998;29:986–91.
24. Schellinger PD, Bath PM, Lees KR, Bornstein NM, Uriel E, Eisert W, et al.; European Stroke Organisation Outcomes Working Group. Assessment of additional endpoints for trials in acute stroke—what, when, where, in who? Int J Stroke. 2012;7:227–30.
25. Al-Khindi T, Macdonald RL, Schweizer TA. Cognitive and functional outcome after aneurysmal subarachnoid hemorrhage. Stroke. 2010;41:e519–36.

26. Wolfe CD, Taub NA, Woodrow EJ, Burney PG. Assessment of scales of disability and handicap for stroke patients. Stroke. 1991;22:1242–4.
27. Wilson JT, Hareendran A, Hendry A, Potter J, Bone I, Muir KW. Reliability of the Modified Rankin Scale across multiple raters: benefits of a structured interview. Stroke. 2005;36:777–81.
28. Bruno A, Shah N, Akinwuntan AE, Close B, Switzer JA. J Stroke Cerebrovasc Dis. 2012 Apr 25. [Epub ahead of print] PMID: 22541607.
29. Schaefer PW, Huisman T, Sorensen G, Gonzalez RG, Schwamm LH. Diffusion-weighted MR Imaging in closed head injury: high correlation with initial Glasgow Coma Scale score and score on Modified Rankin Scale at discharge. Neuroradiology. 2004;233:58–66.
30. Wilson JT, Hareendran A, Grant M, Baird T, Schulz UG, Muir KW, et al. Improving the assessment of outcomes in stroke: use of a structured interview to assign grades on the Modified Rankin Scale. Stroke. 2002;33:2243–6.
31. Quinn TJ, Dawson J, Walters MR, Lees KR. Reliability of the modified Rankin Scale: a systematic review. Stroke. 2009;40(10):3393–5.
32. Bruno A, Shah N, Lin C, Close B, Hess DC, Davis K, et al. Improving modified Rankin Scale assessment with a simplified questionnaire. Stroke. 2010;41(5):1048–50.
33. McArthur K, Beagan ML, Degnan A, Howarth RC, Mitchell KA, McQuaige FB, et al. Properties of proxy-derived modified Rankin Scale assessment. Int J Stroke. 2012 Feb 15. doi:10.1111/j.1747-4949.2011.00759.x. [Epub ahead of print].
34. Brott TG, Adams HP, Olinger CP, Marler JR, Barsan WG, Biller J, et al. Measurements of acute cerebral infarction: a clinical examination scale. Stroke. 1989;20:864–70.
35. Kasner SE, Chalela JA, Luciano JM, Cucchiara BL, Raps EC, McGarvey ML, et al. Reliability and validity of estimating the NIH stroke scale score from medical records. Stroke. 1999;30(8):1534–7.
36. Schiemanck SK, Post MW, Witkamp TD, Kappelle LJ, Prevo AJ. Relationship between ischemic lesion volume and functional status in the 2nd week after middle cerebral artery stroke. Neurorehabil Neural Repair. 2005;19:133–8.
37. Saver JL, Johnston KC, Homer D, Wityk R, Koroshetz W, Truskowski LL, et al. Infarct volume as a surrogate or auxiliary outcome measure in ischemic stroke clinical trials. Stroke. 1999;30:293–8.
38. Lyden P, Claesson L, Havstad S, Ashwood T, Lu M. Factor analysis of the National Institutes of Health Stroke Scale in patients with large strokes. Arch Neurol. 2004;61:1677–80.
39. Derex L, Nighoghossian N, Hermier M, Adeleine P, Berthezène Y, Philippeau F, et al. Influence of pretreatment MRI parameters on clinical outcome, recanalization and infarct size in 49 stroke patients treated by intravenous tissue plasminogen activator. J Neurol Sci. 2004;225:3–9.
40. Fink JN, Selim MH, Kumar S, Silver B, Linfante I, Caplan LR, et al. Is the association of National Institutes of Health Stroke Scale scores and acute magnetic resonance imaging stroke volume equal for patients with right- and left-hemisphere ischemic stroke? Stroke. 2002;33:954–8.
41. Woo D, Broderick JP, Kothari RU, Lu M, Brott T, Lyden PD, et al. Does the National Institutes of Health Stroke Scale favor left hemisphere strokes? Stroke. 1999;30:2355–9.
42. Dewey HM, Donnan GA, Freeman EJ, Sharples CM, Macdonell RA, McNeil JJ, et al. Interrater reliability of the National Institutes of Health Stroke Scale: rating by neurologists and nurses in a community-based stroke incidence study. Cerebrovasc Dis. 1999;9(6):323–7.
43. Lyden PD, Lu M, Levine SR, Brott TG, Broderick J, NINDS rtPA Stroke Study Group. A modified National Institutes of Health Stroke Scale for use in stroke clinical trials: preliminary reliability and validity. Stroke. 2001;32(6):1310–7.
44. Meyer BC, Hemmen TM, Jackson CM, Lyden PD. Modified National Institutes of Health Stroke Scale for use in stroke clinical trials: prospective reliability and validity. Stroke. 2002;33(5):1261–6.

45. Kasner SE, Cucchiara BL, McGarvey ML, Luciano JM, Liebeskind DS, Chalela JA. Modified National Institutes of Health Stroke Scale can be estimated from medical records. Stroke. 2003;34(2):568–70.
46. Muir KW, Weir CJ, Murray GD, Povey C, Lees KR. Comparison of neurological scales and scoring systems for acute stroke prognosis. Stroke. 1996;27:1817–20.
47. Williams LS, Weinberger M, Harris LE, Clark DO, Biller J. Development of a Stroke-Specific Quality of Life Scale. Stroke. 1999;30:1362–9.
48. Pickard AS, Johnson JA, Feeny DH. Responsiveness of generic health-related quality of life measures in stroke. Qual Life Res. 2005;14:207–19.
49. Keith RA, Granger CV, Hamilton BB, Sherwin FS. The functional independence measure: a new tool for rehabilitation. Adv Clin Rehabil. 1987;1:6–18.
50. Chau N, Dalter S, Andre JM, Patris A. Inter-rater agreement of two functional independence scales: the Functional Independence Measure (FIM) and a subjective uniform continuous scale. Disabil Rehabil. 1994;16(2):63–71.
51. Segal ME, Ditunno JF, Staas WE. Interinstitutional agreement of individual functional independence measure (FIM) items measured at two sites on one sample of SCI patients. Paraplegia. 1993;31(10):622–31.
52. Kidd D, Stewart G, Baldry J, Johnson J, Rossiter D, Petruckevitch A, et al. The functional independence measure: a comparative validity and reliability study. Disabil Rehabil. 1995;17(1):10–4.
53. Ottenbacher KJ, Hsu Y, Granger CV, Fiedler RC. The reliability of the functional independence measure: a quantitative review. Arch Phys Med Rehabil. 1996;77(12):1226–32.
54. Pollak N, Rheault W, Stoecker JL. Reliability and validity of the FIM for persons aged 80 years and above from a multilevel continuing care retirement community. Arch Phys Med Rehabil. 1996;77(10):1056–61.
55. Black TM, Soltis T, Bartlett C. Using the functional independence measure instrument to predict stroke rehabilitation outcomes. Rehabil Nurs. 1999;24(3):109–14; 121.
56. Oczkowski WJ, Barreca S. The functional independence measure: its use to identify rehabilitation needs in stroke survivors. Arch Phys Med Rehabil. 1993;74(12):1291–4.
57. Segal ME, Gillard M, Schall R. Telephone and in-person proxy agreement between stroke patients and caregivers for the functional independence measure. Am J Phys Med Rehabil. 1996;75(3):208–12.
58. Franchignoni F, Tesio L, Martino MT, Benevolo E, Castagna M. Length of stay of stroke rehabilitation inpatients: prediction through the functional independence measure. Ann Ist Super Sanita. 1998;34(4):463–7.
59. van der Putten JJ, Hobart JC, Freeman JA, Thompson AJ. Measuring change in disability after inpatient rehabilitation: comparison of the responsiveness of the Barthel Index and the Functional Independence Measure. J Neurol Neurosurg Psychiatry. 1999;66:480–4.
60. Hobart JC, Thompson AJ. The five item Barthel index. J Neurol Neurosurg Psychiatry. 2001;71:225–30.
61. Dromerick AW, Edwards DF, Diringer MN. Sensitivity to changes in disability after stroke: a comparison of four scales useful in clinical trials. J Rehabil Res Dev. 2003;40:1–8.
62. Jennett B, Bond M. Assessment of outcome after severe brain damage. Lancet. 1975;1:480–4.
63. Wilson JT, Pettigrew LE, Teasdale GM. Emotional and cognitive consequences of head injury in relation to the Glasgow Outcome Scale. J Neurol Neurosurg Psychiatry. 2000;69(2):204–9.
64. Woischneck D, Firsching R. Efficiency of the Glasgow Outcome Scale (GOS)-Score for the long-term follow-up after severe brain injuries. Acta Neurochir Suppl. 1998;71:138–41.
65. Kosty J, Macyszyn L, Lai K, McCroskery J, Park HR, Stein SC. Relating quality of life to Glasgow outcome scale health states. J Neurotrauma. 2012;29(7):1322–7.
66. Wilson JT, Edwards P, Fiddes H, Stewart E, Teasdale GM. Reliability of postal questionnaires for the Glasgow Outcome Scale. J Neurotrauma. 2002;19(9):999–1005.
67. Gouvier WD, Blanton PD, Kittle KS. Reliability and validity of the Expanded Glasgow Outcome Scale and the Stover-Zieger Scale. Int J Clin Neuropsychol. 1986;8:1–2.

68. Hall K, Cope DN, Rappaport M. Glasgow Outcome Scale and Disability Rating Scale: comparative usefulness in following recovery in traumatic head injury. Arch Phys Med Rehabil. 1985;66:35–7.
69. Hall KM. Overview of functional assessment scales in brain injury rehabilitation. NeuroRehabilitation. 1992;2:98–113.
70. Levin HS, Boake C, Song J, Mccauley S, Contant C, Diaz-Marchan P, et al. Validity and sensitivity to change of the extended Glasgow Outcome Scale in mild to moderate traumatic brain injury. J Neurotrauma. 2001;18(6):575–84.
71. Weir J, Steyerberg EW, Butcher I, Lu J, Lingsma HF, McHugh GS, et al. Does the extended Glasgow Outcome Scale add value to the conventional Glasgow Outcome Scale? J Neurotrauma. 2012;29(1):53–8.
72. Wilson JT, Pettigrew LE, Teasdale GM. Structured interviews for the Glasgow Outcome Scale and the extended Glasgow Outcome Scale: guidelines for their use. J Neurotrauma. 1998;15(8):573–85.
73. Pettigrew LE, Wilson JT, Teasdale GM. Assessing disability after head injury: improved use of the Glasgow Outcome Scale. J Neurosurg. 1998;89(6).939–43.
74. Anon. The EuroQol Group—a new facility for the measurement of health-related quality of life. Health Policy. 1990;16:199.
75. Dorman PJ, Waddell F, Slattery J, Dennis M, Sandercock P. Is the EuroQol a valid measure of health-related quality of life after stroke? Stroke. 1997;28(10):1876–82.
76. Dorman PJ, Waddell F, Slattery J, Dennis M, Sandercock P. Are proxy assessments of health status after stroke with the EuroQol questionnaire feasible, accurate, and unbiased? Stroke. 1997;28(10):1883–7.
77. Pinto EB, Maso I, Vilela RN, Santos LC, Oliveira-Filho J. Validation of the EuroQol quality of life questionnaire on stroke victims. Arq Neuropsiquiatr. 2011;69(2B):320–3.
78. Ronne-Engström E, Enblad P, Lundström E. Outcome after spontaneous subarachnoid hemorrhage measured with the EQ-5D. Stroke. 2011;42(11):3284–6.
79. Dorman P, Slattery J, Farrell B, Dennis M, Sandercock P. Qualitative comparison of the reliability of health status assessments with the EuroQol and SF-36 questionnaires after stroke. United Kingdom Collaborators in the International Stroke Trial. Stroke. 1998;29:63–8.
80. Brazier J, Jones N, Kind P. Testing the validity of the Euroqol and comparing it with the SF-36 health survey questionnaire. Qual Life Res. 1993;2(3):169–80.
81. Badia X, Schiaffino A, Alonso J, Herdman M. Using the EuroQol-5D in the Catalan general population: feasibility and construct validity. Qual Life Res. 1998;7:311–22.
82. Brazier J, Roberts J, Tsuchiya A, Busschbach J. A comparison of the EQ-5D and SF-6D across seven patient groups. Health Econ. 2004;13(9):873–84.
83. Johnson JA, Pickard AS. Comparison of the EQ-5D and SF-12 health surveys in a general population survey in Alberta, Canada. Med Care. 2000;38(1):115–21.
84. Herdman M, Gudex C, Lloyd A, Janssen M, Kind P, Parkin D, et al. Development and preliminary testing of the new five-level version of EQ-5D (EQ-5D-5L). Qual Life Res. 2011;20(10):1727–36.
85. Janssen MF, Pickard AS, Golicki D, Gudex C, Niewada M, Scalone L, et al. Measurement properties of the EQ-5D-5L compared to the EQ-5D-3L across eight patient groups: a multi-country study. Qual Life Res. 2012 Nov 25. [Epub ahead of print].
86. Mahoney FI, Barthel DW. Functional evaluation: the Barthel Index. Md State Med J. 1965;14:61–5.
87. Hsueh IP, Lee MM, Hsieh CL. Psychometric characteristics of the Barthel Activities of Daily Living index in stroke patients. J Formos Med Assoc. 2001;100(8):526–32.
88. Chang KC, Tseng MC, Weng HH, Lin YH, Liou CW, Tan TY. Prediction of length of stay of first-ever ischemic stroke. Stroke. 2002;33(1):2670–4.
89. Granger CV, Greer DS, Liset E, Coulombe J, O'Brien E. Measurement of outcomes of care for stroke patients. Stroke. 1975;6:34–41.
90. Granger CV, Albrecht GL, Hamilton BB. Outcome of comprehensive medical rehabilitation: measurement by PULSES Profile and the Barthel Index. Arch Phys Med Rehabil. 1979;60:145–54.

91. Tilling K, Sterne JA, Rudd AG, Glass TA, Wityk RJ, Wolfe CD. A new method for predicting recovery after stroke. Stroke. 2001;32:2867–73.
92. Della Pietra GL, Savio K, Oddone E, Reggiani M, Monaco F, Leone MA. Validity and reliability of the Barthel Index administered by telephone. Stroke. 2011;42(7):2077–9.
93. Korner-Bitensky N, Wood-Dauphinee S. Barthel Index information elicited over the telephone. Is it reliable? Am J Phys Med Rehabil. 1995;74(1):9–18.
94. Wilkinson PR, Wolfe CD, Warburton FG, Rudd AG, Rudd AG, Howard RS, Ross-Russell RW, et al. Longer term quality of life and outcome in stroke patients: is the Barthel index alone an adequate measure of outcome? Qual Health Care. 1997;6(3):125–30.
95. Duncan PW, Samsa GP, Weinberger M, Goldstein LB, Bonito A, Witter DM, et al. Health status of individuals with mild stroke. Stroke. 1997;28(4):740–5.
96. Salbach N, Mayo N, Higgins J, Ahmed S, Finch LE, Richards CL. Responsiveness and predictability of gait speed and other disability measures in acute stroke. Arch Phys Med Rehabil. 2001;82:1204–12.
97. Quinn TJ, Langhorne P, Stott DJ. Barthel index for stroke trials: development, properties, and application. Stroke. 2011;42(4):1146–51.
98. Uyttenboogaart M, Stewart RE, Vroomen PC, De Keyser J, Luijckx GJ. Optimizing cutoff scores for the Barthel index and the modified Rankin scale for defining outcome in acute stroke trials. Stroke. 2005;36(9):1984–7.
99. Sulter G, Steen C, De Keyser J. Use of the Barthel index and modified Rankin scale in acute stroke trials. Stroke. 1999;30(8):1538–41.
100. Shah S, Vanclay F, Cooper B. Improving the sensitivity of the Barthel Index for stroke rehabilitation. J Clin Epidemiol. 1989;42(8):703–9.
101. Glässel A, Kirchberger I, Linseisen E, Stamm T, Cieza A, Stucki G. Content validation of the International Classification of Functioning, Disability and Health (ICF) Core Set for stroke: the perspective of occupational therapists. Can J Occup Ther. 2010;77(5):289–302.
102. Starrost K, Geyh S, Trautwein A, Grunow J, Ceballos-Baumann A, Prosiegel M, et al. Interrater reliability of the extended ICF core set for stroke applied by physical therapists. Phys Ther. 2008;88(7):841–51.
103. Duncan PW, Wallace D, Lai SM, Johnson D, Embretson S, Laster LJ. The Stroke Impact Scale version 2.0: evaluation of reliability, validity, and sensitivity to change. Stroke. 1999;30:2131–40.
104. Duncan PW, Bode RK, Lai SM, Perera S, Glycine Antagonist in Neuroprotection Americans Investigators. Rasch analysis of a new stroke-specific outcome scale: the Stroke Impact Scale. Arch Phys Med Rehabil. 2003;84:950–63.
105. Edwards B, O'Connell B. Internal consistency and validity of the Stroke Impact Scale 2.0 (SIS 2.0) and SIS-16 in an Australian sample. Qual Life Res. 2003;12:1127–35.
106. Duncan PW, Lai SM, Tyler D, Perera S, Reker DM, Studenski S. Evaluation of proxy responses to the Stroke Impact Scale. Stroke. 2002;33(11):2593–9.
107. Carod-Artal FJ, Ferreira Coral L, Stieven Trizotto D, Menezes MC. Self- and proxy-report agreement on the Stroke Impact Scale. Stroke. 2009;40(10):3308–14.
108. Lin KC, Fu T, Wu CY, Hsieh YW, Chen CL, Lee PC. Psychometric comparisons of the Stroke Impact Scale 3.0 and Stroke-Specific Quality of Life Scale. Qual Life Res. 2010;19(3):435–43.
109. Duncan P, Reker D, Kwon S, Lai SM, Studenski S, Perera S, et al. Measuring stroke impact with the stroke impact scale: telephone versus mail administration in veterans with stroke. Med Care. 2005;43(5):507–15.
110. Kelly-Hayes M, Robertson JT, Broderick JP, Duncan PW, Hershey LA, Roth EJ, et al. The American Heart Association Stroke Outcome Classification. Stroke. 1998;29(6):1274–80.
111. Lai SM, Duncan PW. Evaluation of the American Heart Association Stroke Outcome Classification. Stroke. 1999;30(9):1840–3.
112. Gowland C, Stratford P, Ward M, Moreland J, Torresin W, Van Hullenaar S, et al. Measuring physical impairment and disability with the Chedoke-McMaster Stroke Assessment. Stroke. 1993;24:58–63.
113. Crowe J, Harmer D, Sharpe D. Reliability of the Chedoke-McMaster Disability Inventory in acquired brain injury. Physiother Can. 1996;48(1):25.

114. Finch E, Brooks D, Stratford PW, Mayo NE. Physical rehabilitations outcome measures. A guide to enhanced clinical decision-making. 2nd ed. Toronto: Canadian Physiotherapy Association; 2002.
115. Holbrook M, Skilbeck CE. An activities index for use with stroke patients. Age Ageing. 1983;12(2):166–70.
116. Tooth LR, McKenna KT, Smith M, O'Rourke P. Further evidence for the agreement between patients with stroke and their proxies on the Frenchay Activities Index. Clin Rehabil. 2003;17:656–65.
117. Segal ME, Schall RR. Determining functional/health status and its relation to disability in stroke survivors. Stroke. 1994;25:2391–7.
118. Schuling J, de Haan R, Limburg M, Groenier KH. The Frenchay Activities Index. Assessment of functional status in stroke patients. Stroke. 1993;24(8):1173–7.
119. Lin KC, Chen HF, Wu CY, Yu TY, Ouyang P. Multidimensional Rasch validation of the Frenchay Activities Index in stroke patients receiving rehabilitation. J Rehabil Med. 2012;44(1):58–64.
120. Appelros P. Characteristics of the Frenchay Activities Index one year after a stroke: a population-based study. Disabil Rehabil. 2007;29(10):785–90.
121. Carter J, Mant F, Mant J, Wade D, Winner S. Comparison of postal version of the Frenchay Activities Index with interviewer-administered version for use in people with stroke. Clin Rehabil. 1997;11(2):131–8.
122. Sveen U, Bautz-Holter E, Sødring KM, Wyller TB, Laake K. Association between impairments, self-care ability and social activities 1 year after stroke. Disabil Rehabil. 1999;21(8):372–7.
123. Wade DT, Legh-Smith J, Langton HR. Social activities after stroke: measurement and natural history using the Frenchay Activities Index. Int Rehabil Med. 1985;7(4):176–81.
124. Post MW, de Witte LP. Good inter-rater reliability of the Frenchay Activities Index in stroke patients. Clin Rehabil. 2003;17(5):548–52.
125. Piercy M, Carter J, Mant J, Wade DT. Inter-rater reliability of the Frenchay activities index in patients with stroke and their careers. Clin Rehabil. 2000;14(4):433–40.
126. van Straten A, de Haan RJ, Limburg M, Schuling J, Bossuyt PM, van den Bos GA. A stroke-adapted 30-item version of the Sickness Impact Profile to assess quality of life (SA-SIP30). Stroke. 1997;28(11):2155–61.
127. van Straten A, de Haan RJ, Limburg M, van den Bos GA. Clinical meaning of the Stroke-Adapted Sickness Impact Profile-30 and the Sickness Impact Profile-136. Stroke. 2000;31:2610–5.
128. Buck D, Jacoby A, Massey A, Ford G. Evaluation of measures used to assess quality of life after stroke. Stroke. 2000;31(8):2004–10.
129. Ware Jr JE, Sherbourne CD. The MOS 36-item short-form health survey (SF-36). I. Conceptual framework and item selection. Med Care. 1992;30:473–83.
130. Almborg AH, Berg S. Quality of life among Swedish patients after stroke: psychometric evaluation of SF-36. J Rehabil Med. 2009;41(1):48–53.
131. Unalan D, Soyuer F, Ozturk A. Should the Nottingham Health Profile or the Short Form-36 be given preference in stroke? Neurosciences (Riyadh). 2009;14(1):45–52.
132. Lotus Shyu YI, Lu JF, Chen ST. Psychometric testing of the SF-36 Taiwan version on older stroke patients. J Clin Nurs. 2009;18(10):1451–9.
133. Hagen S, Bugge C, Alexander H. Psychometric properties of the SF-36 in the early post-stroke phase. J Adv Nurs. 2003;44(5):461–8.
134. Anderson C, Laubscher S, Burns R. Validation of the Short Form 36 (SF-36) health survey questionnaire among stroke patients. Stroke. 1996;27(10):1812–6.
135. Owolabi MO. Which is more valid for stroke patients: generic or stroke-specific quality of life measures? Neuroepidemiology. 2010;34(1):8–12.
136. Hobart JC, Williams LS, Moran K, Thompson AJ. Quality of life measurement after stroke: uses and abuses of the SF-36. Stroke. 2002;33(5):1348–56.
137. McDowell I. Measuring health: a guide to rating scales and questionnaires. 3rd ed. New York: Oxford University Press; 2005.

138. House A, Dennis M, Mogridge L, Warlow C, Hawton K, Jones L. Mood disorders in the year after first stroke. Br J Psychiatry. 1991;158:83–92.
139. Svanborg P, Asberg M. A comparison between the Beck Depression Inventory (BDI) and the self-rating version of the Montgomery Asberg Depression Rating Scale (MADRS). J Affect Disord. 2001;64(2–3):203–16.
140. Aben I, Verhey F, Lousberg R, Lodder J, Honig A. Validity of the beck depression inventory, hospital anxiety and depression scale, SCL-90, and Hamilton depression rating scale as screening instruments for depression in stroke patients. Psychosomatics. 2002;43(5):386–93.
141. Turner-Stokes L, Hassan N. Depression after stroke: a review of the evidence base to inform the development of an integrated care pathway. Part 1: Diagnosis, frequency and impact. Clin Rehabil. 2002;16(3):231–47.
142. Berg A, Lönnqvist J, Palomäki H, Kaste M. Assessment of depression after stroke: a comparison of different screening instruments. Stroke. 2009;40(2):523–9.
143. Lincoln NB, Nicholl CR, Flannaghan T, Leonard M, Van der Gucht E. The validity of questionnaire measures for assessing depression after stroke. Clin Rehabil. 2003;17(8):840–6.
144. Folstein MF, Folstein SE, McHugh PR. "Mini-mental state". A practical method for grading the cognitive state of patients for the clinician. J Psychiatr Res. 1975;12(3):189–98.
145. Grace J, Nadler JD, White DA, Guilmette TJ, Giuliano AJ, Monsch AU, et al. Folstein vs modified Mini-Mental State Examination in geriatric stroke. stability, validity, and screening utility. Arch Neurol. 1995;52(5):477–84.
146. Zwecker M, Levenkrohn S, Fleisig Y, Zeilig G, Ohry A, Adunsky A. Mini-Mental State Examination, cognitive FIM instrument, and the Loewenstein Occupational Therapy Cognitive Assessment: relation to functional outcome of stroke patients. Arch Phys Med Rehabil. 2002;83(3):342–5.
147. Adunsky A, Fleissig Y, Levenkrohn S, Arad M, Noy S. Clock drawing task, mini-mental state examination and cognitive-functional independence measure: relation to functional outcome of stroke patients. Arch Gerontol Geriatr. 2002;35(2):153–60.
148. Appelros P, Andersson AG. Changes in Mini Mental State Examination score after stroke: lacunar infarction predicts cognitive decline. Eur J Neurol. 2006;13(5):491–5.
149. Bour A, Rasquin S, Boreas A, Limburg M, Verhey F. How predictive is the MMSE for cognitive performance after stroke? J Neurol. 2010;257(4):630–7.
150. Tombaugh TN, McIntyre NJ. The mini-mental state examination: a comprehensive review. J Am Geriatr Soc. 1992;40(9):922–35.
151. Bleecker ML, Bolla-Wilson K, Kawas C, Agnew J. Age-specific norms for the Mini-Mental State Exam. Neurology. 1988;10:1565–8.
152. Shadlen MF, Larson EB, Gibbons L, McCormick WC, Teri L. Alzheimer's disease symptom severity in Blacks and Whites. J Am Geriatr Soc. 1999;47:482–6.
153. Appelros P. Characteristics of Mini-Mental State Examination 1 year after stroke. Acta Neurol Scand. 2005;112(2):88–92.
154. Nys GM, van Zandvoort MJ, de Kort PL, Jansen BP, Kappelle LJ, de Haan EH. Restrictions of the Mini-Mental State Examination in acute stroke. Arch Clin Neuropsychol. 2005;20(5):623–9.
155. Pendlebury ST, Cuthbertson FC, Welch SJ, Mehta Z, Rothwell PM. Underestimation of cognitive impairment by Mini-Mental State Examination versus the Montreal Cognitive Assessment in patients with transient ischemic attack and stroke: a population-based study. Stroke. 2010;41(6):1290–3.
156. Nasreddine ZS, Phillips NA, Bédirian V, Charbonneau S, Whitehead V, Collin I, et al. The Montreal Cognitive Assessment, MoCA: a brief screening tool for mild cognitive impairment. J Am Geriatr Soc. 2005;53(4):695–9.
157. Damian AM, Jacobson SA, Hentz JG, Belden CM, Shill HA, Sabbagh MN, et al. The Montreal Cognitive Assessment and the mini-mental state examination as screening instruments for cognitive impairment: item analyses and threshold scores. Dement Geriatr Cogn Disord. 2011;31(2):126–31.

158. Toglia J, Fitzgerald KA, O'Dell MW, Mastrogiovanni AR, Lin CD. The Mini-Mental State Examination and Montreal Cognitive Assessment in persons with mild subacute stroke: relationship to functional outcome. Arch Phys Med Rehabil. 2011;92(5):792–8.
159. Larner AJ. Screening utility of the Montreal Cognitive Assessment (MoCA): in place of–or as well as–the MMSE? Int Psychogeriatr. 2012;24(3):391–6.
160. Schweizer TA, Al-Khindi T, Macdonald RL. Mini-Mental State Examination versus Montreal Cognitive Assessment: rapid assessment tools for cognitive and functional outcome after aneurysmal subarachnoid hemorrhage. J Neurol Sci. 2012;316(1–2):137–40.
161. Godefroy O, Fickl A, Roussel M, Auribault C, Bugnicourt JM, Lamy C, et al. Is the Montreal Cognitive Assessment superior to the Mini-Mental State Examination to detect poststroke cognitive impairment? A study with neuropsychological evaluation. Stroke. 2011;42(6):1712–6.
162. Freitas S, Simões MR, Alves L, Santana I. Montreal Cognitive Assessment: influence of sociodemographic and health variables. Arch Clin Neuropsychol. 2012;27(2):165–75.
163. Wong GK, Lam SW, Ngai K, Wong A, Poon WS, Mok V. Validation of the Stroke-specific Quality of Life for patients after aneurysmal subarachnoid hemorrhage and proposed summary subscores. J Neurol Sci. 2012;320(1–2):97–101.
164. Boosman H, Passier PE, Visser-Meily JM, Rinkel GJ, Post MW. Validation of the Stroke Specific Quality of Life scale in patients with aneurysmal subarachnoid haemorrhage. J Neurol Neurosurg Psychiatry. 2010;81(5):485–9.
165. Hsueh IP, Jeng JS, Lee Y, Sheu CF, Hsieh CL. Construct validity of the stroke-specific quality of life questionnaire in ischemic stroke patients. Arch Phys Med Rehabil. 2011;92(7):1113–8.
166. Williams LS, Redmon G, Saul DC, Weinberger M. Reliability and telephone validity of the Stroke-specific Quality of Life (SS-QOL) scale. Stroke. 2000;32:339-b.
167. Chen HF, Wu CY, Lin KC, Li MW, Yu HW. Validity, reliability and responsiveness of a short version of the Stroke-Specific Quality of Life Scale in patients receiving rehabilitation. J Rehabil Med. 2012;44(8):629–36.
168. Post MW, Boosman H, van Zandvoort MM, Passier PE, Rinkel GJ, Visser-Meily JM. Development and validation of a short version of the Stroke Specific Quality of Life Scale. J Neurol Neurosurg Psychiatry. 2011;82(3):283–6.
169. Williams LS, Bakas T, Brizendine E, Plue L, Tu W, Hendrie H, et al. How valid are family proxy assessments of stroke patients' health-related quality of life? Stroke. 2006;37:2081–5.
170. Vrdoljak D, Rumboldt M. Quality of life after stroke in Croatian patients. Coll Antropol. 2008;32(2):355–9.
171. Czechowsky D, Hill MD. Neurological outcome and quality of life after stroke due to vertebral artery dissection. Cerebrovasc Dis. 2002;13:192–7.
172. Ginsberg MD. Neuroprotection for ischemic stroke: past, present and future. Neuropharmacology. 2008;55:363–89.
173. www.Strokecenter.org. Accessed 19 July 2012.
174. www.clinicaltrials.gov. Accessed 19 July 2012.
175. Anon. International stroke trial collaborative group. International Stroke Trial. A randomized trial of aspirin, subcutaneous heparin, both, or neither among 19435 patients with acute ischemic stroke. Lancet. 1997;349:1569–81.
176. Pickard J, Murray G, Illingworth R, Shaw MD, Teasdale GM, Foy PM, et al. British Aneurysm Nimodipine Trial. Effect of oral nimodipine on cerebral infarction and outcome after subarachnoid hemorrhage. BMJ. 1989;298:636–42.
177. Molyneux A, Kerr R, Yu L, Clarke M, Sneade M, Yarnold JA, et al. International Subarachnoid Aneurysm Trial (ISAT) Collaborative Group. International Subarachnoid Aneurysm Trial of neurosurgical clipping versus endovascular coiling in 2143 patients with ruptured intracranial aneurysms: a randomized comparison of effects on survival, dependency, seizures, rebleeding, subgroups and aneurysm occlusion. Lancet. 2005;366:809–17.

178. Scott RB, Eccles F, Molyneux AJ, Kerr RS, Rothwell PM, Carpenter K. Improved cognitive outcomes with endovascular coiling of ruptured intracranial aneurysms: neuropsychological outcomes from the International Subarachnoid Aneurysm Trial (ISAT). Stroke. 2010;41:1743–7.
179. Anon. North American Symptomatic Carotid Endarterectomy Trial Collaborators. Beneficial effect of carotid endarterectomy in symptomatic patients with high-grade carotid stenosis. N Engl J Med. 1991;325:445–53.
180. Halliday A, Mansfield A, Marro J, Peto C, Peto R, Potter J, et al.; MRC Asymptomatic CarotidSurgery Trial (ACST) Collaborative Group. Prevention of disabling and fatal strokes by successful carotid endarterectomy in patients without recent neurological symptoms. Lancet. 2004;363:1491–502.
181. Anon. Carotid endarterectomy for patients with asymptomatic internal carotid artery stenosis. National Institute of Neurological Disorders and Stroke. J Neurol Sci. 1995;129:76–7.
182. Anon. Randomised trial of endarterectomy for recently symptomatic carotid stenosis: final results of the MRC European Carotid Surgery Trial (ECST). Lancet. 1991;351:1379–87.
183. Altinbas A, van Zandvoort MJ, van den Berg E, Jongen LM, Algra A, Moll FL, et al. Cognition after carotid endarterectomy or stenting: a randomized comparison. Neurology. 2011;77:1084–90.
184. Mendelow A, Gregson B, Fernandes H, Murray GD, Teasdale GM, Hope DT, et al.; STICH Investigators. Early surgery versus initial conservative treatment in patients with spontaneous supratentorial intracerebralhaematomas in the International Surgical Trial in Intracerebral Haemorrhage (STICH): a randomised trial. Lancet. 2005;365:387–97.
185. Gregson BA, Broderick JP, Auer LM, Batjer H, Chen XC, Juvela S, et al. Individual patient data subgroup meta-analysis of surgery for spontaneous supratentorial intracerebral hemorrhage. Stroke. 2012;43:1496–504.
186. Lees KR, Zivin JA, Ashwood T, Davalos A, Davis SM, Diener H-C, et al.; for the Stroke–Acute Ischemic NXY Treatment (SAINT I) Trial Investigators. NXY-059 for acute ischemic stroke. N Engl J Med. 2006;354:588–600.
187. Shuaib A, Lees KR, Lyden P, Grotta J, Davalos A, Davis SM, et al.; SAINT II Trial Investigators. NXY-059 for the treatment of acute ischemic stroke. N Engl J Med. 2007;357:562–71.
188. Jüttler E, Schwab S, Schmiedek P, Unterberg A, Hennerici M, Woitzik J, et al.; DESTINY Study Group. Decompressive surgery for the treatment of malignant infarction of the middle cerebral artery (DESTINY): a randomized controlled trial. Stroke. 2007;38:2518–25.
189. Vahedi K, Vicaut E, Mateo J, Kurtz A, Orabi M, Guichard JP, et al.; DECIMAL Investigators. Sequential-design, multicenter, randomized, controlled trial of early decompressive Craniectomyin malignant cerebral artery infarction (DECIMAL Trial). Stroke. 2007;38:2506–17.
190. Hofmeijer J, Kappelle L, Algra A, Amelink GJ, van Gijn J, van der Worp HB, HAMLET Investigators. Surgical decompression for space-occupying cerebral infarction (the Hemicranectomy After Middle Cerebral Artery infarction with Life-threatening Edema Trial [HAMLET]): a multicentre, open randomized trial. Lancet Neurol. 2009;8(4):326–33.

Chapter 16
Cognitive Rehabilitation and Recovery After Stroke

Audrey Bowen and Emma Patchick

Introduction

As the previous chapters have described in detail, many of those fortunate to survive their stroke do so with detrimental alterations to their cognitive and psychological well-being. These impairments impact the affected individual's ability to participate in, and benefit from, multidisciplinary stroke rehabilitation, to safely and independently carry out activities of everyday living, and to resume pre-morbid personal, social, and vocational roles [1–4]. Previously automatic and effortless tasks require exhausting levels of concentration and, despite the efforts invested, often end in perplexing and de-motivating failure. Uncertainty in one's own abilities and reliance on others makes people with cognitive problems vulnerable to frustration, humiliation, worry, and feelings of hopelessness. These topics are covered elsewhere in this book. The current chapter focuses on cognitive rehabilitation by exploring the evidence base from the perspective of informing clinical service improvements and strives to root cognitive recovery firmly within a broader psychological context.

> I couldn't understand why things were so much harder…I couldn't follow things. I worked before my stroke and was…am…an intelligent man, but didn't feel that way anymore. The tests were interesting for me…some bits were so easy, other bits just made me unravel… things I knew I should be able to do. It really helped me and my wife that the girls explained why this was happening…that it was the stroke, not me. I guess I felt it gave me some control to understand it…. Quote from person with stroke. Reprinted with permission from NHS Improvement -Stroke [31].

A. Bowen, B.A. Psychology, M.Sc. Neuropsychology, Ph.D. (✉)
Vascular and Stroke Research Centre, University of Manchester (MAHSC),
Salford Royal Foundation NHS Trust, Salford, UK
e-mail: audrey.bowen@manchester.ac.uk

E. Patchick, B.Sc.
School of Psychological Sciences, University of Manchester, Manchester, Lancashire, UK

T.A. Schweizer and R.L. Macdonald (eds.), *The Behavioral Consequences of Stroke*,
DOI 10.1007/978-1-4614-7672-6_16, © Springer Science+Business Media New York 2014

Consensus on Prioritizing Psychological Problems

Stroke survivors often seek to express that they feel like a different person, their essence has changed, and their self-identity as well as esteem has been threatened, not necessarily by their hemiplegia or their hemianopia but by changes to the cognitive functions underlying their capacity for language, attention, spatial awareness, memory, and so on [5]. Families notice a difference too, although as they anecdotally report, it is the dysexecutive impairments altering social behavior that cause the greatest concern about having "lost" the person they knew. It is therefore not surprising that there is a consensus amongst people with stroke, their health service providers, and stroke rehabilitation researchers regarding the importance of the behavioral consequences of stroke.

Research into psychological problems was raised as a priority area by the National Stroke Strategy for England [6] despite, or perhaps because of, uncertainties regarding the most effective rehabilitation interventions. When stroke survivors were recently asked about their unmet needs following stroke, almost half of the 799 respondents reported problems with their mood and cognition [7]. Of those, a high proportion felt that issues such as memory and concentration had not been addressed appropriately, especially when compared with other issues such as mobility and pain. Similarly, the James Lind Alliance took a comprehensive and rigorous approach to identifying research priorities relating to life after stroke by consulting with stroke survivors, caregivers, and health professionals as well as searching relevant literature. They concluded that the number one research priority was investigating the best ways to improve cognition after stroke [8].

Quality of the Evidence Base for Cognitive Rehabilitation

One conclusion that might be drawn from the above is that there is very little existing research in cognitive rehabilitation. However, there is in fact an abundance of literature on the topic, and cognitive rehabilitation research is now well established with contributions from several fields including neuropsychology, cognitive psychology, clinical psychology, neurorehabilitation, occupational and speech and language therapy, and acquired brain injury. The full gamut of research designs are employed from qualitative methods exploring survivors' perspectives and priorities through the whole range of quantitative methodologies. The latter consist of single case designs and case series, cohort and case–control observational studies, experimental group designs (within and between subject controls) up to and including randomized controlled trials, and the recent emergence of health economic evaluations. Readers interested in the topic of research design for the evaluation of complex interventions such as cognitive rehabilitation are referred to the framework proposed by the Medical Research Council [9, 10].

Perhaps this abundance of evidence is the problem. How do those charged with improving national and local clinical services extract the most relevant and reliable research, especially where it appears contradictory? The two most internationally accepted methods of evidence synthesis for clinical service development are the Cochrane Collaboration's established systematic review and meta-analysis, disseminated widely throughout the world via the web-based Cochrane Library [11]; and the national clinical guidelines/recommendations for stroke now produced and regularly updated by a growing number of countries, e.g., Australia [12], Canada [13], the UK (except Scotland) [14], and a separate guideline for Scotland [15]. Cochrane reviews employ a tried and tested formula for systematic searching to extract and include published and unpublished data that meet agreed quality standards, thereby reducing the risk of bias. This usually restricts the review to evidence collected from well-conducted randomized controlled trials.

From Cochrane Reviews to National Clinical Stroke Guidelines

Cochrane reviews of cognitive rehabilitation that focus on dysfunctions such as neglect, apraxia, memory, perception, and attention problems exist, and others—such as those concerned with executive dysfunction—are close to publication. The Cooksey review of UK healthcare research highlighted two problematic "gaps" that hold back clinical service development in healthcare generally [16]. One of the gaps is relevant to cognitive rehabilitation and is specifically concerned with how we transfer research evidence into clinical knowledge or clinical practice.

Assumptions that data/evidence and knowledge are one and the same are naïve, as is the expectation that clinicians can and will automatically implement published evidence and evidence syntheses into practice. National clinical guidelines seek to address this gap [12–15]. They perform the essential translator role, producing recommendations for implementation into clinical practice based on high quality searching, evidence appraisal, and consensus level agreement. Where evidence is missing, recommendations are formulated around expert opinion and good practice points. Often they also complete the loop by conducting national audits of adherence to the recommendations [17, 18]. This can help by highlighting areas of practice in need of greatest improvement such as the area of psychological needs, including cognitive rehabilitation, in England [7]. The Canadian guideline (i.e., their Stroke Strategy: Best Practice Recommendations) explicitly includes helpful links to "Implementation Resources and Knowledge Transfer Tools" for each topic within stroke care [13, 19].

Aims of This Chapter

There are now several excellent textbooks [20, 21], journal review papers [22, 23], and Cochrane reviews on post-stroke cognitive rehabilitation that can be referred to for detailed descriptions of both the interventions and the studies that evaluate their efficacy [24–30]. The current chapter describes and compares the recommendations for cognitive rehabilitation currently advocated in various National guideline, which themselves are heavily influenced by the Cochrane reviews and randomized controlled trials. We review each cognitive area and conclude with what has been termed "comprehensive holistic neuropsychological rehabilitation." The evidence base for this borrows heavily from the traumatic brain injury literature but suggests a pragmatic way forward for stroke rehabilitation services. The final issues considered will be service organization and the workforce needed to deliver effective cognitive rehabilitation, with reference to the recent National Health Service (NHS) Improvement Program's useful stepped care model of improving stroke services for people with cognitive and mood problems in England [31].

Cognitive Rehabilitation: Screening and Assessment

The most striking common feature of the clinical guidelines is their emphasis on screening and assessment to elicit underlying cognitive impairments and determine the likely functional and personal impact for each individual with stroke. In some guidelines a larger proportion of the recommendations focus on assessment compared to restorative or compensatory interventions (e.g., Scottish). Providing explanations to demystify patients and caregivers is often a core recommendation and the rationale for this is illustrated in the previous quote from a person with stroke [31]. The following definition of cognitive rehabilitation from the Scottish guideline places this message up front. It also highlights the current paucity of evidence for the benefits of assessment [15]. Although the Scottish guideline writers raise a valid methodological concern with the one existing study, the practical and cost implications of using qualified psychologists rather than assistants would need careful consideration.

> Cognitive rehabilitation concerns efforts to help patients understand their impairment and to restore function or to compensate for lost function (e.g., by teaching strategies) in order to assist adaptation and facilitate independence….When cognitive problems are suspected and relatives report personality change, the patient can be referred to a clinical psychologist to provide assessment and where appropriate, psychological intervention which may include career education and support. One [randomized controlled trial] found a trend only toward reduced [caregiver] strain when this service was provided. Assistant psychologists, not fully trained clinical psychologists, were used in this study. Reprinted with permission from Scottish Intercollegiate Guidelines Network [15]

Key recommendations on the topic of screening and assessment have been extracted and presented in Table 16.1. These include the reminder that assessment should determine a person's cognitive strengths and not just their impairments. The stroke team needs to be informed regarding the person's learning potential and how best to maximize that, not just for the rehabilitation of their cognitive difficulties, but as an "integral part of the [multidisciplinary] rehabilitation plan" [15]. Other recommendations common amongst guidelines concern balancing the utility of

Table 16.1 Recommendations from National Clinical Guidelines: screening and assessment for cognitive problems (selected extracts)

Australia	a) All patients should be screened for cognitive and perceptual deficits using validated and reliable screening tools. b) Patients identified during screening as having cognitive deficits should be referred for comprehensive clinical neuropsychological investigations.
UK[a]	A. Interventions or patient management should be organised so that people with cognitive difficulties can participate in the treatments and regularly reviewed and evaluated. B. Every patient seen after a stroke should be considered to have at least some cognitive losses in the early phase. Routine screening should be undertaken to identify the patient's broad level of functioning, using simple standardised measures (e.g. Montreal Cognitive Assessment MOCA). C. Any patient not progressing as expected in rehabilitation should have a more detailed cognitive assessment to determine whether cognitive losses are causing specific problems or hindering progress. D. Care should be taken when assessing patients who have a communication impairment. The advice from a speech and language therapist should be sought where there is any uncertainty about these individuals… E. The patient's cognitive status should be taken into account by all members of the multidisciplinary team when planning and delivering treatment. F. Planning for discharge from hospital should include an assessment of any safety risks from persisting cognitive impairments. G. People returning to cognitively demanding activities (e.g. some work, driving) should have their cognition assessed formally beforehand.
Scotland	A full understanding of the patient's cognitive strengths and weaknesses should be an integral part of the rehabilitation plan. *Screening* Short, standardised cognitive screening measures can be used by a health professional with knowledge and experience of the presentations of cognitive functioning and factors influencing it. They can be used as a broad screen to reduce the possibility that problems will be missed and as a measure of progress. It is important for staff to understand that these screening measures will miss some of the cognitive problems which can be most important for rehabilitation and eventual functioning. These are varied but can include such issues as poor awareness of deficits or their implications, slowing of information processing, and the ability to cope with distraction. Care needs to be taken in selecting measures for use with people who have communication difficulties and, ideally, the selection should be made in collaboration with a speech and language therapist. *Assessment* Screening measures do not provide information about the depth and nature of the patient's problems or strengths and therefore do not constitute an assessment sufficient for rehabilitation planning or for establishing suitability for a particular work role (e.g. operating machinery). Administering and interpreting full assessment results requires specialist training and should be carried out in the context of clinical interviews with access to background information. Stroke patients should have a full assessment of their cognitive strengths and weaknesses when undergoing rehabilitation or when returning to cognitively demanding activities such as driving or work. Cognitive assessment may be carried out by occupational therapists with expertise in neurological care, although some patients with more complex needs will require access to specialist neuropsychological expertise.

(continued)

Table 16.1 (continued)

Canada	1. All high-risk patients should be screened for cognitive impairment using a validated screening tool. 2. Screening to investigate a person's cognitive status should address arousal, alertness, attention, orientation, memory, language, agnosia, visuospatial/perceptual function, praxis and executive functions such as insight, judgment, social cognition, problem solving, abstract reasoning, initiation, planning and organization. 3. The Montreal Cognitive Assessment is considered more sensitive to cognitive impairment than the Mini Mental Status Exam in patients with vascular cognitive impairment. Its use is recommended when vascular cognitive impairment is suspected. Additional validation is needed for the Montreal Cognitive Assessment as well as other potential screening instruments such as the 5-min protocol from the Vascular Cognitive Impairment Harmonization recommendations. 4. Post-stroke patients should also be screened for depression, since depression has been found to contribute to cognitive impairment in stroke patients. A validated screening tool for depression should be used. 5. Post-stroke patients who have cognitive impairment detected on a screening test should receive additional cognitive and/or neuropsychologic assessments as appropriate to further guide management.

[a]Covers all of the UK except Scotland, which has a separate guideline

Selected extracts reprinted with permission from:

- National Stroke Foundation. Clinical Guidelines for Stroke Management; 2010. Melbourne, Australia [12]
- Intercollegiate Stroke Working Party. National clinical guideline for stroke, 4th edition. London: Royal College of Physicians, 2012 [14]
- Scottish Intercollegiate Guidelines Network (SIGN). Management of patients with stroke: Rehabilitation, prevention and management of complications and discharge planning. A national clinical guideline. Edinburgh: SIGN, 2010 [15]
- Lindsay MP, Gubitz G, Bayley M, Hill MD, Davies-Schinkel C, Singh S, and Phillips S. Canadian Best Practice Recommendations for Stroke Care (Update 2010). Prepared by the Canadian Stroke Strategy Best Practices and Standards Writing Group, on behalf of the Canadian Stroke Strategy (a joint initiative of the Canadian Stroke Network and the Heart and Stroke Foundation of Canada). 2010; Ottawa, Ontario, Canada: Canadian Stroke Network [13]

brief screening tools against consideration of their limitations, when to refer for more detailed assessment and by whom. Examples of useful tools are given in some guidelines. The Canadian and the recent update of the UK (except Scotland) guidelines suggest the Montreal Cognitive Assessment as a simple, standardized screening tool. The latter suggests more detailed assessments within later sections covering specific cognitive impairments (Table 16.1).

Timing and Workforce Mobilization: Cognitive Screening and Assessment

Workforce competencies for cognitive screening and assessment require careful planning as does the timing of these activities, which should influence clinical decision-making and outcomes for people with stroke, without using valuable

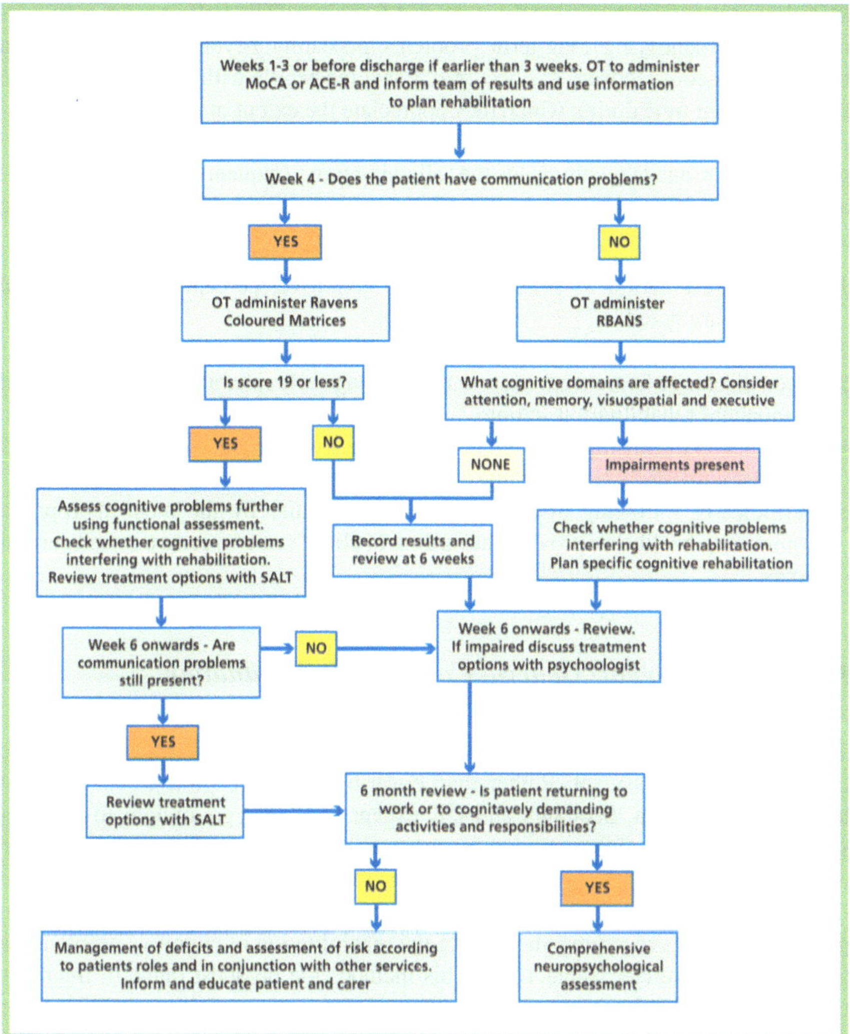

Fig. 16.1 Pathway for assessing cognitive problems. Reprinted with permission from Gillham S, Clark L. Psychological care after stroke—improving stroke services for people with cognitive and mood disorders. NHS Improvement—Stroke, 2011. http://www.improvement.nhs.uk/stroke/Psychologicalcareafterstroke/tabid/177/Default.aspx

resources to simply confirm the obvious (i.e., most acute stroke patients will have some cognitive impairment). Investigations should provide more information than a simple "cognitive impairment absent/present" tick box. Guidelines emphasize the roles of occupational therapists and psychologists. A recent document from the NHS Improvement Stroke program for England [31] suggests a pathway for assessing cognitive problems by way of the first step towards cognitive rehabilitation (Fig. 16.1). As shown, key time points in the UK model are: pre-transfer of care from hospital to community at 6 weeks and 6 months. The latter review is recommended for identifying long-term problems persisting beyond the period when much spontaneous recovery has occurred. For some people with stroke, this can also be a significant time during which they appreciate the extent of their residual cognitive difficulties and the need to adjust and accept compensatory rehabilitation strategies and aids. Canada recommends the following more frequent cognitive screening/assessment regime (and extends this to those who have had a transient ischemic attack) "at various transition points throughout the continuum of stroke care [13]":

1. During presentation to emergency when cognitive, perceptual, or functional concerns are noted.
2. Upon admission to acute care, particularly if any evidence of delirium is noted.
3. Upon discharge home from acute care or during early rehabilitation if transferred to inpatient rehabilitation setting.
4. Periodically during inpatient rehabilitation stage according to client progress and to assist with discharge planning.
5. Periodically following discharge to the community by the most appropriate community healthcare provider according to client's needs, progress, and current goals.

Beyond Assessment: General Cognitive Rehabilitation

The National guideline differ slightly in how they treat the management of cognitive problems after assessment. Rather than covering general cognitive rehabilitation most (e.g., Australia, Scotland, and UK except Scotland) go straight to domain-specific advice (e.g., interventions for memory and neglect). These often include recommendations of assessment tools specific to that impairment but the point here is that they also cover restorative and compensatory techniques. The Canadian guideline includes recommendations for the rehabilitation of cognitive problems as a single collective (see Table 16.2). This includes the broadest range of interventions including psychopharmacology (not reprinted here, see full report [13]) since this guideline covers "vascular cognitive impairment and dementia."

Domain-Specific Recommendations

The Australian, Scottish, and UK (except Scotland) guidelines take the approach of dividing cognition into specific impairments. Recommendations for attention, memory, neglect, and aphasia are covered by all. Apraxia and executive functions are included in the UK (except Scotland) and Australian guidelines. Agnosia is specifically covered by the Australian guideline whilst the most recent guideline (UK with the exception of Scotland) makes recommendations more broadly on perception. Space does not permit detailed coverage of all eight domains. The approach taken

Table 16.2 Canadian recommendations: interventions for general cognitive problems (extracts)

Patients who demonstrate cognitive impairments in the screening process should be referred to a healthcare professional with specific expertise in this area for additional cognitive, perceptual and/or functional assessments. • Additional assessments should be undertaken to determine the severity of impairment and impact of deficits on function and safety in activities of daily living and instrumental activities of daily living, and to implement appropriate remedial, compensatory and/or adaptive intervention strategies. • A team approach is recommended, and healthcare professionals may include an occupational therapist, neuropsychologist, psychiatrist, neurologist, geriatrician, speech–language pathologist or social worker.
An individualized, patient-centered approach should be considered to facilitate resumption of desired activities such as return to work, leisure, driving, volunteer participation, financial management, home management and other instrumental activities of daily living.
Intervention strategies including rehabilitation should be tailored according to the cognitive impairments and functional limitations as well as remaining cognitive abilities, as identified through in-depth assessment and developed in relation to patients' and caregivers' needs and goals.
Strategy training provides individuals who have limitations in activities of daily living with compensatory strategies to promote independence and should be offered to patients with cognitive challenges. The evidence for the effectiveness of specific interventions for cognitive impairment in stroke is limited and requires more research. • Attention training may have a positive effect on specific, targeted outcomes and should be implemented with appropriate patients.
Compensatory strategies can be used to improve memory outcomes.

Extracts reprinted with permission from Lindsay MP, Gubitz G, Bayley M, Hill MD, Davies-Schinkel C, Singh S, and Phillips S. Canadian Best Practice Recommendations for Stroke Care (Update 2010). Prepared by the Canadian Stroke Strategy Best Practices and Standards Writing Group, on behalf of the Canadian Stroke Strategy (a joint initiative of the Canadian Stroke Network and the Heart and Stroke Foundation of Canada). 2010; Ottawa, Ontario, Canada: Canadian Stroke Network [13]

has been to extract the relevant information into tables to enable comparisons between guidelines. The reader is referred to the original documents for specifics on the studies on which these recommendations were made.

Although this modularized approach to cognitive rehabilitation is an oversimplification intended to aid clarity, it is also a true reflection of the design of the majority of the rehabilitation studies, which focus on a single impairment (e.g., neglect). In clinical practice, rehabilitation acknowledges that each cognitive domain, such as perception, attention, and memory, cannot be considered in isolation, as most everyday activities draw on a range and interaction of cognitive abilities.

Attention/Concentration

Each of the four guidelines mentions the pivotal role played by attention and the impact of attentional impairments. The ability to select and concentrate on relevant information or events is fundamental to everyday life. When this ability is impaired, other cognitive skills will be affected. Attention can therefore be considered a "mediator" or starting point for many aspects of cognition. Attentional deficits have an acute negative impact on functional ability [32–34].

Trials of rehabilitation of attention involve a number of different approaches. Computerized rehabilitation has been used; this allows repetition of tasks that draw on attention [35–37]. Approaches also focus on practice and development of specific strategies for time pressure management (TPM) [38, 39]. TPM is an intervention directly aimed at behavioral and cognitive change in treatment situations that are designed to mirror real-life situations. The goal is to develop alternative cognitive strategies to compensate for mental slowness. Attention process training (APT) has also been used [40, 41]. APT is "a theoretically based, hierarchical, multilevel treatment, including sustained, selective, alternating, and divided attention" [40].

A Cochrane systematic review of attention [24] concluded that there was no evidence to refute or support the use of specific rehabilitation techniques for attentional impairments that improve functional independence after stroke. An update to this review is in progress. The latest update to Cicerone's review of cognitive rehabilitation for attention impairments [23] made practice standard recommendations for interventions for traumatic brain injury but this may well be applicable to stroke. The UK (except Scotland) guidelines, the most recently updated of all the guidelines, make recommendations based mainly on consensus opinion and a recent underpowered randomized controlled trial [39] of TPM (see Table 16.3). Although inconclusive, the latter trial suggests that TPM shows promise with younger, more physically independent stroke survivors and that it is feasible to train staff to deliver TPM in hospital or community stroke services.

Overall, there is a lack of high quality trials to inform selection of specific interventions and much of the evidence is at consensus level. Adequately powered randomized controlled trials of TPM and other interventions (e.g., APT) would greatly improve the evidence base for these commonly disabling impairments (Table 16.3).

Memory

Memory impairments (see Chap. 8) are related to a general reduction in functional ability for everyday tasks, even after factors such as age and stroke severity are taken into consideration [42]. Memory impairments also are upsetting for family members who cope with the consequences of forgetfulness; caregiver well-being correlates negatively with a patient's memory problems [43]. The following simple three-step model has been advocated as useful for explaining and offering interventions to rehabilitate the effects of memory impairments:

1. Encoding—organizing and processing information for later recall. Encoding may happen consciously or unconsciously.
2. Consolidation—the process by which a piece of information becomes stored in memory in a more permanent way.
3. Retrieval and recognition—recalling previously encoded and consolidated information in a meaningful way [44].

Table 16.3 Recommendations from National Clinical Guidelines: Attention (extracts)

Australia	Cognitive rehabilitation can be used in stroke survivors with attention and concentration deficits
Canada	The evidence for the effectiveness of specific interventions for cognitive impairment in stroke is limited and requires more research • Attention training may have a positive effect on specific, targeted outcomes and should be implemented with appropriate patients
Scotland	There is not yet sufficient evidence to support or refute the benefits of cognitive rehabilitation for patients with problems of attention
UK[a]	A. Any person after stroke who appears easily distracted or unable to concentrate should have their attentional abilities (e.g. focused, sustained and divided) formally assessed B. Any person with impaired attention should have cognitive demands reduced through: – having shorter treatment sessions – taking planned rests – reducing background distractions – avoiding work when tired. C. Any person with impaired attention should: – be offered an attentional intervention (e.g. Time Pressure Management, Attention Process Training, environmental manipulation), ideally in the context of a clinical trial – receive repeated practice of activities they are learning.

[a]Covers all of the UK except Scotland, which has a separate guideline
Selected extracts reprinted with permission from:

- National Stroke Foundation. Clinical Guidelines for Stroke Management; 2010. Melbourne, Australia [12]
- Intercollegiate Stroke Working Party. National clinical guideline for stroke, 4th edition. London: Royal College of Physicians, 2012 [14]
- Scottish Intercollegiate Guidelines Network (SIGN). Management of patients with stroke: Rehabilitation, prevention and management of complications and discharge planning. A national clinical guideline. Edinburgh: SIGN, 2010 [15]
- Lindsay MP, Gubitz G, Bayley M, Hill MD, Davies-Schinkel C, Singh S, and Phillips S. Canadian Best Practice Recommendations for Stroke Care (Update 2010). Prepared by the Canadian Stroke Strategy Best Practices and Standards Writing Group, on behalf of the Canadian Stroke Strategy (a joint initiative of the Canadian Stroke Network and the Heart and Stroke Foundation of Canada). 2010; Ottawa, Ontario, Canada: Canadian Stroke Network [13]

As suggested in Table 16.4, there are two main methods used in memory rehabilitation: (1) approaches to help encode, store, and retrieve new information (e.g., deep [semantic] encoding of material); and (2) teaching compensatory techniques to reduce disabilities (e.g., diaries, electronic organizers, and audio alarms). The Cochrane review for memory impairments post-stroke [26] concluded that there was "no evidence to support or refute the effectiveness of memory rehabilitation on functional outcomes, and objective, subjective, and observer-rated memory measures." The more recent guidelines' conclusions regarding the effectiveness of memory rehabilitation note there are serious limitations in the evidence base. The Australian and UK (except Scotland) recommendations are the most detailed and are very similar. There is widespread agreement between Cochrane reviewers and guideline writers that research is needed to establish both the clinical effectiveness

Table 16.4 Recommendations from National Clinical Guidelines: Memory (extracts)

Scotland	There is not yet sufficient evidence to support or refute the benefits of cognitive rehabilitation for patients with problems of attention or memory.
Canada	The evidence for the effectiveness of specific interventions for cognitive impairment in stroke is limited and requires more research. • compensatory strategies can be used to improve memory outcomes
Australia	Any patient found to have memory impairment causing difficulties in rehabilitation or adaptive functioning should: • be referred for a more comprehensive assessment of their memory abilities • have their nursing and therapy sessions tailored to use techniques which capitalise on preserved memory abilities • be assessed to see if compensatory techniques to reduce their disabilities, such as notebooks, diaries, audiotapes, electronic organisers and audio alarms, are useful • be taught approaches aimed at directly improving their memory • have therapy delivered in an environment as like the patient's usual environment as possible to encourage generalisation.
UK[a]	A. Patients who complain of memory impairment and those clinically considered to have difficulty in learning and remembering should have their memory assessed using a standardised measure such as the Rivermead Behavioural Memory Test (RBMT). B. Any patient found to have memory impairment causing difficulties in rehabilitation or undertaking activities should: • be assessed medically to check that there is not another treatable cause or contributing factor (e.g. hypothyroidism) • have their profile of impaired and preserved memory abilities determined (as well as the impact of any other cognitive deficits on memory performance for example, attentional impairment) • have nursing and therapy sessions altered to capitalise on preserved abilities • be taught approaches that help them to encode, store and retrieve new information for example, spaced retrieval (increasing time intervals between review of information) or deep encoding of material (emphasizing semantic features) • be taught compensatory techniques to reduce their prospective memory problems, such as using notebooks, diaries, electronic organisers, pager systems and audio alarms • have therapy delivered in an environment that is as similar to the usual environment for that patient as possible.

[a]Covers all of the UK except Scotland, which has a separate guideline

Selected extracts reprinted with permission from:

- National Stroke Foundation. Clinical Guidelines for Stroke Management; 2010. Melbourne, Australia [12]
- Intercollegiate Stroke Working Party. National clinical guideline for stroke, 4th edition. London: Royal College of Physicians, 2012 [14]
- Scottish Intercollegiate Guidelines Network (SIGN). Management of patients with stroke: Rehabilitation, prevention and management of complications and discharge planning. A national clinical guideline. Edinburgh: SIGN, 2010 [15]
- Lindsay MP, Gubitz G, Bayley M, Hill MD, Davies-Schinkel C, Singh S, and Phillips S. Canadian Best Practice Recommendations for Stroke Care (Update 2010). Prepared by the Canadian Stroke Strategy Best Practices and Standards Writing Group, on behalf of the Canadian Stroke Strategy (a joint initiative of the Canadian Stroke Network and the Heart and Stroke Foundation of Canada). 2010; Ottawa, Ontario, Canada: Canadian Stroke Network [13]

(particularly at an activity rather than impairment level of outcome measurement) and the patient acceptability of different memory rehabilitation approaches, recruiting larger, more representative, groups of stroke patients (Table 16.4).

Neglect

Unilateral spatial neglect was originally classified as a perceptual impairment, before being widely accepted as an attentional disorder. It tends to stand alone these days perhaps because neglect is the most frequently researched topic within cognitive rehabilitation for stroke. The disabling effects of neglect have been well documented [45] (see Chap. 4). Although severe neglect is rather easily recognized, diagnosing milder neglect can be less obvious and only become apparent when observing higher-level activities such as driving, preparing a meal, and interacting in real-world social situations [46]. These difficulties obviously impact patient function and safety on transfer of care from hospital to community.

There is a relative wealth of research evidence in this field. Twelve randomized controlled trials were included in the Cochrane review of the cognitive rehabilitation of neglect [25]. A recent update of this review (in press) has included a further 11 trials [47–57]. Providing visual scanning training remains a popular intervention in neglect trials, as is the use of prisms. The latter is sometimes prescribed as an aid to be routinely worn on glasses but recent pilot trials have succeeded in determining the feasibility (but not yet the effectiveness) of prism adaptation training, a short therapist-led intervention using prisms during a specific computerized training activity [54].

The original review [25] concluded that cognitive rehabilitation can improve performance on impairment level tests but there is insufficient evidence to support or refute its effectiveness at reducing disability, one of the main aims of rehabilitation. This gap in the evidence base is due to limitations in the quality of the research studies, especially around the reduction of bias and the choice of appropriate outcome measures. The updated review will provide a systematic determination of whether the evidence base has been strengthened recently but for now the National guideline recommendations remain mostly at the consensus level and stress the need to invite people with neglect to participate in clinical trials (Table 16.5).

Aphasia

Aphasia (see Chap. 6) rehabilitation is a topic that has generated considerable research interest for decades and yet controversies regarding the quality of the evidence base remain. Clinical uncertainty persists around the most clinically and cost-effective method of supporting people with aphasia. Several major trials [55–58] and an update to the existing Cochrane review [28] that are likely to impact on National guideline were recently published. The new trials primarily concern impairment-focused intervention delivered at varying rates of intensity in the acute phase of the stroke pathway. Overall, the recent evidence does not support this

Table 16.5 Recommendations from National Clinical Guidelines: Neglect (extracts)

Canada	No specific recommendation beyond assessment
Scotland	Patients with visuospatial neglect should be assessed and taught compensatory strategies.
Australia	a) Any patient with suspected or actual neglect or impairment of spatial awareness should have a full assessment using validated assessment tools. b) Patients with unilateral neglect can be trialled with one or more of the following interventions: • simple cues to draw attention to the affected side • visual scanning training in addition to sensory stimulation • prism adaptation • eye patching • mental imagery training or structured feedback.
UK[a]	A. Any patient with a stroke affecting the right cerebral hemisphere should be considered at risk of reduced awareness on the left side and should be tested formally if this is suspected clinically. B. Due to the fluctuating presentation of neglect a standardised test battery such as the Behavioural Inattention Test should be used in preference to a single subtest, and the effect on functional tasks such as dressing and mobility should be determined. C. Any patient shown to have impaired attention to one side should be: – given a clear explanation of the impairment – taught compensatory strategies to help reduce impact on functional activities such as reading – given cues to draw attention to the affected side during therapy and nursing procedures – monitored to ensure that they do not eat too little through missing food on one side of the plate – offered interventions aimed at reducing the functional impact of the neglect (eg visual scanning training, limb activation, sensory stimulation, eye patching, prism wearing, prism adaptation training), ideally within the context of a clinical trial.

[a]Covers all of the UK except Scotland, which has a separate guideline
Selected extracts reprinted with permission from:

- National Stroke Foundation. Clinical Guidelines for Stroke Management; 2010. Melbourne, Australia [12]
- Intercollegiate Stroke Working Party. National clinical guideline for stroke, 4th edition. London: Royal College of Physicians, 2012 [14]
- Scottish Intercollegiate Guidelines Network (SIGN). Management of patients with stroke: Rehabilitation, prevention and management of complications and discharge planning. A national clinical guideline. Edinburgh: SIGN, 2010 [15]
- Lindsay MP, Gubitz G, Bayley M, Hill MD, Davies-Schinkel C, Singh S, and Phillips S. Canadian Best Practice Recommendations for Stroke Care (Update 2010). Prepared by the Canadian Stroke Strategy Best Practices and Standards Writing Group, on behalf of the Canadian Stroke Strategy (a joint initiative of the Canadian Stroke Network and the Heart and Stroke Foundation of Canada). 2010; Ottawa, Ontario, Canada: Canadian Stroke Network [13]

approach at this time point. A qualitative study of patients' perspectives, nested within our own trial, suggested service reorganization to provide a more psychosocial approach to early aphasia rehabilitation, perhaps shifting the cognitive neuropsychological model approach to later [59–63]. In addition to rehabilitation directed at the language impairment, emerging evidence supports the effectiveness of structured behavioral interventions in reducing low mood in people with aphasia [64] right across the pathway.

This flurry of recent research interest in aphasia is welcome news for people with aphasia and their caregivers but makes it difficult to compare the latest recommendations from guidelines as several have yet to be updated (see Table 16.6). Interested readers are referred directly to the studies referenced previously and to the recent Cochrane review and UK (except Scotland) guideline. There remains a striking need for research into interventions for people with chronic aphasia and to supporting caregivers and other communication partners.

Table 16.6 Recommendations from National Clinical Guidelines: Aphasia (extracts)

UK[a]	A. All patients with communication problems following stroke should have an initial assessment by a speech and language therapist to diagnose the communication problem and to explain the nature and implications to the patient, family and multidisciplinary team. Routine reassessment of the impairment or diagnosis in the early stages of stroke (immediate and up to four months) should not be performed unless there is a specific purpose eg to assess mental capacity. B. In the early stages of stroke (immediate and up to four months) patients identified as having aphasia as the cause of the impairment should be given the opportunity to practise their language and communication skills as tolerated by the patient. C. Beyond the early stages of stroke (immediate and up to four months), patients with communications problems caused by aphasia should be reassessed to determine if they are more suitable for more intensive treatment with the aim of developing greater participation in social activities. This may include a range of approaches such as using an assistant or volunteer, family member or communication partner guided by the speech and language therapist, computer-based practice programmes and other functional methods. D. Patients with impaired communication should be considered for assistive technology and communication aids by an appropriately trained clinician. E. Patients with aphasia whose first language is not English should be offered assessment and communication practice in their preferred language. F. Education and training of health/social care staff, carers and relatives regarding the stroke patient's communication impairments should be provided by a speech and language therapist. Any education and training should enable communication partners to use appropriate communication strategies to optimise patient engagement and choice, and the delivery of other rehabilitation programmes. G. Any person with stroke at home who has continuing communication difficulty due to aphasia and whose social interactions are limited by it should be provided with information about any local or national groups for people with long-term aphasia, and referred to the group as appropriate.
Canada	Patients with aphasia should be taught supportive conversation techniques. Access to training for care providers in programs that facilitate communication with stroke survivors with aphasia.
Scotland	Aphasic stroke patients should be referred for speech and language therapy. Where the patient is sufficiently well and motivated, a minimum of two hours per week should be provided. Where appropriate, treatments for aphasia may require a minimum period of six months to be fully effective. Referral to the volunteer stroke service .... should be considered as an adjunct.

(continued)

Table 16.6 (continued)

Australia	a) All patients should be screened for communication deficits using a screening tool that is valid and reliable. b) Those patients with suspected communication difficulties should receive formal, comprehensive assessment by a specialist clinician. c) Where a patient is found to have aphasia, the clinician should: • document the provisional diagnosis • explain and discuss the nature of the impairment with the patient, family/carers and treating team, and discuss and teach strategies or techniques which may enhance communication • in collaboration with the patient and family/carer, identify goals for therapy and develop and initiate a tailored intervention plan. The goals and plans should be reassessed at appropriate intervals over time. d) All written information on health, aphasia, social and community supports (such as that available from the Australian Aphasia Association or local agencies) should be available in an aphasia-friendly format. e) Alternative means of communication (such as gesture, drawing, writing, use of augmentative and alternative communication devices) should be used as appropriate. f) Interventions should be individually tailored but can include: • treatment of aspects of language (including phonological and semantic deficits, sentence level processing, reading and writing) following models derived from cognitive neuropsychology • constraint-induced language therapy • the use of gesture • supported conversation techniques • delivery of therapy programs via computer. g) The routine use of piracetam is NOT recommended. h) Group therapy and conversation groups can be used for people with aphasia and should be available in the longer term for those with chronic and persisting aphasia. i) People with chronic and persisting aphasia should have their mood monitored. j) Environmental barriers facing people with aphasia should be addressed through training communication partners, raising awareness of and educating about aphasia in order to reduce negative attitudes, and promoting access and inclusion by providing aphasia-friendly formats or other environmental adaptations. People with aphasia from culturally and linguistically diverse backgrounds may need special attention, for example, from trained healthcare interpreters. k) The impact of aphasia on functional activities, participation and quality of life, including the impact upon relationships, vocation and leisure, should be assessed and addressed as appropriate from early post-onset and over time for those chronically affected.

[a]Covers all of the UK except Scotland, which has a separate guideline
Selected extracts reprinted with permission from:

- National Stroke Foundation. Clinical Guidelines for Stroke Management; 2010. Melbourne, Australia [12]
- Intercollegiate Stroke Working Party. National clinical guideline for stroke, 4th edition. London: Royal College of Physicians, 2012 [14]
- Scottish Intercollegiate Guidelines Network (SIGN). Management of patients with stroke: Rehabilitation, prevention and management of complications and discharge planning. A national clinical guideline. Edinburgh: SIGN, 2010 [15]
- Lindsay MP, Gubitz G, Bayley M, Hill MD, Davies-Schinkel C, Singh S, and Phillips S. Canadian Best Practice Recommendations for Stroke Care (Update 2010). Prepared by the Canadian Stroke Strategy Best Practices and Standards Writing Group, on behalf of the Canadian Stroke Strategy (a joint initiative of the Canadian Stroke Network and the Heart and Stroke Foundation of Canada). 2010; Ottawa, Ontario, Canada: Canadian Stroke Network [13]

Other Cognitive Domains: Apraxia, Perception, Agnosia, and Executive Functions

As mentioned previously, not all the guidelines address each of these topics so, where available, they are simply listed in a single table (see Table 16.7). Cochrane reviews exist for apraxia [27] and perception [29] and one on executive function has been submitted for publication [30]. The apraxia review is now out of date but relevant rehabilitation trials published since that review are included in the recent UK (except Scotland) guideline (see the guideline's evidence tables). Generally these topics lack a clear evidence base (in the case of apraxia of speech [65] there are no trials at all) and implications for future research are discussed in the reviews. The Australian guideline selects the management of agnosia as a research priority, although they are alone in this (Table 16.7).

Table 16.7 Recommendations from National Clinical Guidelines: other cognitive domains

Apraxia: Australia	a) People with suspected difficulties executing tasks but who have adequate limb movement should be screened for apraxia and, if indicated, complete a comprehensive assessment. b) For people with confirmed apraxia, tailored interventions (e.g. strategy training) can be used to improve ADL.
Apraxia: UK[a]	A. Any person who has difficulties in executing tasks despite apparently adequate limb movement should be assessed formally for the presence of apraxia. B. Any person found to have apraxia should: – have their profile of impaired and preserved action abilities determined using a standardised approach (e.g. Test of Upper Limb Apraxia TULIA) – have the impairment and the impact on function explained to them, their family, and their treating team. – be given therapies and/or taught compensatory strategies specific to the deficits identified ideally in the context of a trial
Executive functions: Australia	a) Patients considered to have problems associated with executive functioning deficits should be formally assessed using reliable and valid tools that include measures of behavioural symptoms. b) External cues, such as a pager, can be used to initiate everyday activities in stroke survivors with impaired executive functioning. c) In stroke survivors with impaired executive functioning, the way in which information is provided should be considered.
Executive functions: UK[a]	A. Any person who appears to have adequate skills to perform complex activities but who fails to organise the tasks needed should be formally assessed for the dysexecutive syndrome, for example using the Behavioural Assessment of the Dysexecutive Syndrome (BADS). B. Any person with an executive disorder and activity limitation should be taught compensatory techniques. This may include internal strategies (eg self-awareness and goal setting) and/or external strategies (eg use of electronic organizers or pagers, or use of written checklists) ideally in the context of a clinical trial. C. When a patient's activities are affected by an executive disorder, the nature and effects of the impairment and ways of supporting and helping the patient should be discussed with others involved (eg family and staff).

(continued)

Table 16.7 (continued)

Agnosia: Australia	The presence of agnosia should be assessed by appropriately trained personnel and communicated to the stroke team.
Perception: UK[a]	A. Any person who appears to have perceptual difficulties should have a formal perceptual assessment (eg using the Visual Object and Space Perception battery (VOSP)) B. Any person found to have agnosia should: – have the impairment explained to them, their carers and their treating team – be offered a perceptual intervention, ideally within the context of a clinical trial

[a]Covers all of the UK except Scotland, which has a separate guideline
Selected extracts reprinted with permission from:

- National Stroke Foundation. Clinical Guidelines for Stroke Management; 2010. Melbourne, Australia [12]
- Intercollegiate Stroke Working Party. National clinical guideline for stroke, 4th edition. London: Royal College of Physicians, 2012 [14]
- Scottish Intercollegiate Guidelines Network (SIGN). Management of patients with stroke: Rehabilitation, prevention and management of complications and discharge planning. A national clinical guideline. Edinburgh: SIGN, 2010 [15]
- Lindsay MP, Gubitz G, Bayley M, Hill MD, Davies-Schinkel C, Singh S, and Phillips S. Canadian Best Practice Recommendations for Stroke Care (Update 2010). Prepared by the Canadian Stroke Strategy Best Practices and Standards Writing Group, on behalf of the Canadian Stroke Strategy (a joint initiative of the Canadian Stroke Network and the Heart and Stroke Foundation of Canada). 2010; Ottawa, Ontario, Canada: Canadian Stroke Network [13]

Models of Comprehensive Neuropsychological Rehabilitation

It is clinically intuitive that for maximum efficacy a program of cognitive rehabilitation must be delivered as part of a comprehensive neuropsychological approach and within a clear pathway specifying different levels of involvement by differently skilled professionals. Comprehensive programs are sometimes referred to, especially within the US traumatic brain injury rehabilitation literature, as "holistic" [22] although in Europe the term holistic usually relates to alternative medicine.

The inclusion of recommendations on a comprehensive neuropsychological approach is very new in national stroke guidelines, appearing for the first time in 2012 [14]. It is based on a biopsychosocial model of illness for the organization and delivery of psychological care after stroke. As stated in the preamble to the forthcoming UK (except Scotland) guideline:

> The comprehensive model was developed because domain-specific cognitive rehabilitation interventions (e.g. memory rehabilitation) tend not to address the complexity of life after stroke. The same limitation applies to interventions that focus on a specific mood disorder and this may lead to ineffective treatment (e.g., cognitive problems misdiagnosed as depression).

Comprehensive-holistic rehabilitation programmes integrate evaluations of cognition, behaviour and mood to formulate the individual's difficulties. They then assist in the development of alternative or compensatory expectations and behaviours, leading towards independent self-management. They acknowledge that people with stroke may have limited awareness of their impairments or their impact (anosognosia), and that many therapies require motivation for engagement. [14]

The evidence base for comprehensive rehabilitation is mostly at the level of case series or cohort studies and largely focused on rehabilitation after acquired brain injury. There have also been two randomized controlled trials, the findings from which support the integration of cognitive, interpersonal, and functional skills [66, 67]. However, there is no unequivocal evidence that benefits are long-lasting (i.e., beyond the end of the treatment), which is a key requirement of an effective rehabilitation program. Interested readers are referred to two recent reviews of this topic [22, 23]. The UK (except Scotland) guideline is therefore largely at the level of consensus and based on extrapolation from promising research with younger, traumatically brain injured samples. The main recommendation concerns how multidisciplinary team (MDT) services are delivered, by whom and when, advocating a dynamic, rather than linear, stepped care approach, whereby patients move up and down the following steps of the model as required:

- *Step 1* comprises the routine assessments conducted within the MDT of all admitted patients, and the more detailed assessment of patients exhibiting symptoms of psychological disorder at any time after stroke.
- *Step 2* comprises the management of mild or moderate problems by MDT members who have been appropriately trained and where possible working under specialist supervision.
- *Step 3* comprises the management of more severe or persistent disorder, usually by a specialist.

The model in Fig. 16.2 illustrates the approach recommended by the NHS Stroke Improvement Program for England [31] and was developed from the stepped care model for adults with depression described by the National Institute for Health and Clinical Excellence (NICE) [68]. The latter defines stepped care as providing "a framework in which to organize the provision of services supporting patients, [caregivers] and healthcare professionals in identifying and accessing the most effective interventions." The NHS Improvement publication includes more details on operationalizing the stepped care model for people with stroke, including cognitive problems [31]. One of the core aspects of the model concerns skill mix and the employment of trained non-psychologists at certain steps of the model. This is a specific issue in the UK where difficulty accessing clinical psychologists has been a common and persisting finding from national audits [17, 69].

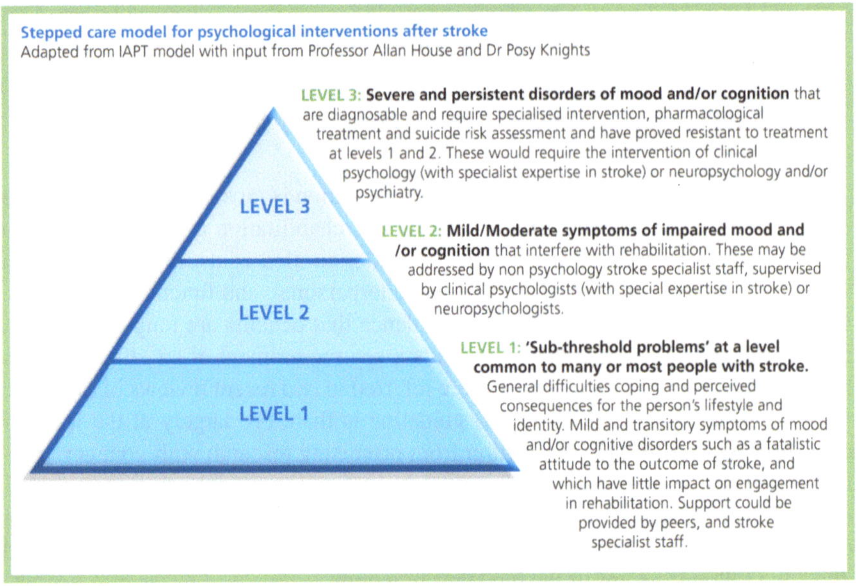

Fig. 16.2 Stepped care model for psychological interventions after stroke. Reprinted with permission from Gillham S, Clark L. Psychological care after stroke—improving stroke services for people with cognitive and mood disorders. NHS Improvement—Stroke, 2011. http://www.improvement.nhs.uk/stroke/Psychologicalcareafterstroke/tabid/177/Default.aspx

Summary

There is much to celebrate in the achievements of those working to develop an evidence-based approach for the rehabilitation of people with cognitive problems after stroke. Certain cognitive domains (e.g., neglect and aphasia) have attracted considerable research interest resulting in a range of interventions, many trials, and other levels of evidence. These feed into Cochrane systematic reviews and inform national clinical guidelines. These are exciting times with great potential for significant service improvement through emerging evidence for comprehensive neuropsychological rehabilitation approaches. In addition, practical recommendations for service delivery and organization are beginning to appear such as through recent modifications to the English stepped care model of psychological services.

On the other hand, even within heavily researched topics such as aphasia and neglect, there is still considerable uncertainty about which interventions to use, for which subgroup, when in the stroke pathway, and at what intensity. These are important questions. Furthermore, it is not clear why some topics (e.g., apraxia and memory) are, relatively speaking, under-researched and it certainly does not appear to be linked to either low prevalence or minimal impact on activity or social role

Implications for Research

Future studies should:

1. Provide a sufficiently detailed theoretical rationale for, and description of, the interventions including type and amount to allow implementation into clinical practice and research replication.
2. Provide a standard care control group, carefully documenting the content and amount of standard care, which can be highly variable.
3. Include detailed diagnostic information on individuals' perceptual problems given the heterogeneity in perceptual problems in terms of type, severity, and likely impact on everyday function.
4. Ensure low risk of study bias through rigorous methodological development and reporting, e.g., ensure allocation concealment, attempt to blind outcome assessors and report the success or failure, report all loss to follow-up, report results from all outcome measures, and control for other possible sources of bias.
5. Be of sufficient size to have adequate statistical power to answer clinically important questions about long-term functional outcomes.
6. Specify a primary endpoint and include analysis of other key outcomes such as adverse events, psychosocial benefits, and other outcomes deemed important by service users.
7. Adopt an intention-to-treat approach to measurement of outcomes in all individuals as well as to analysis of measured outcomes by treatment group.
8. Include a health economic assessment.

participation. Nor is it certain that simply producing "more of the same" research is the most productive way forward. As suggested in several of the Cochrane reviews of cognitive rehabilitation (see following for a recent example from the perception review [29]), future research could greatly improve clinical care through certain methodological and reporting changes:

Several countries now produce and audit against national clinical guidelines. In terms of cognitive rehabilitation there is reasonable consistency between the nations. Sometimes their differences are simply due to their publication date, with less evidence available to the older guidelines. The Scottish, Canadian, and Australian publications were in 2010, whereas the UK (excluding Scotland) guideline from the Royal College of Physicians London was updated for publication in 2012. Other differences result from the choice of either a wide or more focused breadth of topics and of course judgments about the standards set for accepting a piece of evidence, the criteria for which are described within each guideline.

Finally, the oft-repeated conclusion when examining the evidence is that we need more evidence! However there is also a need—and indeed it is already being met—for a paradigm shift in how we think about rehabilitation for people with cognitive problems. We need to reach a balance between domain-specific research (essential for helping us understand specific impairments and mechanisms for recovery) and research into broad-based comprehensive approaches (that treat the person's cognitive deficits within the broader perspective of impact on everyday life and well-being). We must also engage in implementation research, so that the emerging evidence is translated into clinical practice.

Acknowledgements We are grateful to the authors of the national clinical guidelines (Australia, Canada, Scotland, RCP London) and the NHS Improvement—Stroke program for England and to Wiley for allowing us to reprint sections of their work. Our ideas are the result of collaborative working with the RCP London's Intercollegiate Stroke Working Party, and especially the members of the cognitive psychology subgroup: Janet Cockburn, David Gillespie, Andrew Bateman, and Peter Knapp whose input we wish to acknowledge.

Potential Conflict of Interest: Audrey Bowen is a member of the RCP London Intercollegiate Working Party for Stroke that produces the National Clinical Guideline referred to as UK (except Scotland), and author of some of the studies referred to in this chapter.

References

1. British Psychological Society. Psychological services for stroke survivors and their families: briefing paper no 19. Leicester: British Psychological Society; 2002.
2. Claesson L, Lindén T, Skoog I, Blomstrand C. Cognitive impairment after stroke—impact on activities of daily living and costs of care for elderly people. The Goteborg 70+ Stroke Study. Cerebrovasc Dis. 2005;19(2):102–9.
3. Patel MD, Coshall C, Rudd AG, Wolfe CD. Cognitive impairment after stroke: clinical determinants and its associations with long-term stroke outcomes. J Am Geriatr Soc. 2002;50(4): 700–6.
4. Tatemichi TK, Desmond DW, Stern Y, Paik M, Sano M, Bagiella E. Cognitive impairment after stroke: frequency, patterns, and relationship to functional abilities. J Neurol Neurosurg Psychiatry. 1994;57(2):202–7.
5. Barker-Collo SL. Depression and anxiety 3 months post stroke: prevalence and correlates. Arch Clin Neuropsychol. 2007;22(4):519–31.
6. Department of Health. National stroke strategy. London: Department of Health; 2007.
7. McKevitt C. The Stroke Association: UK stroke survivors needs survey; 2010. ISBN 0901548650.
8. Pollock A, Brady M, Langhorne P. DORIS: Database of Research in Stroke—current best evidence, ongoing research and priorities for future research relating to stroke rehabilitation. Int J Stroke. 2011;6 Suppl 2:50–1.
9. Medical Research Council. A Framework for development and evaluation of RCTs for complex interventions to improve health. London: MRC; 2000. http://www.mrc.ac.uk/Utilities/Documentrecord/index.htm?d=MRC003372. Accessed 6 Feb 2012.
10. Medical Research Council. Developing and evaluating complex interventions: new guidance. London: MRC; 2008. http://www.mrc.ac.uk/Utilities/Documentrecord/index.htm?d=MRC004871. Accessed 8 Feb 2012.
11. Cochrane Library. http://www.thecochranelibrary.com/view/0/index.html. Accessed 6 Feb 2012.
12. National Stroke Foundation. Clinical guidelines for stroke management, Melbourne, Australia; 2010. http://www.strokefoundation.com.au/. Accessed 8 Feb 2012.

13. Lindsay MP, Gubitz G, Bayley M, Hill MD, Davies-Schinkel C, Singh S, Phillips S. Canadian Best Practice Recommendations for Stroke Care (Update 2010). Prepared by the Canadian Stroke Strategy Best Practices and Standards Writing Group, on behalf of the Canadian Stroke Strategy (a joint initiative of the Canadian Stroke Network and the Heart and Stroke Foundation of Canada). Ottawa, ON: Canadian Stroke Network; 2010. http://www.canadianstrokenetwork.ca/. Accessed 8 Feb 2012.
14. Intercollegiate Stroke Working Party. National clinical guideline for stroke. 4th ed. London: Royal College of Physicians; 2012.
15. Scottish Intercollegiate Guidelines Network (SIGN). 118. Management of patients with stroke: rehabilitation, prevention and management of complications and discharge planning. A national clinical guideline. Edinburgh: SIGN; 2010. http://www.sign.ac.uk/guidelines/published/numlist.html. Accessed 8 Feb 2012.
16. Cooksey, Sir David. A review of UK health research funding. London: HMSO; 2006. http://www.hm-treasury.gov.uk/pre_budget_report/prebud_pbr06/other_docs/prebud_pbr06_odcooksey.cfm. Accessed 31 Jan 2012
17. Intercollegiate Stroke Working Party. National sentinel stroke audit 2010—public report. London: Royal College of Physicians; 2011.
18. National Stroke Foundation. National Stroke Audit—acute services organisational survey report. Melbourne, Australia: National Stroke Foundation; 2011.
19. Canadian Stroke Network. The quality of stroke care in Canada 2011. http://www.canadianstroke network.ca/. Accessed 6 Feb 2012
20. Lincoln NB, Kneebone II, Macniven JAB, Morris RC. Psychological management of stroke. West Sussex: Wiley-Blackwell; 2012.
21. Johnstone B, Stonnington HH, editors. Rehabilitation of neuropsychological disorders: a practical guide for rehabilitation professionals. Philadelphia, PA: Psychology Press; 2001.
22. Cattelani R, Zettin M, Zoccolotti P. Rehabilitation treatments for adults with behavioral and psychosocial disorders following acquired brain injury: a systematic review. Neuropsychol Rev. 2010;20:52–85.
23. Cicerone KD, Langenbahn DM, Braden C, Malec JF, Kalmar K, Fraas M, et al. Evidence-based cognitive rehabilitation: updated review of the literature from 2003 through 2008. Arch Phys Med Rehabil. 2011;92:519–30.
24. Lincoln N, Majid M, Weyman N. Cognitive rehabilitation for attention deficits following stroke. Cochrane Database Syst Rev. 2000;(4):CD002842. doi:10.1002/14651858.CD002842.
25. Bowen A, Lincoln N. Cognitive rehabilitation for spatial neglect following stroke. Cochrane Database Syst Rev. 2007;(2):CD003586. doi:10.1002/14651858.CD003586.pub2.
26. das Nair R, Lincoln N. Cognitive rehabilitation for memory deficits following stroke. Cochrane Database Syst Rev. 2007;(3):CD002293. doi:10.1002/14651858.CD002293.pub2.
27. West C, Bowen A, Hesketh A, Vail A. Interventions for motor apraxia following stroke. Cochrane Database Syst Rev. 2008;(1):CD004132. doi:10.1002/14651858.CD004132.pub2.
28. Brady MC, Kelly H, Godwin J, Enderby P. Speech and language therapy for aphasia following stroke. Cochrane Database Syst Rev. 2012;(5):CD000425. doi:10.1002/14651858.CD000425.pub3.
29. Bowen A, Knapp P, Gillespie D, Nicolson DJ, Vail A. Non-pharmacological interventions for perceptual disorders following stroke and other adult-acquired, non-progressive brain injury. Cochrane Database Syst Rev. 2011;(4):CD007039. doi:10.1002/14651858.CD007039.pub2.
30. Chung CSY, Pollock A, Campbell T, Durward BR, Hagen S. Cognitive rehabilitation for executive dysfunction in patients with stroke or other adult non-progressive acquired brain damage (Protocol). Cochrane Database Syst Rev. 2010;(3):CD008391. doi:10.1002/14651858.CD008391 (The Review has now been submitted for publication).
31. Gillham S, Clark L. Psychological care after stroke—improving stroke services for people with cognitive and mood disorders. NHS improvement—stroke; 2011. http://www.improvement.nhs.uk/stroke/Psychologicalcareafterstroke/tabid/177/Default.aspx. Accessed 6 Feb 2012.
32. Hyndman D, Pickering RM, Ashburn A. The influence of attention deficits on functional recovery post stroke during the first 12 months after discharge from hospital. J Neurol Neurosurg Psychiatry. 2008;79(6):656–63.

33. Levitt T, Johnstone B. The assessment and rehabilitation of attention disorders. In: Johnstone B, Stonnington HH, editors. Rehabilitation of neuropsychological disorders: a practical guide for rehabilitation professionals. Philadelphia, PA: Psychology Press; 2001. p. 27–52.
34. Robertson IH, Ridgeway V, Greenfield E, Parr A. Motor recovery after stroke depends on intact sustained attention: a 2-year follow-up study. Neuropsychology. 1997;11(2):290–5.
35. Gray J, Robertson I, Pentland B, Anderson S. Microcomputer-based attentional retraining after brain damage: a randomised group controlled trial. Neuropsychol Rehabil. 1992;2(2):97–115.
36. Schoettke H. Rehabilitation von Aufmerksamkeits storungen nach einem Schlagenfall—Effectivitat eines verhaltensmedizinisch–neuropsychologischen Aufmerksamkeitstrainings. Verhaltenstherapie. 1997;7:3.
37. Sturm W, Willmes K. Efficacy of a reaction training on various attentional and cognitive functions in stroke patients. Neuropsychol Rehabil. 1991;1:259–80.
38. Fasotti L, Kovacs F, Eling PATM, Brouwer WH. Time pressure management as a compensatory strategy training after closed head injury. Neuropsychol Rehabil. 2000;10(1):47–65.
39. Winkens I, Van Heugten CM, Wade DT, Habets EJ, Fasotti L. Efficacy of time pressure management in stroke patients with slowed information processing: a randomized controlled trial. Arch Phys Med Rehabil. 2009;90(10):1672–9.
40. Barker-Collo SL, Feigin VL, Lawes CM, Parag V, Senior H, Rodgers A. Reducing attention deficits after stroke using attention process training: a randomized controlled trial. Stroke. 2009;40(10):3293–8.
41. Sohlberg MM, McLaughlin KA, Pavese A, Heidrich A, Posner MI. Evaluation of attention process training and brain injury education in persons with acquired brain injury. J Clin Exp Neuropsychol. 2000;22(5):656–76.
42. Wade DT, Parker V, Langton HR. Memory disturbance after stroke: frequency and associated losses. Int Rehabil Med. 1986;8(2):60–4.
43. Clark PC, Dunbar SB, Shields CG, Viswanathan B, Aycock DM, Wolf SL. Influence of stroke survivor characteristics and family conflict surrounding recovery on caregivers' mental and physical health. Nurs Res. 2004;53(6):406–13.
44. Skeel R, Edwards S. The assessment and rehabilitation of memory impairments. In: Johnstone B, Stonnington HH, editors. Rehabilitation of neuropsychological disorders: a practical guide for rehabilitation professionals. Philadelphia, PA: Psychology Press; 2001. p. 27–52.
45. Appelros P, Karlsson GM, Seiger A, Nydevik I. Neglect and anosognosia after first-ever stroke: incidence and relationship to disability. J Rehabil Med. 2002;34(5):215–20.
46. Teasell R, Salter K, Bitensky J, Bhogal SK, Foley N, Anita Menon A, et al. EBRSR: evidence-based review of stroke rehabilitation: perceptual disorders. http://www.ebrsr.com/uploads/Module-13_Perceptual-Disorders.pdf. Accessed 15 Oct 2011.
47. Ferreira HP, Leite Lopes MA, Luiz RR, Cardoso L, André C. Is visual scanning better than mental practice in hemispatial neglect? Results from a pilot study. Top Stroke Rehabil. 2011;18(2):155–61.
48. Fong KN, Chan MK, Ng PP, Tsang MH, Chow KK, Lau CW, et al. The effect of voluntary trunk rotation and half-field eye-patching for patients with unilateral neglect in stroke: a randomized controlled trial. Clin Rehabil. 2007;21(8):729–41.
49. Luukkainen-Markkula R, Tarkka IM, Pitkänen K, Sivenius J, Hämäläinen H. Rehabilitation of hemispatial neglect: a randomized study using either arm activation or visual scanning training. Restor Neurol Neurosci. 2009;27(6):663–72.
50. Nys GM, de Haan EH, Kunneman A, de Kort PL, Dijkerman HC. Acute neglect rehabilitation using repetitive prism adaptation: a randomized placebo-controlled trial. Restor Neurol Neurosci. 2008;26(1):1–12.
51. Polanowska K, Seniów J, Paprot E, Leśniak M, Członkowska A. Left-hand somatosensory stimulation combined with visual scanning training in rehabilitation for post-stroke hemineglect: a randomised, double-blind study. Neuropsychol Rehabil. 2009;19(3):364–82.
52. Schroder A, Wist ER, Homberg V. TENS and optokinetic stimulation in neglect therapy after cerebrovascular accident: a randomized controlled study. Eur J Neurol. 2008;15(9):922–7.

53. Tsang MH, Sze KH, Fong KN. Occupational therapy treatment with right half-field eye-patching for patients with subacute stroke and unilateral neglect: a randomised controlled trial. Disabil Rehabil. 2009;31(8):630–7.
54. Turton AJ, O'Leary K, Gabb J, Woodward R, Gilchrist ID. A single blinded randomised controlled pilot trial of prism adaptation for improving self-care in stroke patients with neglect. Neuropsychol Rehabil. 2010;20(2):180–96.
55. Kerkhoff G, Keller I, Artinger F, Hildebrandt H, Marquardt C, Reinhart S, Ziegler W. Recovery from auditory and visual neglect after optokinetic stimulation with pursuit eye movements—transient modulation and enduring treatment effects. Neuropsychologia. 2012;50(6):1164–77.
56. Mizuno K, Tsuji T, Takebayashi T, Fujiwara T, Hase K, Liu M. Prism adaptation therapy enhances rehabilitation of stroke patients with unilateral spatial neglect: a randomized, controlled trial. Neurorehabil Neural Repair. 2011;25(8):711–20.
57. Welfringer A, Leifert-Fiebach G, Babinsky R, Brandt T. Visuomotor imagery as a new tool in the rehabilitation of neglect: a randomised controlled study of feasibility and efficacy. Disabil Rehabil. 2011;33:2033–43.
58. Bowen A, Hesketh A, Patchick E, Young A, Davies L, Vail A, et al. Effectiveness of enhanced communication therapy in the first four months after stroke for aphasia and dysarthria: a randomised controlled trial. BMJ. 2012;345:e4407.
59. Bowen A, Hesketh A, Patchick E, Young A, Davies L, Vail A, et al.; The ACT NoW investigators. Clinical effectiveness, cost effectiveness and service users' perceptions of early, well-resourced communication therapy following a stroke, a randomised controlled trial (The ACT NoW Study). Health Technol Assess. 2012;16(26):1–160.
60. Laska AC, Kahan T, Hellblom A, Murray V, von Arbin M. A randomized controlled trial on very early speech and language therapy in acute stroke patients with aphasia. Cerebrovasc Dis. 2011;1:66–74.
61. Godecke E, Hird K, Lalor EE, Rai T, Phillips MR. Very early poststroke aphasia therapy: a pilotrandomizedcontrolledefficacytrial.IntJStroke.2011.doi:10.1111/j.1747-4949.2011.00631.
62. de Jong-Hagelstein M, van de Sandt-Koenderman WM, Prins ND, Dippel DW, Koudstaal PJ, Visch-Brink EG. Efficacy of early cognitive linguistic treatment and communicative treatment in aphasia after stroke: a randomised controlled trial (RATS-2). J Neurol Neurosurg Psychiatry. 2011;82:399–404.
63. Young A, Gomersall T, Bowen A. Trial participants' experiences of early, enhanced speech and language therapy after stroke compared with employed visitor support: a qualitative study nested within a RCT. Clin Rehabil. (Online First 26/7/12).
64. Thomas SA, Lincoln NB, Walker MF, Macniven J, Haworth H. Communication and Low Mood (CALM) study: a randomised controlled trial evaluating behaviour therapy for low mood in people with aphasia after stroke. Int J Stroke. 2011;6 Suppl 2:27–8.
65. West C, Hesketh A, Vail A, Bowen A. Interventions for apraxia of speech following stroke. Cochrane Database Syst Rev. 2005;(4):CD004298. doi:10.1002/14651858.CD004298.pub2.
66. Salazar AM, Warden DL, Schwab K, Spector J, Braverman S, Walter J, et al. Cognitive rehabilitation for traumatic brain injury: a randomized trial. Defense and Veterans Head Injury Program (DVHIP) Study Group. JAMA. 2000;283(23):3075–81.
67. Cicerone KD, Mott T, Azulay J, Sharlow-Galella MA, Ellmo WJ, Paradise S, et al. A randomized controlled trial of holistic neuropsychologic rehabilitation after traumatic brain injury. Arch Phys Med Rehabil. 2008;89(12):2239–49.
68. The National Institute for Health and Clinical Excellence. Depression the NICE guideline on the treatment and management of depression in adults (updated edition) CG90. The British Psychological Society & The Royal College of Psychiatrists; 2010. http://guidance.nice.org.uk/CG90/Guidance. Accessed 8 Feb 2012.
69. National Audit Office. Reducing brain damage: faster access to better stroke care. Norwich: Stationery Office; 2005. http://www.nao.org.uk/publications/0506/reducing_brain_damage.aspx. Accessed 9 Feb 2012.

Index

A
Acetaminophen, 17
Activities of daily living (ADL), 186–187, 229
Acute intracerebral hemorrhage, 202–203
Acute Stroke Ischemic NXY-059 Treatment (SAINT), 304
Agnosia, 322, 332
Alzheimer's disease (AD), 169, 170
American Heart Association (AHA) guidelines, 16, 17
American Heart Association Stroke Outcome Classification (AHA.SOC), 290
Aneurysmal subarachnoid hemorrhage (aSAH)
  clipping *vs.* coiling, 188
    encephalomalacia, 189
    pSAH survivors, 190
    ruptured aneurysms, 189
  cognitive outcome
    attention, 179–180
    executive function, 182–185
    language, 185
    memory, 181–182
    visuospatial function, 185–186
  CT angiography, 178
  functional outcome
    daily living activities, 186–187
    instrumental activities of daily living, 187
    return to work, 187–188
  incidence, 178
  neuroimaging, 190–191
  prognostic factors, 284
  risk factors, 177
  thirty-day case fatality rate, 178, 179
Anomic aphasia, 101
Anterior cerebral artery infarcts, 141
Anterior communicating artery (ACoA), 182, 183
Aphasia, 122, 123, 327–330. *See also* Post-stroke aphasia (PSA)
Apoplexy, 1
Apraxia
  classification
    errors in performance, 54
    representational/nonrepresentational gesture, 53–54
    single-gesture performance, 52
    task and input modality, 52
    tool/object use, 52
  clinical assessment implications, 61–62
  cognitive rehabilitation and recovery, 322, 331
  functional performance
    after intervention, 64
    conceptual system, 64
    functional dependency, 62–63
    gesture production, 65
    observational perspective, 62
    severity, 63
  models
    conceptual, 60
    production systems, 60–61
    routes to action, 61
    sensory-perceptual, 60
  neural correlation
    callosal lesions, 57
    hand gestures, 59
    hemispheric asymmetries, 55–56
    left hemisphere, 57–59
    neuroimaging, 58, 59
    stroke, 57
    voxel-based lesion, 56

T.A. Schweizer and R.L. Macdonald (eds.), *The Behavioral Consequences of Stroke*, DOI 10.1007/978-1-4614-7672-6, © Springer Science+Business Media New York 2014

Arcuate fasciculus, 98, 100, 101, 104, 106
Arterial spin labeling (ASL)
  Alzheimer disease, 274
  arterial transit time, 274
  BOLD signal, 273
  CBF-weighted, 272, 273
  RF pulses, 272

B

Behavioral disorder. *See* Dysexecutive syndrome
Beta-amyloid, 209, 210
Bloodletting, 2
Blood oxygenation level-dependent fMRI
  phases, 267
  resting-state, 271–272
  task-based, 269–271
Boston naming test (BNT), 154, 156
Brain microbleeds
  CAA, 208
  hemosiderin, 207
  magnetic resonance imaging, 264
  MRI T2*-weighted gradient-recalled echo, 207
British Aneurysm Nimodipine Trial, 302
Broca's aphasia, 99–100, 124

C

CAA. *See* Cerebral amyloid angiopathy (CAA)
CADASIL. *See* Cerebral autosomal dominant arteriopathy with subcortical ischemic leukoencephalopathy (CADASIL)
Callosal apraxia, 53
Cardio-embolic stroke, diagnosis, 6–7
Cardiogenic brain embolism, sources of, 9
Carotid endarterectomy, 10
Cerebellar stroke, 26
Cerebral amyloid angiopathy (CAA)
  abeta, 209
  Alzheimer's disease, 209
  amyloid, 208
  Boston criteria, 210
  brain autopsy, 211
  factors, 212
  lobar microbleeds, 210
Cerebral autosomal dominant arteriopathy with subcortical ischemic leukoencephalopathy (CADASIL), 212–213
Cerebral small vessel disease, 7
Cerebral venous thrombosis, 218
Chedoke-McMaster stroke assessment scale, 290
CNS vasculitis, 213–214
Cognitive disorder. *See* Dysexecutive syndrome
Cognitive rehabilitation and recovery
  aims, 317–318
  Cochrane reviews, 317, 325, 327, 331, 335
  comprehensive neuropsychological rehabilitation models
    evidence base, 333
    inclusions, 332
    multidisciplinary team services, 333
    stepped care model, 333, 334
    UK guideline, 332, 333
  definition of, 318
  domain-specific recommendations
    agnosia, 322, 332
    aphasia, 327–330
    apraxia, 322, 331
    attention, 323–325
    Canadian recommendations, 323
    concentration, 323–325
    executive functions, 322, 331
    memory impairments, 324–327
    neglect, 327, 328
    perception, 332
  impacts, 315
  psychological problems, 316
  quality of, 316–317
  randomized controlled trials, 316, 324, 327, 333
  screening and assessment
    general, 322
    national clinical stroke guidelines, 317–320
    Scottish guidelines, 318
    timing and workforce mobilization, 320–322
Conduction aphasia, 100
Constraint induced movement therapy, 37
Contrast-enhanced magnetic resonance angiography (CE-MRA), 263
Covert brain infarcts, 6

D

Decompressive craniectomy, 24–26, 304–305
Desmoteplase, 19
Diffusion tensor imaging (DTI)
  EPI techniques, 266
  FA value, 266
  tractography, 267, 268

Diffusion-weighted imaging (DWI), 261
Disconnection syndrome, 84–85
Double-step saccade task, 80
Drug therapy, PSA, 110–112
DTI. *See* Diffusion tensor imaging (DTI)
Dural arteriovenous fistula, 219
Dysexecutive syndrome
  clinical implications
    attention, 144
    cognitive assessment, 144, 145
    gaze abnormalities, 144
    monitoring and regulation, 145
    motor sequence, 144
    neuropsychological examination, 144
    NIH scale score, 143
  functions and disorders
    abulia-apathy, 141, 142
    action speed and slowing, 137, 138
    akinetic mutism, 141, 143
    anterior cerebral artery infarcts, 141
    apathy, 138
    aspects of, 135, 136
    behavioral and cognitive disorders, 136, 137
    cerebral infarction, 140, 141
    cerebral venous thrombosis, 138, 143
    confusional states, 141
    deficit profile, 138
    definition, 135
    determinants of, 140
    facilitatory paratonia, 144
    frontal lobe damage, 136, 137
    GREFEX study, 136, 137
    heteroactivation, 141
    hypoactivity, 138, 141–143
    impairment, 139
    infarcts, 140, 141
    intracerebral hemorrhage, 143
    lacunar infarcts, 142
    oppositional paratonia, 144
    psychomotor speed, 143
    reaction time test, 137
    subarachnoid hemorrhage, 142–143
    thalamic dementia, 142
    thalamic infarcts, 141–142
    verbal fluency test, 138

E

Echo planar imaging (EPI), 260
Electroconvulsive therapy (ECT), 234–235
Electrodermal activity (EDA), 270, 271
Elevator counting task, 76
Embolic stroke, mechanisms of, 8
Emotional communication disorders
  classification and domains, 119
  facial emotion
    anterior *vs.* posterior lesions, 128
    expressions, 128–129
    hemispheric strokes, 125–127
    iconic hypothesis, 127
    LHD patients, 127, 128
    mirror-somatosensory hypothesis, 128
    neutral facial discrimination, 126
    nonverbal emotions, 127
    RHD patients, 126–128
    ventral temporal-occipital lesion, 128
    visuospatial facial perceptual disorder, 126, 127
  speech (*see also* Speech emotion disorders)
    expression, 123–125
    prosody, 120–122
    verbal stimuli, 122–123
Episodic memory
  familiarity, 152
  hippocampus
    anterograde and retrograde amnesia, 160, 161
    familiarity-based memory, 158
    hypoxic ischemic encephalopathy, 159
    left *vs.* right, 160
    medial temporal lobe, 160
    multiple trace theory, 158
    recollection, 158, 159
  and posterior association areas, 167–169
  prefrontal cortex
    aneurysm ruptures, 166
    confabulation, 165
    impairment, 164
    left *vs.* right, 167
    top-down deficit, 165
  recollection, 152
  and thalamus (*see* Thalamus)
EuroQol, 289, 303
Explicit memory, 152
EXPRESS study, 9

F

Facial action scoring system (FACS), 128
Facial emotion disorders. *See* Emotional communication disorders
Fisher grade, 180
Fluid-attenuated inversion recovery (FLAIR), 261
Fluoxetine, 230
Frenchay activities index, 290
Frontal lobe, 182–184

G
Gadolinium diethyl-triamine-penta-acetic acid (Gd-DTPA), 263
Glasgow Outcome Scale (GOS), 184, 185
Global aphasia, 100–101
GREFEX study, dysexecutive syndrome, 136–138

H
Health-related quality of life (HRQoL)
  primary caregivers, 248–250
  SIP and NHP, 245
  standard-gamble method, 243
  time-tradeoff method, 243
  utility index, 244
Hematoma, 29, 30
Hemispheric strokes, 123, 125–127
Hemorrhagic stroke
  acute medical management, 28
  core area, 15
  ICH, 24, 26, 27
  IVH, 28
  SAH, 24, 27
  surgical treatment, 29–30
Hippocampus
  anterograde and retrograde amnesia, 160, 161
  familiarity-based memory, 158
  hypoxic ischemic encephalopathy, 159
  left *vs.* right, 160
  medial temporal lobe, 160
  multiple trace theory, 158
  recollection, 158, 159
HRQoL. *See* Health-related quality of life (HRQoL)
Hunt and Hess grade, 180, 186
Hyperglycemia, 17, 285
Hypertension, 16, 17
Hypoglycemia, 17
Hypotension, 17
Hypothermia, 17
Hypoxic ischemic encephalopathy, 159

I
ICH. *See* Intracerebral hemorrhage (ICH)
Iconic hypothesis, 127
Ideational apraxia, 52
Ideomotor apraxia (IMA), 52
Imitation, 52
Implicit memory
  basal ganglia and cerebellum, 156
  neuropsychological measures, 154
  priming, 152
  procedural, 152
Infarcts, 140–143
Instrumental activities of daily living (IADL), 187
International Stroke Trial, 302
International Subarachnoid Aneurysm Trial (ISAT), 303
International Surgical Trial in Intracerebral Hemorrhage (STICH), 304
Intracerebral hemorrhage (ICH). *See also* Hemorrhagic stroke
  bleeding into parenchyma, 24–27
  cerebral amyloid angiopathy
    abeta, 209
    Alzheimer's disease, 209
    amyloid, 208
    Boston criteria, 210
    brain autopsy, 211
    factors, 212
    lobar microbleeds, 210
  cognitive consequences
    aphasia, 205
    caudate nucleus and putamen, 205
    cerebral amyloid angiopathy, 203
    dementia, 203
    leukoaraiosis, 203, 204
    MoCA, 204
    neuropsychological testing, 204
    striatum, 205
    WCST, 205
  cognitive symptoms/signs
    basal ganglia, 202
    delirium, 203
    thalamus, 202
  diagnosis and medical management, 206–207
  dysexecutive syndrome, 143
  etiology and epidemiology, 26
    cerebral amyloid angiopathy, 200, 201
    incidence, 200
    locations, 202
    primary and secondary causes, 201
  ischemic stroke
    CADASIL, 212–213
    cerebral venous thrombosis, 218
    dural arteriovenous fistula, 219
    hypercoagulability and hyperviscosity states, 218
    intravascular large B cell lymphoma, 214
    moyamoya syndrome, 215–217
    sickle cell disease, 217–218

systemic lupus erythematosis, 215
vasculitis, 213–214
microbleeds
CAA, 208
hemosiderin, 207
MRI T2*-weighted gradient-recalled echo, 207
surgical treatment, 29–30
Intravascular large B cell lymphoma, 214
Intraventricular hemorrhage (IVH). *See* Hemorrhagic stroke
Ischemic stroke
acute medical management
AHA guidelines, 16, 17
fever, 17
hypoglycemia and hyperglycemia, 17
hypotension and hypertension, 16, 17
imaging techniques, 17
arterial occlusion, 16
causes, 15
complications, 11
core area, 15
decompressive craniectomy, 24–26
etiology, 5
interventions, 10
intra-arterial and mechanical thrombolysis, 21
advantages, 20
basilar occlusion, 24
combination therapy, 22
embolectomy, 22, 23
IMS II trial, 22
MELT trial, 20
MERCI trial, 22
pro-urokinase and urokinase, 20
retrievable stent strategies, 22, 23
rtPA, 22
stent-assisted recanalization, 24
use of, 20
intracerebral hemorrhage
CADASIL, 212–213
cerebral venous thrombosis, 218
dural arteriovenous fistula, 219
hypercoagulability and hyperviscosity states, 218
intravascular large B cell lymphoma, 214
moyamoya syndrome, 215–217
sickle cell disease, 217–218
systemic lupus erythematosis, 215
vasculitis, 213–214
intravenous thrombolysis, 18–19
treatment, 15

L

Lacunar infarcts, 142
Large vessel disease, 5
Left-hemisphere damage (LHD). *See* Emotional communication disorders
Leukoaraiosis, 203, 204
Limb apraxia. *See* Apraxia
Limb-kinetic apraxia, 53

M

Magnetic resonance imaging (MRI)
acute stroke
CE-MRA, 263
cerebrospinal fluid, 261
components, 261, 262
3DTOF, 263
Gd-DTPA, 263
recombinant tissue plasminogen activator, 262
T2*-weighted images, 264
aneurysmal subarachnoid hemorrhage, 182, 183, 190
arterial spin labeling
Alzheimer disease, 274
arterial transit time, 274
BOLD signal, 273
CBF-weighted, 272, 273
RF pulses, 272
BOLD fMRI, 267, 269
development, 260
diffusion tensor imaging
EPI techniques, 266
FA value, 266
tractography, 267, 268
intervention and rehabilitation, 277
multi-parametric MRI protocols, 276
nuclear magnetic resonance (*see* Nuclear magnetic resonance (NMR))
physical principles, 258–259
resting-state fMRI, 271–272
in stroke recovery, 264
stroke symptoms, 264–266
task-based fMRI, 269–271
Malignant edema, 25

Memory loss
  Boston naming test, 156
  declarative memory, 153
  episodic (*see also* Episodic memory)
    measurement, 153
    neuronal damage, 157
  explicit memory, 152
  hippocampus (*see also* Hippocampus)
    left *vs.* right, 160
    multiple trace theory, 158
    recollection *vs.* familiarity, 158–160
  implicit memory, 152
  long-term memory, 153
  neuropsychological measures, 154–155
  posterior association areas, 167–169
  post-stroke dementia
    Alzheimer's disease, 170
    large vessel, 169
    mixed, 169
    vascular, 169, 170
  prefrontal cortex
    aneurysm ruptures, 166
    confabulation, 165
    impairment, 164
    left *vs.* right, 167
    top-down deficit, 165
  semantic memory, 153
  short-term memory, 152, 153
  thalamus
    hippocampal damage, 163
    left *vs.* right, 164
    medial dorsal thalamic nuclei, 161
    medial temporal lobe structures, 162
Mini-Mental State Exam (MMSE), 230
Mirror-somatosensory hypothesis, 128
Modified Rankin Scale (mRS), 287
Montreal Cognitive Assessment tool (MoCA), 204
Moyamoya syndrome, 215–217
MRI. *See* Magnetic resonance imaging (MRI)

N
National clinical guidelines, 317–320
National Institutes of Health Stroke Scale (NIHSS), 288, 289
Neglect syndrome. *See also* Spatial neglect syndrome
  cognitive rehabilitation and recovery, 327, 328
  critical lesions, 81–84
  disconnection syndrome, 84–85
  spatial attention-based models, 74, 75
Neuropsychological rehabilitation. *See* Cognitive rehabilitation and recovery
North American Symptomatic Carotid Endarterectomy Trial (NASCET), 303
Nottingham Health Profile (NHP), 245
Nuclear magnetic resonance (NMR), 259

P
Pantomime, 52
Pathophysiology, 10–11
Penumbra, 15
Perimesencephalic subarachnoid hemorrhage (pSAH), 177, 190
PFC. *See* Prefrontal cortex (PFC)
Pharmacotherapy, PSA, 108–109
Phonemes, 120
Posterior fossa decompression, 26
Post-stroke aphasia (PSA)
  aphasia
    acute, 96, 98
    communication, 95
    conceptualization of, 96
    definition, 95
    ischemic and subcortical infarctions, 96
    National Aphasia Association reports, 96
    quality of life, 96, 103
    rehabilitation settings, 96
    thalamic, 97
  assessment of
    aforementioned cognitive impairments, 103
    cognitive neuropsychological approach, 102, 103
    diagnosis, 102
    language deficits evaluation, 102, 104
    screening test, 102, 103
    standardized aphasia test, 102, 103
  classification of
    algorithm of, 98, 99
    anomic, 101
    anterior perisylvian and extrasylvian, 97
    Broca's aphasia, 97–100
    clinical perspective, 98
    conduction, 100
    global, 100–101
    impaired and preserved repetition, 98
    lesion-symptom mapping, 98
    transcortical, 98, 101
    Wernicke's aphasia, 100
  multidimensional approach, 102–104

prognosis and recovery process
cerebral artery, 104, 105
diffusion tensor imaging, 106
environmental contingencies, 105
mechanisms, 104, 105
MRI, 106
patient's characteristics, 105
patterns of, 105
tractography, 107
treatment interventions
brain stimulation, 109
drug therapy, 110–112
epidural stimulation, 109
pharmacotherapy, 108–109
speech-language therapy, 107–108
Post stroke dementia
Alzheimer's disease, 170
intracerebral hemorrhage, 203
large vessel, 169
mixed, 169
vascular, 169, 170
Post-stroke depression (PSD)
antidepressant medication, 230–232
cognitive consequences, 230
diagnosis
aphasia, 229
DSM-IV-TR, 228
rating scales, 228
mortality, 231
non-pharmacological treatments
behavioral interventions, 233–234
electroconvulsive therapy, 234–235
prevention, 235–236
psychotherapy, 234
repetitive transcranial magnetic stimulation, 234
pharmacological treatment, 232–233
physical and functional consequences, 229
psychostimulants, 233
social consequences, 231
Post-traumatic stress disorder (PTSD), 248, 250
Prefrontal cortex (PFC)
aneurysm ruptures, 166
confabulation, 165
impairment, 164
left *vs.* right, 167
sensorimotor control, 41–42
top-down deficit, 165
Prism adaptation process, 85–88
Prosody. *See* Speech emotion disorders
PSA. *See* Post-stroke aphasia (PSA)
PSD. *See* Post-stroke depression (PSD)
Psychostimulants, 233

Q
Quality-adjusted life-year (QALY), 242, 243
Quality of Life (QoL)
burden of care, 250
caregivers, 248–250
coping strategies, 246
depression, 247
drug A and B, 243
HRQoL
SIP and NHP, 245
standard-gamble method, 243
time-tradeoff method, 243
utility index, 244
pain, 241
post-stroke depression, 246, 248
posttraumatic stress disorder, 248
QALY, 242, 243
sexual dysfunction, 246
subarachnoid hemorrhage, 247

R
Recombinant tissue plasminogen activator (rtPA), 18–20, 22, 262
Repetitive transcranial magnetic stimulation (rTMS), 234
Resting-state fMRI (rs-fMRI), 271–272
Rey-Osterrieth Complex Figure-Recall subtest, 181, 186
Right-hemisphere damage (RHD). *See* Emotional communication disorders

S
Saccadic remapping, 80, 81
Scottish guidelines, 318
Selective serotonin reuptake inhibitor (SSRI), 230
Sensorimotor control
constraint induced movement therapy, 37
fMRI scans, 40, 41
MRI scan, 39
N20-P27 amplitudes, 38, 39
perceptual thresholds, 38–40
PFC influences, 41–42
recovery process
aerobic training, 44
bimanual movement training, 43
brain repair, 43
cardiac rehabilitation, 45
CNS, 42, 43
cortical activity, 44
deficits, 43

Sensorimotor control (*cont.*)
determinants of, 44
M1 lesions, 43, 44
neurophysiological changes, 42
persisting disability, 45
PMd activity, 44
standard discipline-specific therapy, 44, 45
walking program, 45
somatosensory system, 37, 38
thalamic lesions, 38
upper limb dysfunction, 37
ventroposterolateral nucleus, 38, 39
Short-term memory, 181
Sickle cell disease, 217–218
Sickness impact profile (SIP), 245
Signal-to-noise ratio (SNR), 258
Silent brain infarcts, 6
Single photon emission computed tomography (SPECT), 180
SLT. *See* Speech-language therapy (SLT)
Spatial errors, 55
Spatial neglect syndrome
anatomy
critical lesions, 81–84
disconnection syndrome, 84–85
attention-based models, 74, 75
classic case of
cancellation task, 72, 73
causes, 71, 72
disorder of consciousness, 71
extinction disorder, 72
line bisection and figure copying, 72, 73
sub-syndromes, 72, 73
vision impairments, 72
deficits of
impaired time perception, 77, 79
nonspatial sustained attention, 75–76
representational updation, 79–82
spatial working memory, 76–78
symptoms constellation, 74
definition, 71
double-step saccade task, 80
elevator counting task, 76
mental models, 79–81, 88
recovery process
limb restriction, 85
neck muscle vibration and caloric stimulation, 85
prism adaptation, 85–88
rehabilitation, 85–88
remapping, 80
saccadic remapping, 80, 81
Spatial working memory (SWM), 76–78
Speech emotion disorders
expression, 123–125
prosody
alternate hypothesis, 121
case study, 120
phonemes, 120
propositional messages, 120
RHD and LHD patients, 121, 122
syntactic-propositional prosody, 121
verbal stimuli, 122–123
Speech-language therapy (SLT), 107–108
Stroke
clinical presentation, 11
complications, 11
definition, 5
epidemiology
age-adjusted mortality rates, 2–4
developed countries, 2
United States, 3
etiology, 5–8
historical perspective, 1–2
imaging investigations, 11–12
management, 12
pathophysiology, 10–11
risk factors, 8–10
Stroke impact scale (SIS), 289
Stroke trials
acetylsalicylic acid, 284
blood pressure control, 284
clinical outcomes and comorbid burden, 285
functional activity measures
Acute Stroke Ischemic NXY-059 Treatment, 304
British Aneurysm Nimodipine Trial, 302
carotid surgery trials, 303–304
decompressive craniectomy trials, 304–305
International Stroke Trial, 302
International Subarachnoid Aneurysm Trial, 303
International Surgical Trial In Intracerebral Hemorrhage, 304
health research methods, 286–287
measurement and outcome scales
AHA.SOC, 290
Chedoke-McMaster stroke assessment scale, 290
Frenchay activities index, 290
impairment (disability) scales, 289
methodology, 287–288
Modified Rankin Scale (mRS), 287
NIHSS, 288

physical impairment and functional activity, 289
quality of life, 290–301
stroke impact scale, 289
WHO international classification, 289
neuroplasticity, 284
prognostic factors, 284–285
tissue plasminogen activator, 302
Subarachnoid hemorrhage (SAH), 142–143. *See also* Hemorrhagic stroke
Sustained-attention to response task (SART), 265
SWM. *See* Spatial working memory (SWM)
Systemic lupus erythematosis (SLE), 215

T
Tenecteplase, 19
Thalamic infarcts, 141–142
Thalamus
hippocampal damage, 163
intracerebral hemorrhage, 202
left *vs.* right, 164
medial dorsal thalamic nuclei, 161
medial temporal lobe structures, 162
Three-dimensional time of flight angiography (3DTOF), 263
Thrombectomy, 20–24
Thrombolysis. *See* Treatment
Tissue plasminogen activator, 302
Transcortical aphasia, 101
Transcortical motor aphasia, 124
Transcortical sensory aphasia, 124
Transient ischemic attacks (TIAs), etiology, 6
Treatment
aim, 15
hemorrhagic stroke
acute medical management, 28
core area, 15
ICH, 24, 26, 27
IVH, 28
SAH, 24, 27
surgical treatment, 29–30
ischemic stroke
acute medical management, 16–17
arterial occlusion, 16
core area, 15
decompressive craniectomy, 24–26
intra-arterial and mechanical thrombolysis, 20–24
intravenous thrombolysis, 18–19
Tricyclic antidepressant (TCA), 230

U
UK guidelines, 332, 333

V
Ventroposterolateral nucleus (VPL), 38, 39
Verbal memory, 181, 182
Visual memory, 181, 182
Visual Reproduction subtest, 181
Visuospatial facial perceptual disorder, 126, 127
Voxel-based lesion symptom mapping (VLSM), 265
Voxel-based morphometry (VBM), 265, 266

W
Wernicke's aphasia, 100, 121, 124
Wisconsin Card Sorting Test (WCST), 144, 184, 205

MIX
Papier aus verantwortungsvollen Quellen
Paper from responsible sources
FSC® C105338

If you have any concerns about our products,
you can contact us on
ProductSafety@springernature.com

In case Publisher is established outside the EU,
the EU authorized representative is:
Springer Nature Customer Service Center GmbH
Europaplatz 3, 69115 Heidelberg, Germany

Printed by Libri Plureos GmbH
in Hamburg, Germany